Advances in Remote Sensing in Coastal Geomorphology II

Advances in Remote Sensing in Coastal Geomorphology II

Editors

José Juan de Sanjosé Blasco
Germán Flor-Blanco
Ramón Blanco Chao

MDPI

Basel • Beijing • Wuhan • Barcelona • Belgrade • Novi Sad • Cluj • Manchester

Editors
José Juan de Sanjosé Blasco
University of Extremadura
Cáceres
Spain

Germán Flor-Blanco
University of Oviedo
Oviedo
Spain

Ramón Blanco Chao
University of Santiago de Compostela
Santiago de Compostela
Spain

Editorial Office
MDPI
St. Alban-Anlage 66
4052 Basel, Switzerland

This is a reprint of articles from the Special Issue published online in the open access journal *Remote Sensing* (ISSN 2072-4292) (available at: https://www.mdpi.com/journal/remotesensing/special_issues/rs_coastal_geomorphology2).

For citation purposes, cite each article independently as indicated on the article page online and as indicated below:

Lastname, A.A.; Lastname, B.B. Article Title. *Journal Name* **Year**, *Volume Number*, Page Range.

ISBN 978-3-7258-1421-3 (Hbk)
ISBN 978-3-7258-1422-0 (PDF)
doi.org/10.3390/books978-3-7258-1422-0

Contents

remote sensing

MDPI

Article

Using Remote Sensing Methods to Study Active Geomorphologic Processes on Cantabrian Coastal Cliffs

María José Domínguez-Cuesta [1,*], Laura Rodríguez-Rodríguez [1], Carlos López-Fernández [1], Luis Pando [1], José Cuervas-Mons [1], Javier Olona [2], Pelayo González-Pumariega [3], Jaime Serrano [1], Pablo Valenzuela [4] and Montserrat Jiménez-Sánchez [1]

1 Department of Geology, University of Oviedo, 33005 Oviedo, Spain
2 TerraDat Geophysics S.L., 33428 Llanera, Spain
3 Department of Mine Exploitation and Prospecting, University of Oviedo, 33600 Mieres, Spain
4 Department of Earth Sciences and Condensed Matter Physics, University of Cantabria, 39005 Santander, Spain
* Correspondence: dominguezmaria@uniovi.es

Abstract: Gravitational processes on inaccessible cliffs, especially in coastal areas, are difficult to study in detail with only in situ techniques. This difficulty can be overcome by the complementary application of remote sensing methods. This work focuses on an active complex landslide affecting the slope of the Tazones Lighthouse (Cantabrian coast, North Iberian Peninsula), which has been monitored since June 2018. The aim of this research is to establish a conceptual model of the internal structure of the slope. A remote multitechnique approach was applied, including landscape deformation analysis from photogrammetric surveys, ground motion detection applying A-DInSAR techniques and Sentinel-1 satellite data, and electrical resistivity tomography. The obtained results showed the great potential of some of the remote techniques, such as UAV photogrammetry and electrical tomography, and the ineffectiveness of others, such as A-DInSAR, which failed to provide adequate results due to the profuse vegetation. This work made it possible to establish a geological model of the functioning of the slope of the Tazones Lighthouse and to deduce the surface extent of the destabilized mass (70,750 m^2), the rupture surface shape (stepped), its in-depth extent (10–50 m), the volume of materials involved (~3,550,000 m^3) and the type of landslide (complex including a predominant translational slide). The combination of field and remote sensing data significantly increased the possibility of reaching a comprehensive geological interpretation of landslides on rocky coasts.

Keywords: cliff; rocky coast; Cantabrian coast; UAV; photogrammetric surveys; A-DInSAR; electrical tomography; geological model

Citation: Domínguez-Cuesta, M.J.; Rodríguez-Rodríguez, L.; López-Fernández, C.; Pando, L.; Cuervas-Mons, J.; Olona, J.; González-Pumariega, P.; Serrano, J.; Valenzuela, P.; Jiménez-Sánchez, M. Using Remote Sensing Methods to Study Active Geomorphologic Processes on Cantabrian Coastal Cliffs. *Remote Sens.* **2022**, *14*, 5139. https://doi.org/10.3390/rs14205139

Academic Editors: José Juan de Sanjosé Blasco, Germán Flor-Blanco and Ramón Blanco Chao

Received: 27 August 2022
Accepted: 8 October 2022
Published: 14 October 2022

1. Introduction

The Earth's surface is constantly undergoing major geomorphological changes, both spatially and temporally, which are mainly linked to the sensitivity of the relief to anthropogenic and climatic changes [1,2]. This is especially relevant in the context of the global change our planet has been facing in recent decades [3]. Many of the landslides caused by natural causes are among the processes associated with global change and involve the movement of land masses from the highest to the lowest part of slopes, sometimes reaching volumes higher than 10^8 m^3, in which case they are termed "Giant Landslides" [4]. The rates of landform erosion due to these processes range between 2.65 and 5.17 mm yr^{-1} at watershed scale [5]. Furthermore, these processes pose a hazard to human settlements, the evaluation of which involves the creation of maps and predictive models. For this purpose, geomorphological criteria and mathematical modelling are used at a cartographic scale, while multidisciplinary studies and continuous monitoring are needed at a detailed scale (discrete landslides) [6]. Remote sensing techniques are crucial for studying landslides and establishing the susceptibility of the terrain to these processes [7] and other gravitational phenomena, such as subsidence [8].

In the case of active slopes, it is critical to establish the geometry of the landslides, determining not only their spatial distribution at the surface and shallow depths, but also the position of the rupture surface at depth, which is a challenging task. It is essential to establish the foreseeable evolution of mass movement in the future and to estimate the volume of material that can be mobilized [9]. With the aid of this information, predictive risk and susceptibility analysis models can be constructed as a basis for taking the necessary measures to prevent damage. On accessible slopes, the combination of monitoring techniques and field work with remote sensing permits systematic predictive analysis [10,11]. However, in inaccessible areas, the use of remote sensing becomes a key tool to establish the spatial distribution, the three-dimensional geometry in depth and the temporal evolution of mass movements.

A particular case of inaccessible slopes affected by active gravity processes are coastal cliffs, where a complex interaction between marine and subaerial processes occurs. This leads to marine scour of the cliff base and different gravity processes, all leading to the progressive inland retreat of the coastline. The retreat rate is conditioned by multiple factors, but global data reveal that the main drivers are associated with the properties of bedrock and its strength. In fact, in different cliffs around the world, average values of erosion reach 2.9 cm yr^{-1} for hard rocks, 10 cm yr^{-1} for medium rocks and 23 cm yr^{-1} for weak rocks [12]. Coastline retreat along cliff coasts is one of the major global problems affecting densely populated coastal areas. Moreover, recent works show that cliff retreat has experienced an acceleration in recent years: for example, in Great Britain, values between 2–6 cm yr^{-1} prevailed for most of the Holocene and contrast dramatically with historical records of rapid retreat at 22–32 cm yr^{-1} at the same sites during the last 150 years [13]. However, cliff retreat rates are not necessarily regular over time. They could experience sudden recession episodes due to the occurrence of voluminous landslides, potentially able to produce tsunamis [14].

Considering that topographic and accessibility factors usually limit the development of systematic in situ studies of unstable coastal cliffs, remote sensing techniques are essential for monitoring cliff retreat rates and evaluating their stability and potential impact on threatened coastal communities. The Cantabrian coast (North Iberian Peninsula) is eminently rocky [15], and, in some specific stretches, retreat rates of more than 2 m yr^{-1} have been recorded over the last 15 years [16]. This work focuses on the unstable Tazones Lighthouse cliff, located on the Cantabrian Coast where an active c. 70,000 m^2 complex landslide has been systematically monitored since June 2018 [17]. Slope movement presumably takes place through a combination of translation and sliding mechanisms strongly influenced by bedrock discontinuities, but some important questions remain unsolved. These include the internal structure and the in-depth location of the rupture surface, which are crucial to estimate the volume of displaced material. In this work, we aim to establish a conceptual model of the internal structure of the Tazones Lighthouse slope by applying a combination of remote sensing techniques to a previously known environment, studied on the surface and characterized by means of geomorphological mapping and geomatic criteria.

As we will show, our main conclusions suggest that (1) unmanned aerial vehicle (UAV) photogrammetry and geophysics provide good results, unlike A-DInSAR techniques, which require the installation of passive reflectors; and (2) the integration of remotely sensed data and those previously collected through in situ fieldwork make it possible to obtain a coherent model of the functioning of the slope at depth.

Settings of Study Area

The study area is a 700 m long stretch of the Cantabrian coast facing north and east (Figure 1). In this area, geomorphological activity was detected due to gravity processes that caused the appearance of several cracks inland. This led to the destruction of a restaurant located 150 m from the coast and increased activity of falling debris and blocks on the front of the cliff. The area was monitored by field work and UAV photogrammetric surveys between June 2018 and May 2020 [17]. A set of complex movements have developed affecting the Jurassic bedrock of the Lastres Formation [18,19], a multilayer sandstone and marl dipping 14–17° north (towards the sea), which is affected by two families of joints (J_1: 166/75; J_2: 85/89; [17]) and multiple fault systems (NE-SW, NW-SE, E-W and NNW-SSE; [20]). This deltaic formation displays frequent changes in lateral facies and strata thickness. Previous works refer to old jet mining works in the surroundings [21–23], but no clear signs of mining were found in the field before, despite the fact that manual cores were extracted in one of the areas where archaeologists referred to the presence of old mining activity [17,24].

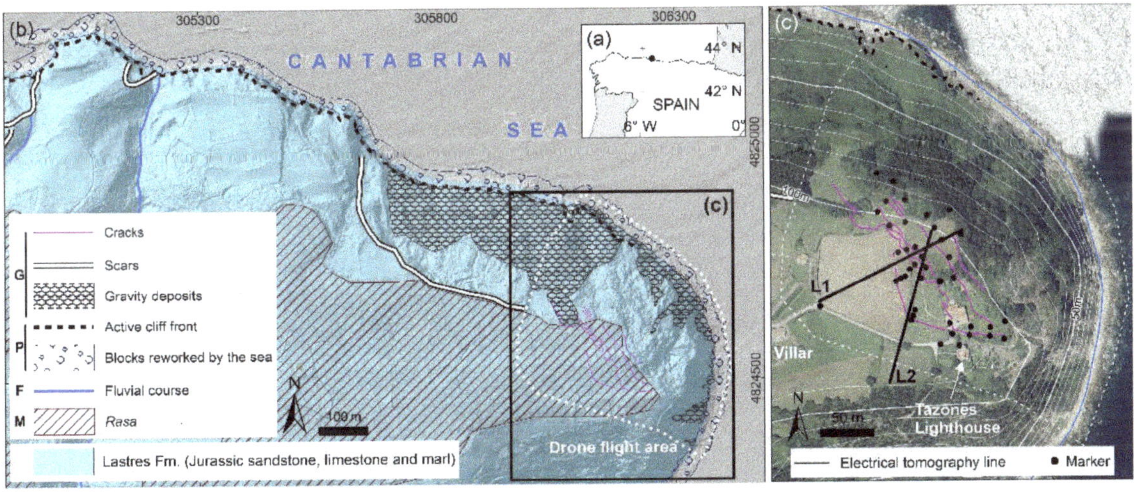

Figure 1. (**a**) Location of the study area in the North of Spain. (**b**) Lithology [25] and main geomorphological aspects of the study area (M, marine; F, fluvial; P, polygenic; G, gravity). Terrain model: LIDAR-PNOA 2012 CC-BY 4.0scne.es. UTM coordinates ETRS89 30T. (**c**) Location of the electrical tomography lines, markers and main cracks in the surroundings of the Tazones Lighthouse. Background image: OrtoPNOA 2017 CC-BY 4.0 scne.es.

Various research techniques have been applied in the area, including structural and geomatic measurements by total station and remote sensing, which have made it possible to establish a 4D model of the evolution of the slope [17]. The first two years of monitoring of the 38 installed marker points (Figure 1) showed that, spatially, the movement is strongly influenced by lithology/rock structure (S_0, J_1, and J_2) and, temporally, by rainfall/soil moisture and/or storm swell [17].

However, one of the remaining challenges in the characterization of the Tazones Lighthouse slope is the analysis of the internal structure and the exact depth of the rupture surface. As shown below, the combination of different remote sensing techniques has made it possible to achieve this objective, providing a solid basis for assessing both the degree of stability of the slope and the volume of material that can be mobilized if, in extreme circumstances, a sudden landslide event occurs.

2. Materials and Methods

The methodology of this study is based on a multitechnique approach and include the following tasks, which are described in detail below: (i) analysis of landscape deformations from photogrammetric surveys; (ii) ground motion detection, applying A-DInSAR techniques and Sentinel-1 satellite data; (iii) realization of electrical resistivity tomography lines; and (iv) geological and geomorphological interpretation of the Tazones Lighthouse slope.

2.1. Analysis of Landscape Deformations from Photogrammetric Surveys

The photogrammetric surveys were performed with a quadcopter Icom3D Cárabo S3, equipped with a Sony α5100 camera (24.3 Mpx CMOS sensor and pixel size of 4 μm) and Sony E 20 mm f/2.8 prime lens. The APS-C size sensor (23.5 × 15.6 mm) involves a 30 mm equivalent focal length for a full-frame sensor. The aerial surveys took place on 17 November 2018, and 29 November 2019, and flight planning was carried out using UAV-GeoFlip software. The coordinates of the ground control points (GCP), nine permanent sites distributed throughout the study area (16 hm^2), were measured using Topcon GR-3 receivers linked to the Spanish GNSS Reference Stations Network (ERGNSS) of the Spanish National Geographic Institute (permanent station XIX1; 5°41′54.15187″W/43°33′28.94305″N coordinates) [26]. The observation method was approached with a fixed receiver at the AUX301 control point (static GNSS permanent station XIX1, 6 h), and a mobile receiver at the other GCP (rapid static GNSS) (Tables S1–S4; Figures S1 and S2). Achieving the correct exposure, image quality and sharpness during photographic acquisition involved a fixed shutter speed (1/800 s), medium apertures (f/5.6 and 6.3), and a maximum ISO limited to 800. The average flight height for both surveys was 117 m above ground level. The drone flights provided 167 images in 2018 and 162 in 2019 (6000 × 4000 px), ensuring a ground resolution close to 2.2 cm/px, and almost 1.6 hm^2 of average ground coverage per photograph.

Photogrammetric processing was addressed using Agisoft Metashape Professional (v. 1.5.2), a well-known software that is also widely used in coastal studies e.g., [27–29]. This produced for each aerial survey a 3D textured point cloud (RGB) with an average of 34 million points, providing a data density of 220 points/m^2; a digital surface model (DSM) with a 9 cm/px resolution; and an orthophoto with 2.5 cm of ground sample distance (GSD) (Figures S3 and S4). The GNSS positioning of checkpoints using the receivers provided the estimation of errors. The root-mean-squared error remained below 2.9 cm in both flights (in X, Y), and averaged 2.48 cm for Z (Table 1).

Table 1. Estimated RMSE (cm) at ground control points and check points.

	2018		**2019**	
	GCP (9)	**CP (12)**	**GCP (8)**	**CP (7)**
X	0.028	1.445	0.528	2.863
Y	0.061	1.604	0.711	2.761
Z	0.014	3.438	0.341	1.538

The investigation on ground deformations was based on the comparative analysis of the geomatic products available at different dates. In 3D, it involved calculating the distances between clouds using the Multiscale Model to Model Cloud Comparison algorithm (M3C2), a plugin implemented in Cloud Compare software (v. 2.12). This tool offers a robust way of obtaining distances using the local roughness of surfaces to estimate uncertainty and makes it possible to differentiate significant changes [30], which demonstrates its usefulness studying coastal areas e.g., [31,32].

2.2. Ground Motion Detection Applying A-DInSAR Techniques and Sentinel-1 Satellite Data

Advanced differential synthetic aperture radar interferometry (A-DInSAR) techniques have been applied to detect ground motion. Available SAR data consisted of 113 IW-SLC SAR images acquired by Sentinel-1A/B satellite in C-Band (5.5 cm wavelength) and descending orbit, from 4 January 2018, to 5 February 2020 (Table 2). Two different approaches have been used for A-DInSAR processing: (i) PSIG chain [33,34], which is based on the method of Persistent Scatterer Interferometry [35,36] and (ii) P-SBAS processing service [37–39], available at the GEP [40,41]. Resolution of PSIG results is 14 × 4 m, while P-SBAS resolution is 90 × 90 m.

Table 2. Main acquisition parameters of the SAR satellite dataset used.

Satellite	Sentinel-1
Sensor	A/B
Band	C
Wavelength	5.55 cm
Acquisition mode	Interferometric Wide
Polarization	VV
SAR product	Single Look Complex
Acquisition orbit	Descending
Temporal period	1 April 2018–2 May 2020
Revisit period	6–12 days
Resolution	14 × 4 m
Incidence angle	39°
Track	154
Number of SAR images	113

A-DInSAR procedure consisted of the usual next steps: (i) co-registration of SAR images and generation of wrapped interferograms; (ii) estimation of residual topographic errors and line-of-sight (LOS) mean velocity over a network of points (PS) according to amplitude or coherence threshold criteria; (iii) interferogram phase unwrapping and assessment of deformation time series; (iv) estimation and filtering of atmospheric components; and (v) geocoding and implementation of LOS mean deformation velocity map (mm yr^{-1}) and deformation time series (mm) in a geographic information system (QGIS software, v. 3.16).

2.3. Electrical Resistivity Tomography Lines

The resistivity study was developed on 13 August 2019, in the surroundings of the Tazones Lighthouse, along two lines covering the area with the highest density of cracks (Figure 1c): line 1 (L1), with a length of 210 m and an orientation of N55° E, and line 2 (L2), with a length of 207 m and an orientation of N10° E. The positions of both lines, including both terrain with evidence of instability and, a priori, stable terrain, were perpendicular to the main detected cracks. The lengths of the lines were conditioned by the inaccessibility due to the steep terrain and the high density of vegetation. Electrical resistivity tomography (ERT) acquisition was performed with an IRIS Syscal Junior Switch 72 system and electrodes spaced 3 m apart. The dipole–dipole and Wenner–Schlumberger arrays were developed on both lines. Both types of data were merged and inverted (with an algorithm based on a robust inversion) at the same time to improve the accuracy and stability of the resistivity sections. Resistivity sections were calculated considering the most reliable data acquired in the field. Calculations on lines 1 and 2 were based on 1190 and 1312 apparent resistivity measures, respectively. Root-mean-squared error between observed and calculated apparent resistivity was 12.05% for L1 and 16.39% for L2.

2.4. Geomorphological Mapping and Geological Interpretation of the Tazones Lighthouse Slope

Geomorphological mapping was performed through fieldwork and photointerpretation of the two photogrammetric surveys with UAV. The technical details of the aircraft are given in Section 2.1. Elevation point clouds of both flights were used to derive digital surface and digital hillshade models of the terrain surface (no vegetation filtering was applied). Using the information of both photogrammetric surveys, we mapped the location of cracks, the limits of the landslide area and the trace of four strata which can be easily identified in both flights to track if they remain still, or if they shift with time.

Two topographic sections were completed using the digital elevation model information based on the UAV point cloud taken in 2019. Both sections follow the geophysical lines surveyed in 2018, L1 and L2, extending them to the cliff foot. The geophysical resistivity models obtained for both lines were projected into their respective positions in the 2019 topographic profiles to support the geological interpretation of the structure. Additionally, the three families of discontinuities surveyed in the field (S_0: 2/14; J_1: 166/75; J_2: 85/89 [17]) were projected in the area not covered by the geophysical survey, considering a random spacing based on the data taken from the geomechanical stations [17]. As sections are not necessarily normal to the strike of these discontinuities, the apparent dip angles were calculated using an alignment diagram [42].

3. Results

3.1. Analysis of Landscape Deformations from Photogrammetric Surveys

The comparison of the point clouds using the M3C2 algorithm reveals that 94% of the studied area underwent changes that the software considers significant (Figure 2a). The remaining 6% corresponds mainly to the area with the tallest trees. It also includes the N end of the mass movement, a small land mobilization in the central-eastern part of the study area, and isolated patches of the N–S oriented cliff located in the SE. However, the forested area, with trees between 10 and 18 m high, is not representative for studying the mass movement as it does not provide direct information on the ground surface. This area with trees and shrubs was mapped from the orthoimages and digital surface models in order to extract it from the point cloud comparison. Thus, the total comparison area is 5.5 hm^2.

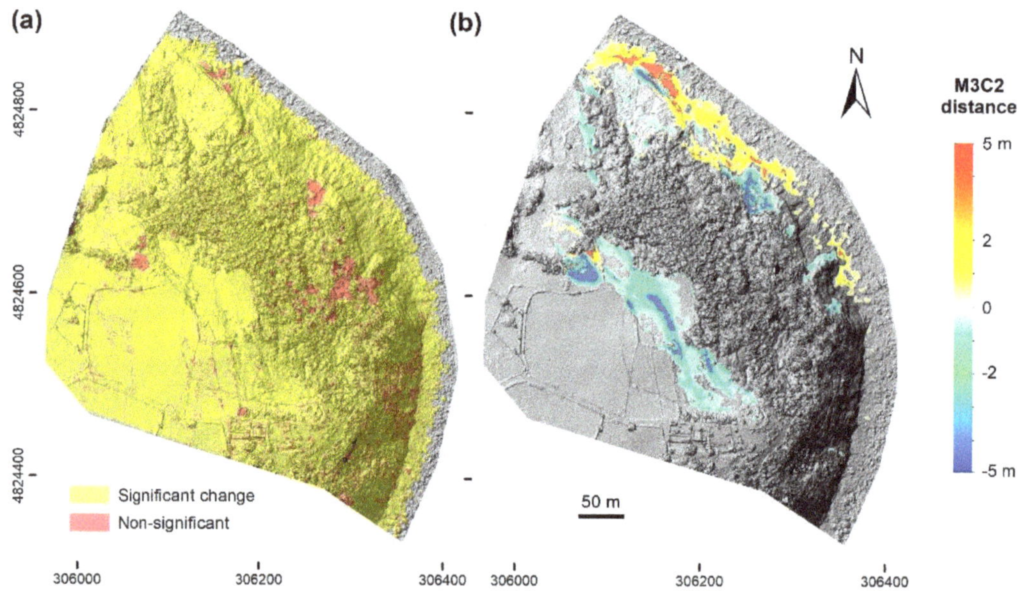

Figure 2. Results of cloud-to-cloud comparison. UTM coordinates referenced to ETRS89 30T. (**a**) Significant change. (**b**) Distance (M3C2).

The M3C2 distance measurements (calculated prioritizing the z-axis), detect both rises and falls in topographic elevation between the 2018 and 2019 flights (Figure 2b). The extreme values in each case are +4.8 and −4.7 m, and the average is −0.1 m. However, through field observations, it was estimated that a range of ±30 cm may be due to changes in height of herbaceous vegetation, and that their movement may cause uncertainties for photogrammetric reconstruction. This represents 60% of the investigated area, from which, therefore, no conclusions on slope dynamics can be drawn. The most relevant changes between the two point clouds lead to differentiating two zones according to the predominance of positive and negative distances.

The distances associated with topographic collapse reach an extension of 1.26 hm^2 and are mainly concentrated along the southern edge of the wooded area. The range from −1 to −3 m represents 11.6% of the total area studied, and mostly corresponds to two elongated depressions in a NNW–SSE direction, with M3C2 distances varying between −0.8 and

−2.7 m along their axes. Downward displacements greater than −3 m were only obtained in 0.9% of the territory. On the other hand, the main positive distance values are found in the northern perimeter, bordering the mass movement front. Very occasionally (0.5% of the total area), at the northern end of the slope where a rocky block collapsed, they exceed 3 m but most of the calculated positive distances are between 30 cm and 2 m (15% of the studied area).

3.2. Ground Motion Detection Applying A-DInSAR Techniques and Sentinel-1 Satellite Data

We obtained two LOS mean deformation velocity maps (mm yr^{-1}) by means of PSIG software and P-SBAS processing from GEP (Figure 3). The PSIG approach allowed us to obtain 331 PS, with LOS ground deformation velocities between −5.3 and 8.5 mm yr^{-1}, 265 of which have a velocity between −2.5 and 2.5 mm yr^{-1} (80.1%) (Figure 3a). In the case of P-SBAS, only 20 PS were measured, for which the LOS mean velocities range between −3.7 and 2.9 mm yr^{-1} (Figure 3b). All detected PS are outside the Tazones landslide area; thus, the area cannot be monitored by A-DInSAR techniques. This can be due to (i) the high complexity of this predominantly forested area and the absence of adequate natural reflectors, and (ii) a mass movement velocity that is too fast to be detected by means of A-DInSAR techniques. Even so, PSIG procedure substantially improves the obtained results with respect to P-SBAS processing, due to the high degree of manipulation and control of processing parameters along the whole PSIG chain and the large difference of spatial resolution between the two software.

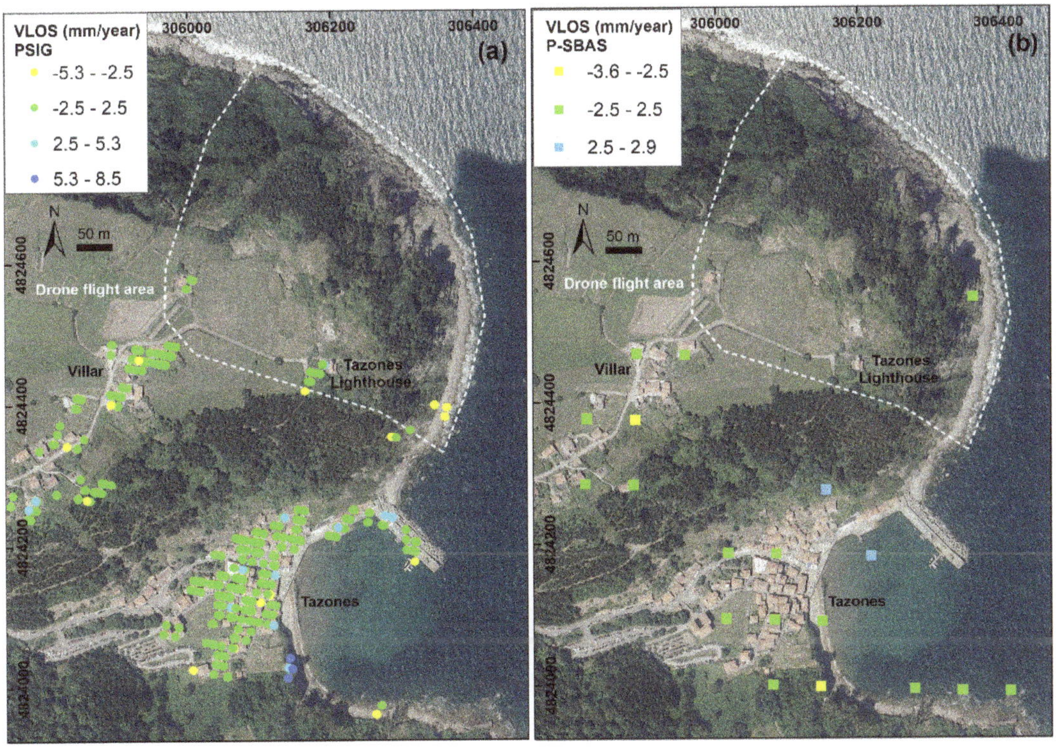

Figure 3. Tazones Lighthouse area LOS mean deformation velocity maps (mm yr^{-1}). UTM coordinates referenced to ETRS89 30T. (**a**) LOS velocity map obtained by PSIG approach (pixel resolution: 14 × 4 m); (**b**) LOS velocity map obtained by P-SBAS processing of GEP service (pixel resolution: 90 × 90 m). OrtoPNOA 2017 CC-BY 4.0 scne.es.

The PSIG LOS mean deformation velocity map (Figure 3a) revealed that Villar village and the Tazones Lighthouse building area are both stable zones, with LOS rate velocities varying between −5.2 and 4.4 mm yr^{-1}. Hence, these areas can be considered outside the limits of the Tazones landslide area. The PS with deformations, with LOS velocities below −2.5 and over 2.5 mm yr^{-1}, could be related to different artefacts, such as uncorrected atmospheric components or topographic and orbital errors. However, south of Tazones village, there are four PS with LOS velocities of 5 to 8.5 mm yr^{-1}, which could be related to a landslide incidence (Figure 3a).

3.3. Interpretation of the Electrical Resistivity Tomography Lines

Line 1 displays a 5 to 10 m thick surface level, with relatively high electrical resistivity values (>400 Ωm), tilted about 9° to the northeast (Figure 4a). This level is interrupted by a high-resistivity anomaly (>9000 Ωm) 125 m away from the origin of the line, which affects the entire thickness of the level without displacing it. In contrast, between 155 and 165 m, a second interruption is observed, coinciding with a small topographic trench area, where the high resistivity level achieves 4 to 5 m vertical displacement. At greater depth and below the high-resistivity uppermost level, a parallel low-resistivity interval (<35 Ωm) is observed. This level is interrupted by two areas of slightly higher resistivity values (<40 Ωm) at 115 m and 155 m, in a consistent fashion with the discontinuities described for the overlying high resistivity level.

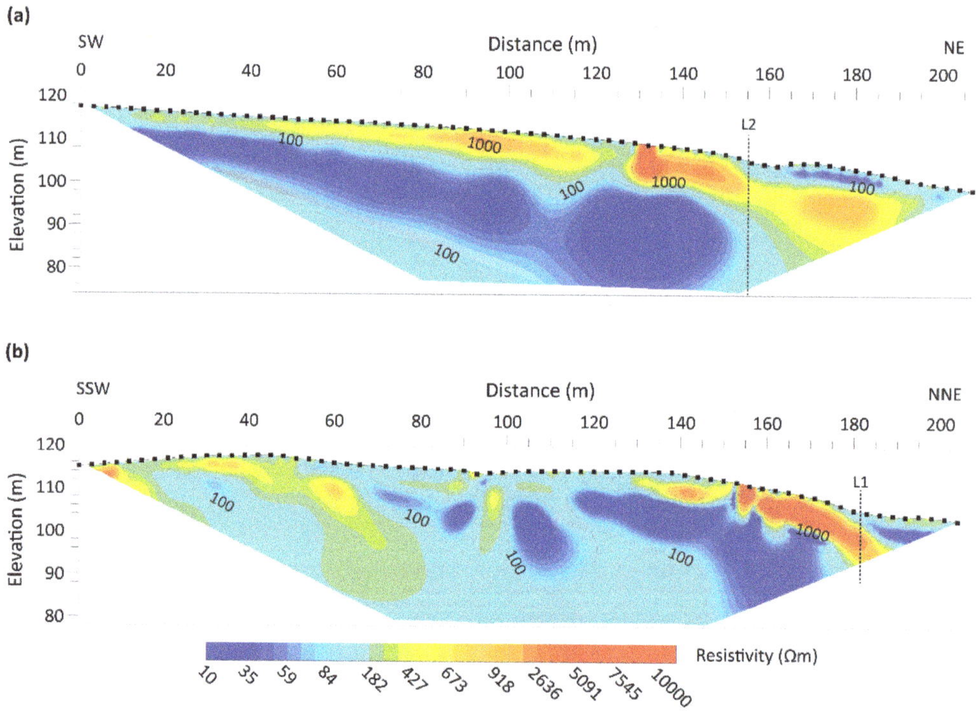

Figure 4. Electrical resistivity tomography profiles. (**a**) Line 1; (**b**) Line 2. Locations are shown in Figure 1.

Line 2 shows a distribution of electrical resistivity zones of greater complexity than Line 1. A high-resistivity (>3000 Ωm) interval of about 10 m thickness is observed between 130 to 195 m from the origin, tilted around 14° to the northeast and with increasing resistivity values (>9000 Ωm) from 155 m onward. As in Line 1, a low-resistivity unit (<100 Ωm) is observed below the upper high-resistivity unit, which progressively thins out southwards. This low-resistivity interval is interrupted by a narrow corridor of higher resistivity values at 130 m from the origin. To the south, at 95 m from the origin, a vertical anomaly is observed, recording intermediate resistivity values (180–400 Ωm) that reach a depth of around 16 m, coinciding with a small topographic trench at the surface. This discontinuity apparently does not shift the low-resistivity anomaly. Finally, close to the origin of Line 2, another interval of high resistivity values (>3000 Ωm) can be observed overlying on top of a low-resistivity interval (<180 Ωm).

Units showing intermediate-to-high resistivity values could correspond to sandstone intervals of the Lastres Formation, reaching the highest resistivity values (>9000 Ωm) in zones of enhanced porosity due to intense weathering of sandstone layers [43,44], open cracks or widespread fractures. In contrast, low-resistivity anomalies are likely related to either a higher content in clay materials (e.g., marly intervals of the Lastres Formation) or in humidity. Regarding the subvertical resistivity anomalies that intercept zones of high and low resistivity values, they could correspond to the development of incipient cracks preceding the formation of open cracks, since they are consistent with recently formed topographic depressions. For example, an open crack appeared two years after the geophysical survey, precisely at the vertical anomaly located at 155 m from the origin in Line 1, showing a vertical displacement of 2.5 m.

3.4. Geological and Geomorphological Interpretation of the Tazones Lighthouse Slope

Up to 16 cracks were identified based on the hillshade model derived from the UAV flight of 2018 (Figure 5a). The longest and innermost one is a pair of parallel cracks running in a NNW–SSE to NW–SE direction, which flank a topographic trench passing right in between the Tazones lighthouse and the old restaurant. Based on these cracks and the trench zone, we estimated the limits of the potentially unstable part of the cliff. Therefore, our working hypothesis assumed that the unstable cliff portion extended from the semi-circular trench area down to the cliff foot, covering a surface extent of c. 70,922 m^2.

The hillshade model derived from the UAV flight of 2019 (Figure 5b) reveals that new cracks were developing in the uppermost part of the slope. At the western end of the study area, the new cracks broadened the extent of the unstable zone further inland than expected in our working hypothesis. To the east, the topographic trench zone remained stable, and the new cracks are clearly developing closer to the cliff than expected. Towards the foot of the cliff, we noticed that three of the four stratification traces, identified in the field and also recognizable in the 2018 and 2019 hillshade models, remain stable (a, b and c in Figure 5b), while the material on top of layer c undergoes translational sliding according to Varnes [45]. Thus, the rupture surface is presumably placed at the top of layer c or even at a higher elevation in the cliff. In contrast, to the north, the rupture surface intersects the topographic surface at greater depth (closer to the foot of the cliff) since the trace of layer d slides downslope and even suffers toppling instability processes. Therefore, the unstable portion of the cliff actually covers a surface extent of 70,750 m^2. The gif animation provided in the Supplementary Material (Video S1) shows that this portion of the slope undergoes translational sliding, transitioning to flows and topples towards the foot of the cliff.

Regarding the internal structure, Figure 6 shows two topographic sections following the geophysical tomography of lines 1 and 2, extending both profiles to the north until they reach the foot of the cliff. The contact between the subhorizontal high- and low-resistivity anomalies observed in line 1 is consistent with the apparent dipping of the stratification following the direction of line 1 (about 9°). The highest resistivity values coincide with an open crack of 3 m wide and 10 m deep. The discontinuity observed in the high-resistivity uppermost level, which causes an apparent displacement of c. 5 m, could be explained as the result of a rupture surface that propagates downward following the J_2 joints, for which the expected apparent dip along Line 1 is 84°N. At depth, the rupture surface could be following a combination of S_0 and J_2 discontinuities until it intercepts the topographic surface close to the top of layer c, at ~35 m above sea level. (Figures 5b and 6a). Regarding the J_1 joints, their expected apparent dip along Line 1 is 51°S, which is consistent with the presence of toppling processes in layer d (Figure 5b).

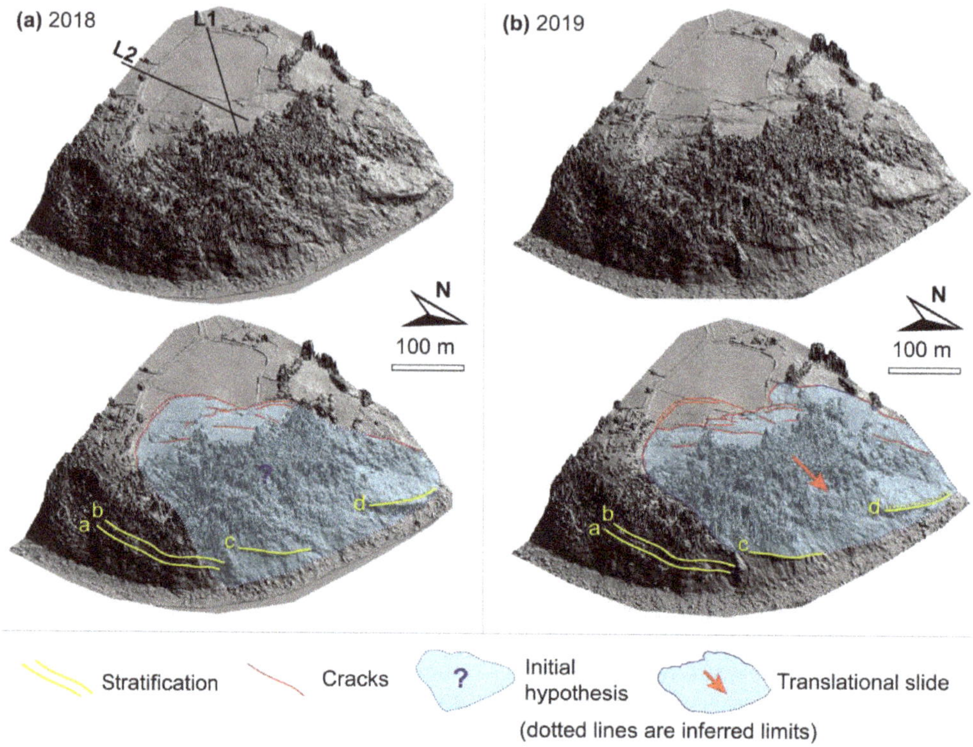

Figure 5. (**a**) Hillshade digital terrain model created from the 2018 UAV, showing the position of electrical tomography lines 1 and 2 (**top**) and the geomorphologic interpretation of the slope, delimiting the potential extent of the unstable portion of the Tazones Lighthouse slope that defines our working hypothesis (**bottom**). The yellow lines indicate the trace of four identifiable sandstone layers of the Lastres Formation to compare their position over time. (**b**) Hillshade digital terrain model created from the 2019 UAV (**top**) showing its geomorphologic interpretation and the limits of the area affected by the complex landslide one year later (**bottom**). Only the trace of layer d shifted downslope, indicating that the rupture surface daylights close to the foot of the cliff to the west, while it intercepts the topographic surface at higher elevations to the east (atop layer c).

The second topographic profile follows Line 2 (Figure 6b) and is almost normal to the strike of the stratification, hence showing its true dip (14°N). The geologic structure seems more complex with alternative high- and low-resistivity anomalies that could correspond to sandstone and marl alternations of the Lastres Formation, and which are locally interrupted by some discontinuities that cause the vertical displacement of anomalies and do not belong to the two joint families known in the area (J_1 and J_2). The slope instability starts 150 m away from the origin of this profile, where the highest resistivity level is interrupted by some discontinuities that fit the expected apparent dip of the J_1 joints (~73.5°S) and J_2 joints (~86°N). Similar to Line 1, the rupture surface seems to be controlled by the S_0 and J_2 discontinuities, intercepting the topographic surface close to the foot of the cliff.

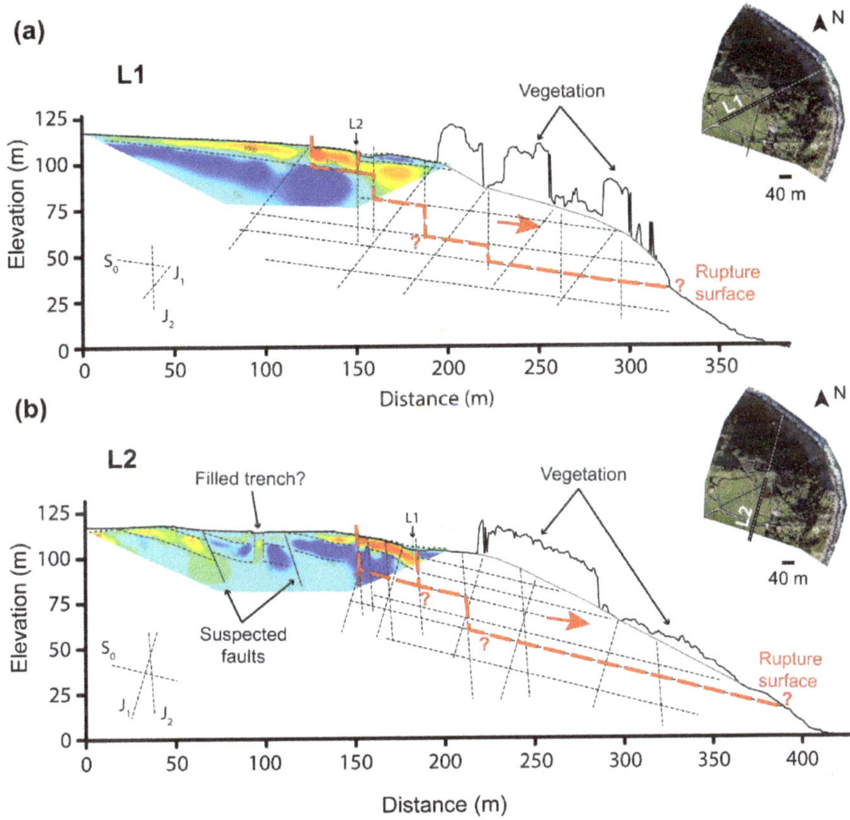

Figure 6. (**a**) In-depth interpretation of the section running along Line 1 of the electrical tomography and reaching the sea. (**b**) In-depth interpretation of the section running along Line 2 of the electrical tomography and reaching the sea.

4. Discussion

Comprehensive monitoring of rocky coastal cliffs has been revealed as one of the comprehensive ways to understand their medium- to long-term evolution. The different monitoring methods we applied in the Tazones Lighthouse surroundings uncovered the topographic changes that took place since June 2018, providing a better understanding of the limits and internal structure of the unstable area.

Photogrammetric surveys made it possible to establish the areas where the main deformations of the landscape occurred. Our initial hypothesis was proved to be wrong in the south-western part of the landslide, as this is an area that did not show significant displacement during the studied period (Figures 2 and 5). With respect to the results of cloud-to-cloud comparison, the extreme values are not only very low in percentage, but also spatially diffuse (Figure 2). They are not concentrated in specific areas, but rather define a scattered mottling throughout the study area, especially in the wooded area (with no notable groupings of pixels). Therefore, they are not considered representative values.

The application of A-DInSAR techniques provided inconclusive results, as LOS mean deformation velocity maps obtained by PSIG and P-SBAS approaches (Figure 3) do not show the real ground deformation in the Tazones landslide area, which has been revealed by other in situ and remotely sensed methods. For this reason, we made new tests in March–May 2020 to improve A-DInSAR results by means of different approaches into the PSIG chain, such as densification of points and regional processing. None of these attempts have improved the results. In urban and densely populated areas, buildings and infrastructure are suitable reflectors. However, in mountainous, rural and vegetated areas, it would be necessary to install artificial PCR (passive corner reflectors) or AR (active reflectors), which would ensure a good response from the SAR signal [46,47]. In the Tazones Lighthouse Landslide, the installation of this type of device would improve the results of A-DInSAR and help evaluate the deformation data of the terrain in the landslide area. This system, boosted with artificial reflectors, has been successfully applied to forested landslides [48,49].

The geomorphologic interpretation of derived UAV hillshade models allowed us to refine the actual limits of the Tazones Lighthouse complex landslide and its classification as a combination of translational slide with minor flow and topple processes near the base of the cliff, where the rupture surface intercepts the topographic surface between 0 and 35 m above sea level. The geophysical survey provided new insights regarding the internal structure of the slope instability, which displays low-tilting resistivity anomalies that are interpreted as sandstone and marl rich strata. These subhorizontal anomalies are interrupted by low-resistivity vertical anomalies that fit the expected apparent dip of the two joint families, J_1 and J_2, which have been controlling the generation of new open cracks at the surface since 2018 [17]. We interpret that the stratification and J_2 are the two main discontinuities favoring the in-depth propagation of the rupture surface, most likely of stepped geometry, while J_1 clearly favors the topple processes observed at the cliff base. Therefore, considering its surface extent (70,750 m^2) and average thickness (~50 m), the volume of the Tazones landslide could approach ~3,550,000 m^3. In the stable part of the slope, the area called "filled trench" in Figure 6 shows a local anomaly with high-resistivity values matching a topographic low that was considered the landslide scar in our initial hypothesis (Figure 5a). This anomaly does not seem to be associated with any geological structure, since no vertical displacement is observed in S_0. Although there is evidence of surface change in this area since 1984 [50], the cloud-to-cloud comparison and Video S1 show that this zone has remained stable in recent years. Moreover, the electrical tomography results could be consistent with anthropogenic origin of this structure, as suggested in previous works [24], which link this structure to old jet-mining labors. However, the sediment cores extracted in the filled trench do not provide conclusive evidence to confirm the jet mining theory [17]. Finally, we observed a couple of additional subvertical anomalies in Line 2 outside the complex landslide influence that apparently cause a displacement of the subhorizontal anomalies and do not fit the two families of joints described in the area. We interpret these as potential faults affecting the Jurassic bedrock. A previous work grouped faults into two main families, the orientation of which changes depending on the zone [20]: a NW-SE system that can vary between N040W and N070W and dips over 70° mainly to the NE; and a NE-SW system that ranges from N020E to N060E and dips between 70° and 85° to the NW.

5. Conclusions

In this work, a combination of remote sensing techniques (UAV, A-DInSAR and electrical tomography) was applied to understand the internal structure of the active complex landslide of the Tazones Lighthouse cliff on the Cantabrian coast. These techniques have proved to be essential to complement the information previously available on the activity of this landslide obtained by in situ techniques (geomatic monitoring, manual soundings, hydrogeological, pluviometric, soil moisture and swell data).

UAV and electrical tomography provided important data to reconstruct the internal geometry of the landslide and to delimit its spatial distribution and depth. However, the A-DInSAR technique did not yield good results for this site, as it is a sparsely urbanized area with profuse vegetation and few reflectors.

The conceptual model for the functioning of the slope developed in this work corroborates the previous findings regarding the influence of the lithological (marl and sandstone multilayer) and structural (stratification, two systems of joints and faults) factors in the genesis and advance of the mass movement. Furthermore, the possible geometry of the movement was revealed. In particular, the rupture surface is likely placed at 10–50 m depth and shows a stepped geometry following both the stratification and the two joint families, daylighting at 0 to 35 m above sea level. This made it possible to classify the movement from a genetic point of view as a complex movement (with a predominantly translational slip component, in addition to flow and topple) that mobilizes an estimated volume of 3,550,000 m^3.

Electrical resistivity tomography did not only allow us to define the rupture surface at the top of the slope and to estimate its position at the bottom, but also to find some anomalies that could be compatible with structures of anthropic origin. Nevertheless, the most remarkable result is that it revealed the existence of possible new rupture surfaces consistent with the in-depth extension of the active cracks observed on the surface.

Remote sensing techniques have proven to be crucial to improve the understanding of the functioning of active geomorphological processes in a coastal area with difficult access. Their effectiveness was found to be particularly increased when previous data from detailed in situ studies are available.

Supplementary Materials: The following supporting information can be downloaded at: https://www.mdpi.com/article/10.3390/rs14205139/s1, Figure S1. Position of Ground Control Points (red points) and Check Points (yellow points) on the 2018 flight point cloud. Figure S2. Position of Ground Control Points (red points) and Check Points (yellow points) on the 2019 flight point cloud. Figure S3. Orthophoto obtained by 2018 flight. Figure S4. Orthophoto obtained by 2019 flight. Table S1. Coordinates of the Ground Control Points used in the November-2018 flight (GNSS processing results). Reference system: ETRS89 30N. Orthometric altitude calculated with Geoid EGM08-REDNAP. Table S2. Coordinates of the Check Points used in the November-2018 flight. Reference system: ETRS89 30N. Table S3. Coordinates of the Ground Control Points used in the November-2019 flight (GNSS processing results). Reference system: ETRS89 30N. Orthometric altitude calculated with Geoid EGM08-REDNAP. Table S4. Coordinates of the Check Points used in the November-2019 flight. Reference system: ETRS89 30N. Video S1: Gif animation built from the hillshade digital terrain models obtained from the 2018 and 2019 UAV flights. Its visualisation allows to observe the changes experienced by the Tazones Lighthouse slope, predominantly translational sliding, transitioning to flows and topples towards the foot of the cliff.

Author Contributions: Conceptualization, M.J.D.-C. and M.J.-S.; methodology, all authors; investigation, all authors; writing—original draft preparation, M.J.D.-C., M.J.-S., L.R.-R., J.C.-M., L.P. and J.O.; review, M.J.D.-C., L.R.-R., J.C.-M., L.P., C.L.-F. and P.G.-P.; project administration, M.J.D.-C. and M.J.-S.; funding acquisition, M.J.D.-C. and M.J.-S. All authors have read and agreed to the published version of the manuscript.

Funding: This research is part of (1) the "COSINES" Project (CGL2017-83909-R), Call 2017 for RETOS Projects funded by the Spanish Economy, Industry and Competitiveness Ministry—*Ministerio de Economía, Industria y Competitividad* (MINECO), the Spanish Research Agency—*Agencia Estatal de Investigación* (AEI) and the European Regional Development Found (FEDER), (2) the GEOCANCOSTA research group, supported by the Asturian Regional Government (Spain) (grant number GRUPIN-IDI-2018-184) and (3) 2021 Research Support and Promotion Program-Programa de Apoyo y Promoción de la Investigación 2021 at University of Oviedo [Ref. PAPI-21-PF-22].

Acknowledgments: Authors are grateful to Oriol Monserrat and Anna Barra from *Centro Tecnológico de Telecomunicaciones de Cataluña* (CTTC) for their help and comments during the A-DInSAR processing, to Carlos Espadas from INGECOR GEOMÁTICA for the drone flights and to the City Council and Civil Protection of the Villaviciosa municipality for making it possible to analyse the cliff from the sea.

Conflicts of Interest: The authors declare no conflict of interest.

References

1. Cendrero, A.; Remondo, J.; Bonachea, J.; Rivas, V.; Soto, J. Sensitivity of Landscape Evolution and Geomorphic Processes to Direct and Indirect Human Influence. *Geogr. Fis. E Din. Quat.* **2006**, *29*, 125–137.
2. Gariano, S.L.; Guzzetti, F. Landslides in a Changing Climate. *Earth-Sci. Rev.* **2016**, *162*, 227–252. [CrossRef]
3. Steffen, W.; Sanderson, A.; Tyson, P.D.; Jäger, J.; Matson, P.A.; Moore, B., III; Oldfield, F.; Richardson, K.; Schellnhuber, H.J.; Turner, B.L., II; et al. *Global Change and the Earth System: A Planet under Pressure*; Springer: Berlin/Heidelberg, Germany, 2004.
4. Korup, O.; Clague, J.J.; Hermanns, R.L.; Hewitt, K.; Strom, A.L.; Weidinger, J.T. Giant Landslides, Topography, and Erosion. *Earth Planet. Sci. Lett.* **2007**, *261*, 578–589. [CrossRef]
5. Chen, Y.C.; Chang, K.T.; Lee, H.Y.; Chiang, S.H. Average Landslide Erosion Rate at the Watershed Scale in Southern Taiwan Estimated from Magnitude and Frequency of Rainfall. *Geomorphology* **2015**, *228*, 756–764. [CrossRef]
6. Chae, B.G.; Park, H.J.; Catani, F.; Simoni, A.; Berti, M. Landslide Prediction, Monitoring and Early Warning: A Concise Review of State-of-the-Art. *Geosci. J.* **2017**, *21*, 1033–1070. [CrossRef]
7. Youssef, A.M.; Pourghasemi, H.R. Landslide Susceptibility Mapping Using Machine Learning Algorithms and Comparison of Their Performance at Abha Basin, Asir Region, Saudi Arabia. *Geosci. Front.* **2021**, *12*, 639–655. [CrossRef]
8. Zhao, Y.; Zhou, L.; Wang, C.; Li, J.; Qin, J.; Sheng, H.; Huang, L. Analysis of the Spatial and Temporal Evolution of Land Subsidence in Wuhan, China from 2017 to 2021. *Remote Sens.* **2022**, *14*, 3142. [CrossRef]
9. Michel, J.; Dario, C.; Marc-Henri, D.; Thierry, O.; Ivanna Marina, P.; Bejamin, R. A Review of Methods Used to Estimate Initial Landslide Failure Surface Depths and Volumes. *Eng. Geol.* **2020**, *267*, 105478. [CrossRef]
10. Montoya-Montes, I.; Rodríguez-Santalla, I.; Sánchez-García, M.J.; Alcántara-Carrió, J.; Martín-Velázquez, S.; Gómez-Ortiz, D.; Martín-Crespo, T. Mapping of Landslide Susceptibility of Coastal Cliffs: The Mont-Roig Del Camp Case Study. *Geol. Acta* **2012**, *10*, 439–455. [CrossRef]
11. Morales, T.; Clemente, J.A.; Damas Mollá, L.; Izagirre, E.; Uriarte, J.A. Analysis of Instabilities in the Basque Coast Geopark Coastal Cliffs for Its Environmentally Friendly Management (Basque-Cantabrian Basin, Northern Spain). *Eng. Geol.* **2021**, *283*, 106023. [CrossRef]
12. Prémaillon, M.; Regard, V.; Dewez, T.J.B.; Auda, Y. GlobR2C2 (Global Recession Rates of Coastal Cliffs): A Global Relational Database to Investigate Coastal Rocky Cliff Erosion Rate Variations. *Earth Surf. Dyn.* **2018**, *6*, 651–668. [CrossRef]
13. Hurst, M.D.; Rood, D.H.; Ellis, M.A.; Anderson, R.S.; Dornbusch, U. Recent Acceleration in Coastal Cliff Retreat Rates on the South Coast of Great Britain. *Proc. Natl. Acad. Sci. USA* **2016**, *113*, 13336–13341. [CrossRef]
14. Teeuw, R.; Rust, D.; Solana, C.; Dewdney, C.; Robertson, R. Large Coastal Landslides and Tsunami Hazard in the Caribbean. *Eos Trans. Am. Geophys. Union* **2009**, *90*, 81–88. [CrossRef]
15. Domínguez-Cuesta, M.J.; Valenzuela, P.; Rodríguez-Rodríguez, L.; Ballesteros, D.; Jiménez-Sánchez, M.; Piñuela, L.; García-Ramos, J.C. Cliff Coast of Asturias. In *The Spanish Coastal Systems: Dynamic Processes, Sediments and Management*; Morales, J.A., Ed.; Springer: Berlin/Heidelberg, Germany, 2019; pp. 49–77. [CrossRef]
16. Domínguez-Cuesta, M.; Ferrer-Serrano, A.; Rodríguez-Rodríguez, L.; López-Fernández, C.; Jiménez-Sánchez, M. Análisis Del Retroceso de La Costa Cantábrica En El Entorno Del Cabo Peñas. *Geogaceta* **2020**, *68*, 63–66.
17. Domínguez-Cuesta, M.J.; González-Pumariega, P.; Valenzuela, P.; López-Fernández, C.; Rodríguez-Rodríguez, L.; Ballesteros, D.; Mora, M.; Meléndez, M.; Herrera, F.; Marigil, M.A.; et al. Understanding the Retreat of the Jurassic Cantabrian Coast (N. Spain): Comprehensive Monitoring and 4D Evolution Model of the Tazones Lighthouse Landslide. *Mar. Geol.* **2022**, *449*, 106836. [CrossRef]
18. Valenzuela, M.; García-Ramos, J.C.; Suarez de Centi, C. The Jurassic Sedimentation in Asturias (N. Spain). *Trab. Geol.* **1986**, *16*, 121–132. [CrossRef]
19. Aramburu, C.; Bermúdez-Rochas, D.D.; Delvene, G.; Fürsich, F.T.; García-Ramos, J.C.; Piñuela, L.; Ruiz-Omeñaca, J.I.; Werner, W. *Las Sucesiones Litorales y Marinas Restringidas Del Jurásico Superior. Acantilados de Tereñes (Ribadesella) y de La Playa de La Griega (Colunga)*; García-Ramos, J.C., Aramburu, C., Eds.; Universidad de Oviedo: Colunga, Spain, 2010.
20. Uzkeda, H.; Bulnes, M.; Poblet, J.; García-Ramos, J.C.; Piñuela, L. Jurassic Extension and Cenozoic Inversion Tectonics in the Asturian Basin, NW Iberian Peninsula: 3D Structural Model and Kinematic Evolution. *J. Struct. Geol.* **2016**, *90*, 157–176. [CrossRef]
21. Bahamonde, J.; Cossio, J.; Muñoz de la Nava, P.; Cembranos, V. *Posibilidades de Azabaches en Asturias*; IGME: Madrid, Spain, 1986; p. 104.
22. Monte Carreño, V. *El Azabache. Piedra Mágica, Joya, Emblema Jacobeo*; Picu Urriellu: Gijón, Spain, 2004.
23. López, M.T. *Mapa de Rocas y Minerales Industriales de Asturias. Escala 1:200.000*; IGME: Madrid, Spain, 2012.
24. Crespo, S.; Sierra, M.; Fernández, S.; Herrera, D. *Plan Especial de Protección y Rehabilitación de Tazones-Villaviciosa*; Ayuntamiento de Villaviciosa: Villaviciosa, Spain, 2008.
25. Pignatelli, R.; Giannini, G.; Ramírez del Pozo, J.; Beroiz, C.; Barón, A. *Mapa Geológico de España Escala 1:50.000, Nº 15 (14-3) Lastres*; IGME: Madrid, Spain, 1972.
26. Spanish National Geographic Institute. Reseña de Estación Permanente—ERGNSS (XIX1). Available online: https://datos-geodesia.ign.es/ERGNSS/fichas/XIX1.pdf (accessed on 28 July 2022).
27. Taddia, Y.; Stecchi, F.; Pellegrinelli, A. Coastal Mapping Using Dji Phantom 4 RTK in Post-Processing Kinematic Mode. *Drones* **2020**, *4*, 9. [CrossRef]

28. Gonçalves, G.; Gonçalves, D.; Gómez-gutiérrez, Á.; Andriolo, U.; Pérez-alvárez, J.A. 3D Reconstruction of Coastal Cliffs from Fixed-wing and Multi-rotor Uas: Impact of Sfm-mvs Processing Parameters, Image Redundancy and Acquisition Geometry. *Remote Sens.* **2021**, *13*, 1222. [CrossRef]
29. Grottoli, E.; Biausque, M.; Rogers, D.; Jackson, D.W.T.; Cooper, J.A.G. Structure-from-Motion-Derived Digital Surface Models from Historical Aerial Photographs: A New 3D Application for Coastal Dune Monitoring. *Remote Sens.* **2021**, *13*, 95. [CrossRef]
30. Lague, D.; Brodu, N.; Leroux, J. Accurate 3D Comparison of Complex Topography with Terrestrial Laser Scanner: Application to the Rangitikei Canyon (N-Z). *ISPRS J. Photogramm. Remote Sens.* **2013**, *82*, 10–26. [CrossRef]
31. Gómez-Gutiérrez, Á.; Gonçalves, G.R. Surveying Coastal Cliffs Using Two UAV Platforms (Multirotor and Fixed-Wing) and Three Different Approaches for the Estimation of Volumetric Changes. *Int. J. Remote Sens.* **2020**, *41*, 8143–8175. [CrossRef]
32. López, L.; Cellone, F. SfM-MVS and GIS Analysis of Shoreline Changes in a Coastal Wetland, Parque Costero Del Sur Biosphere Reserve, Argentina. *Geocarto Int.* **2022**, 1–17. [CrossRef]
33. Barra, A.; Solari, L.; Béjar-Pizarro, M.; Monserrat, O.; Bianchini, S.; Herrera, G.; Crosetto, M.; Sarro, R.; González-Alonso, E.; Mateos, R.M.; et al. A Methodology to Detect and Update Active Deformation Areas Based on Sentinel-1 SAR Images. *Remote Sens.* **2017**, *9*, 1002. [CrossRef]
34. Devanthéry, N.; Crosetto, M.; Monserrat, O.; Crippa, B.; Mróz, M. Data Analysis Tools for Persistent Scatterer Interferometry Based on Sentinel-1 Data. *Eur. J. Remote Sens.* **2019**, *52* (Suppl. 1), 15–25. [CrossRef]
35. Biescas, E.; Crosetto, M.; Agudo, M.; Monserrat, O.; Crippa, B. Two Radar Interferometric Approaches to Monitor Slow and Fast Land Deformation. *J. Surv. Eng.* **2007**, *133*, 66–71. [CrossRef]
36. Devanthéry, N.; Crosetto, M.; Monserrat, O.; Cuevas-González, M.; Crippa, B. An Approach to Persistent Scatterer Interferometry. *Remote Sens.* **2014**, *6*, 6662–6679. [CrossRef]
37. Casu, F.; Elefante, S.; Imperatore, P.; Zinno, I.; Manunta, M.; De Luca, C.; Lanari, R. SBAS-DInSAR Parallel Processing for Deformation Time-Series Computation. *IEEE J. Sel. Top. Appl. Earth Obs. Remote Sens.* **2014**, *7*, 3285–3296. [CrossRef]
38. De Luca, C.; Cuccu, R.; Elefante, S.; Zinno, I.; Manunta, M.; Casola, V.; Rivolta, G.; Lanari, R.; Casu, F. An On-Demand Web Tool for the Unsupervised Retrieval of Earth's Surface Deformation from SAR Data: The P-SBAS Service within the ESA G-POD Environment. *Remote Sens.* **2015**, *7*, 15630–15650. [CrossRef]
39. Manunta, M.; Bonano, M.; Buonanno, S.; Casu, F.; De Luca, C.; Fusco, A.; Lanari, R.; Manzo, M.; Ojha, C.; Pepe, A.; et al. Unsupervised Parallel SBAS-DInSAR Chain for Massive and Systematic Sentinel-1 Data Processing. *Int. Geosci. Remote Sens. Symp.* **2016**, *2016*, 3890–3893. [CrossRef]
40. Galve, J.P.; Pérez-Peña, J.V.; Azañón, J.M.; Closson, D.; Caló, F.; Reyes-Carmona, C.; Jabaloy, A.; Ruano, P.; Mateos, R.M.; Notti, D.; et al. Evaluation of the SBAS InSAR Service of the European Space Agency's Geohazard Exploitation Platform (GEP). *Remote Sens.* **2017**, *9*, 1291. [CrossRef]
41. Geohazards-TEP. Geohazard Exploitation Platform. Available online: https://geohazards-tep.eu/#! (accessed on 21 September 2021).
42. Ragan, D.M. *Structural Geology: An Introduction to Geometrical Techniques*; John Wiley & Sons: New York, NY, USA, 1968.
43. Ruiz de Argandoña, V.G.; Calleja, L.; Suárez Del Río, L.M.; Rodríguez-Rey, A.; Celorio, C. Durabilidad En Ambientes Húmedos de La Arenisca de La Marina (Formación Lastres, Jurásico Superior de Asturias). *Trab. Geol.* **2005**, *25*, 105–115. [CrossRef]
44. García-Ramos, J.C. El Jurásico de La Costa Centro-Oriental de Asturias. Un Monumento Natural de Alto Interés Patrimonial. *Geo-Temas* **2013**, *14*, 19–29.
45. Varnes, D.J. Slope Movement Types and Processes. In *Landslides: Analysis and Control*; Special Rep., 17; Schuster, R.L., Krizek, R.J., Eds.; The National Academy of Sciences: Washington, DC, USA, 1978; pp. 11–33.
46. Crosetto, M.; Luzi, G.; Monserrat, O.; Barra, A.; Cuevas-González, M.; Palamá, R.; Krishnakumar, V.; Wassie, Y.; Mirmazloumi, S.M.; Espín-López, P.; et al. Deformation Monitoring Using SAR Interferometry and Active and Passive Reflectors. *Int. Arch. Photogramm. Remote Sens. Spat. Inf. Sci. ISPRS Arch.* **2020**, *43*, 287–292. [CrossRef]
47. Luzi, G.; Espín-López, P.F.; Pérez, F.M.; Monserrat, O.; Crosetto, M. A Low-Cost Active Reflector for Interferometric Monitoring Based on Sentinel-1 Sar Images. *Sensors* **2021**, *21*, 2008. [CrossRef]
48. Crosetto, M.; Gili, J.A.; Monserrat, O.; Cuevas-González, M.; Corominas, J.; Serral, D. Interferometric SAR Monitoring of the Vallcebre Landslide (Spain) Using Corner Reflectors. *Nat. Hazards Earth Syst. Sci.* **2013**, *13*, 923–933. [CrossRef]
49. Darvishi, M.; Schlögel, R.; Bruzzone, L.; Cuozzo, G. Integration of PSI, MAI, and Intensity-Based Sub-Pixel Offset Tracking Results for Landslide Monitoring with X-Band Corner Reflectors-Italian Alps (Corvara). *Remote Sens.* **2018**, *10*, 409. [CrossRef]
50. López-Toyos, L.; Domínguez-Cuesta, M.J.; Piñuela, L. Procesos de Gravedad y Hallazgos Paleontológicos En La Costa de Los Dinosaurios (Asturias, N España). *Geogaceta* **2021**, *70*, 15–18.

remote sensing

Article

Spatio-Temporal Heterogeneity and Cumulative Ecological Impacts of Coastal Reclamation in Coastal Waters

Jingfang Lu [1], Xianqing Lv [1] and Honghua Shi [2,3,*]

1 Frontier Science Center for Deep Ocean Multispheres and Earth System (FDOMES) and Physical Oceanography Laboratory, Ocean University of China, Qingdao 266100, China
2 First Institute of Oceanography, Ministry of Natural Resources, Qingdao 266061, China
3 Laoshan Laboratory, Qingdao 266237, China
* Correspondence: shihonghua@fio.org.cn

Abstract: The coastal reclamation, as one of the most extreme transformations of the ocean space by humans, still lacks scientific quantitative evaluating methods to a large extent, compared with the evolution of land use patterns. A cumulative ecological impacts of reclamation (RCEI) was established in our study based on ecological influence characteristics of different reclamation types, and the attenuation effect of reclamation on adjacent areas. It was characterized by spatio-temporal features in decades. Here, we estimated that the cumulative reclamation area in the Bohai Sea from 1985 to 2018 was 5839.5 km^2. Under the influence of human activity, proportions of the industrial and urban boundary, marine construction boundaries (e.g., ports, wharves, and bridges), and protective dams were increased significantly, which led to a sharp increase of the RCEI. In addition, spatio-temporal changes of reclamation were affected by the combination of population growth, economic development, urbanization, industrialization, and marine industry development in coastal cities. These results provided an important historical reference for tracking future development of the Bohai Sea by humans and provided basic data support for the development and protection of the ocean.

Keywords: coastal water; movable water bodies boundary (MWBB); cumulative ecological impact; reclamation; driving factor; spatio-temporal heterogeneity

Citation: Lu, J.; Lv, X.; Shi, H. Spatio-Temporal Heterogeneity and Cumulative Ecological Impacts of Coastal Reclamation in Coastal Waters. *Remote Sens.* **2023**, *15*, 1495. https://doi.org/10.3390/rs15061495

Academic Editors: Andrew Clive Banks, Ramón Blanco Chao, Germán Flor-Blanco and José Juan de Sanjosé Blasco

Received: 17 December 2022
Revised: 27 February 2023
Accepted: 6 March 2023
Published: 8 March 2023

1. Introduction

The Circum-Bohai-Sea region has one of the densest populations and the highest degrees of urbanization and industrialization in China. At the end of 2019, the Circum-Bohai-Sea region had a total population of 0.25 billion, accounting for 17.1% of the population of China, with an annual GDP of 1.8×10^5 billion yuan, 20.8% of the GDP of China [1]. Rapid economic expansion and accelerated population growth in the Bohai Sea have caused a sharp increase in land reclamation [2,3]. Between the 1950s and 1980s, reclamation of cultivated land in succession for the coastal area was focused on agricultural development and food production. Between the 1980s and 2000s, coastal reclamation was aimed at increasing salt and agricultural production. In recent decades, driven by the fast-growing economy from 2000 to 2018, Bohai Rim city aggressively expanded coastal industries and urbanization, which resulted in an increasing demand for reclamation. Along with the rapid development of the Bohai Sea, the high-intensity reclamation activities have already caused severe challenges to its ecological environment and the ecosystem service functions, especially the widely distributed and typical coastal wetlands of the Bohai Sea [4]. According to the 2018 Bulletin of China's Marine Ecological Environment, Ministry of Ecology and Environment of the People's Republic of China (2019), the Bohai Sea—as a part of the Western Pacific Ocean and China's inland sea, with land from three directions, surrounded by Liaoning Province, Hebei Province, Shandong Province, and Tianjin City, and connected with the Yellow Sea through the Bohai Strait—is the sea area with the most prominent marine ecological problems, e.g., pollutant emission, oil exploitation, and coastal reclamation in northern China [5,6]. The ecological degradation problems of the Bohai Sea are mainly within its coastal waters, especially those near the Liaodong Bay, Bohai Bay, and Laizhou Bay [7,8]. Therefore, since the "Action Plan for the Integrated Management of the Bohai Sea" jointly issued by the Ministry of Ecology and Environment, the National Development and Reform Commission, and the Ministry of Natural Resources in November 2018, the most stringent management and control of sea reclamation has been implemented in the Bohai Sea area in order to continually improve its coastal ecological functions and gradually restore its fishery resources.

In order to protect the fragile coastal ecosystem that has been severely affected by human activities and global climate change, recently, the assessment on the spatio-temporal evolution of coastal reclamation and its impacts on marine ecosystem has attracted the attention of marine management departments and has been studied from different perspectives in different regions. For example, Chu et al. [9] investigated how reclamation activities altered the coastal morphology and hydro-oceanography, resulting in the deterioration of the tidal flat in Lingding Bay. Bi et al. [10] showed that reclamation changed the natural properties of sea areas and coastlines with varying degrees of impacts on the health of coastal and marine ecosystems, including ocean hydrodynamic conditions, water quality, sediment environment, noise, food chain, biome structure, and diversity. However, the existing research on reclamation has focused on descriptive statistics and habitats affected (tidal flats, coastal sand dunes, freshwater bodies, sand reclaimed from the seabed, seagrass meadows, rocky reefs, etc.) [11], always considering the entire coastal regions as the research unit; thus, little attention has been paid to the spatial and temporal variation in ecological impact factors.

Research on reclamation change has employed a range of methods and approaches, including remote sensing (RS), geographic information system (GIS) mapping, and field surveys. For example, the remote sensing has been widely used to quantify coastal zone changes, and fine-resolution orbital sensors (e.g., Landsat and Sentinel) are very advantageous [12]. Most scientists use remote sensing images to extract coastlines (the boundary between land and sea) to simulate the coastal erosion-siltation process and the dynamic process of human reclamation development, which can then reveal the effects of the ecological and economic impacts of coastal reclamation [13,14]. Shen et al. [15] established an indicator system to investigate the cumulative impact of reclamation, so that the strategies of ecosystem-based management can be applied in the Bohai Sea of China. Nevertheless, research on cumulative impact of coastal reclamation characterized by spatio-temporal features in decades remain insufficient. The current study provides an integrated approach of the GIS technique and remotely sensed satellites data images for the objectives of coastal topography analysis and ecological impacts of coastal reclamation assessment. The spatial distribution characteristics of coastal reclamation were revealed, and their cumulative impacts on the ecological were investigated.

In this paper, studies were focused on the development process, cumulative ecological effects, and driving mechanism of reclamation in near coastal waters, and the conceptual framework is shown in Figure 1. The goals were: (1) to establish a method to define and identify the boundary of movable water bodies (MWBB) in coastal waters; (2) to draw a diagram of the spatio-temporal changes of near coastal reclamation in typical coastal waters; (3) to reveal the characteristics of the spatio-temporal changes of ecology with the coastal reclamation; and (4) to identify the driving forces of reclamation.

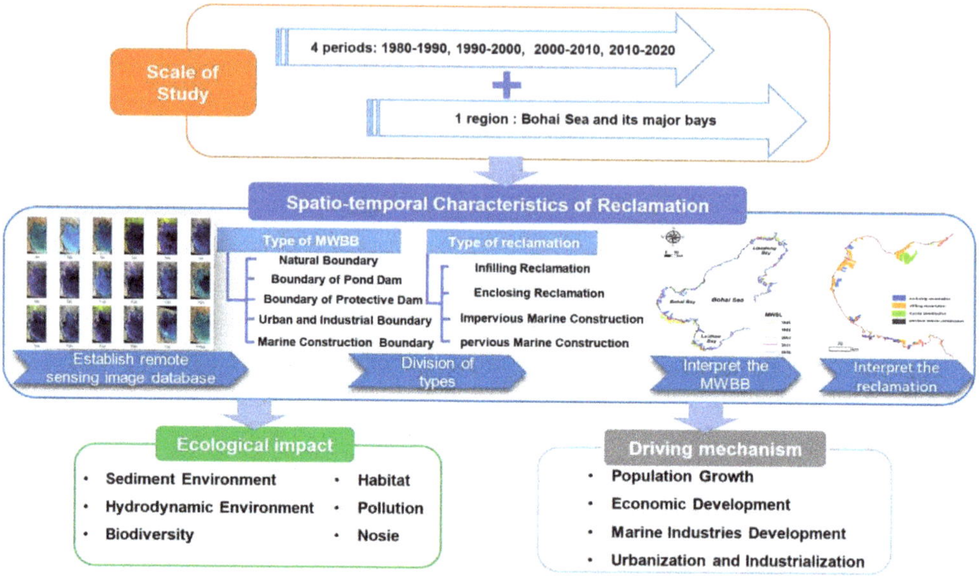

Figure 1. Conceptual framework of study.

2. Materials and Methods

2.1. Study Area

The Bohai Sea is a near-closed continental shelf in the Northwest Pacific between 37–41°N and 117–122°E and is separated from the Yellow Sea with a line between Laotieshanjiao on the Liaodong Peninsula and Penglaijiao on the Shandong Peninsula. Its sea area is approximately 77 × 10^3 km^2, with a coastline of close to 3800 km. It has several bays, including Liaodong Bay (the north of 40°N), Bohai Bay (the west of 118°E), and Laizhou Bay (the south of 37°N), as well as Central Bohai Sea and Bohai Strait. There are 13 cities along its coastline, namely Dalian, Yingkou, Panjin, Jinzhou, Huludao, Qinhuangdao, Tangshan, Cangzhou, Tianjin, Binzhou, Dongying, Weifang, and Yantai (Figure 2).

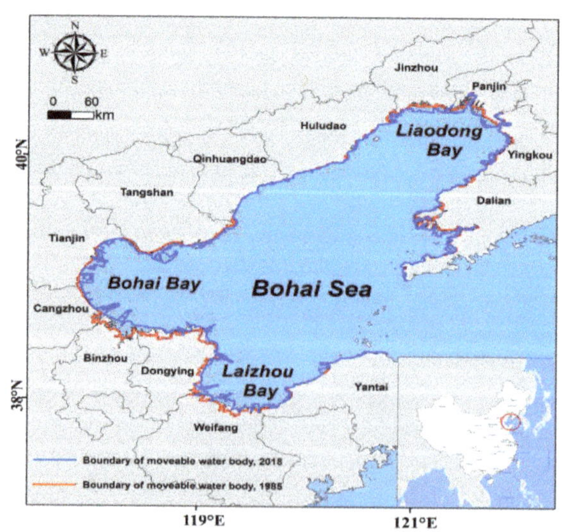

Figure 2. Location and change in boundary of movable water bodies of the study area.

2.2. Data Source

The data used in this study were obtained mainly from the following four resources: remote sensing images from the Landsat 5 satellite in June 1985, May 1994, and July 2003, as well as the Landsat 8 satellite in May 2011 and June 2018, with no or little cloud cover, were used to assess changes in the Bohai Sea's coastal reclamation (Table 1); the vector data, including data extracted from the geographic information system (the length of MWBB, reclamation area, and its utilization types); the statistical data from the China Marine Statistical Yearbook (1995–2018), the statistical yearbooks of Liaoning, Tianjin, Hebei, and Shandong (1998–2018) [16–19]; and Google Earth Pro 7.3 data as a geographic reference for correcting marine constructions such as ports, breakwaters, and cross-sea bridges.

Table 1. List of land satellite image data of Bohai Sea.

Region	Year	Image Sensor	Path/Row	
Bohai Sea	1985, 1994, 2003, 1998–2018	Landsat5 TM, Landsat8 OLI.	120/034, 122/033, 121/032, 120/033,	121/034, 121/033, 120/032, 119/033

2.3. Defining and Identifying Method of the MWBB

2.3.1. A New Definition of the 'Coastline'

The coastline is the boundary between the sea and the land, namely the outline of a coast [20]. In this study, the MWBB was defined as the solid boundary (for example, breakwaters, bridges, and artificial reefs) blocking the movement of the sea and destroying its connectivity at the intersection of land and sea. The main difference between coastline and MWBB is in their natural attributes. The natural attributes of land and sea, which are emphasized by traditional coastlines, are important indicators of ocean management. For example, in the sea surrounding the areas used for mariculture, the sea is still the sea, but it has changed from a naturally moving water body to a basically non-moving one, indicating that its physical and chemical properties have not changed, but its movement characteristics have changed. However, the hydrodynamics, as an important factor affecting the ecological health of the coastal water, is not reflected in the definition of traditional coastline. Therefore, studies based on traditional coastline changes cannot reveal the impact of human activities in the coastal water, which makes ocean management based on ecosystems impossible to implement. On the contrary, with the method of defining and extracting the spatial position of the MWBB in the coastal water, it takes into account whether the sea area reclamation has hydrodynamic conditions and seawater exchange capacity as the foundation to reveal the ecological effect and development of the reclamation and allows for the study of the marine ecological and environmental protection. In this paper, according to Suo et al.'s coastline extraction method, the spatial positions of the two lines were described under the same remote sensing image, and the two lines showed significant spatial heterogeneity, as in Laizhou Bay (from the Yellow River Mouth to Qimujiao), shown in Figure 3 [21].

2.3.2. The Method for Identifying the MWBB

On the basis of the knowledge of the morphological features, vegetation types, and the development and utilization status, we established the relationships between the remote sensing image shape, size, color or tone, shadow, location, structure, texture, and other features and the corresponding interpretation types. Then, the MWBB types were classified as bedrock, sandy, sand–powder silt, lagoon, estuary, pond, industrial and urban land, port, or marine construction through visual interpretation of the remote sensing images. The details are shown in Table 2.

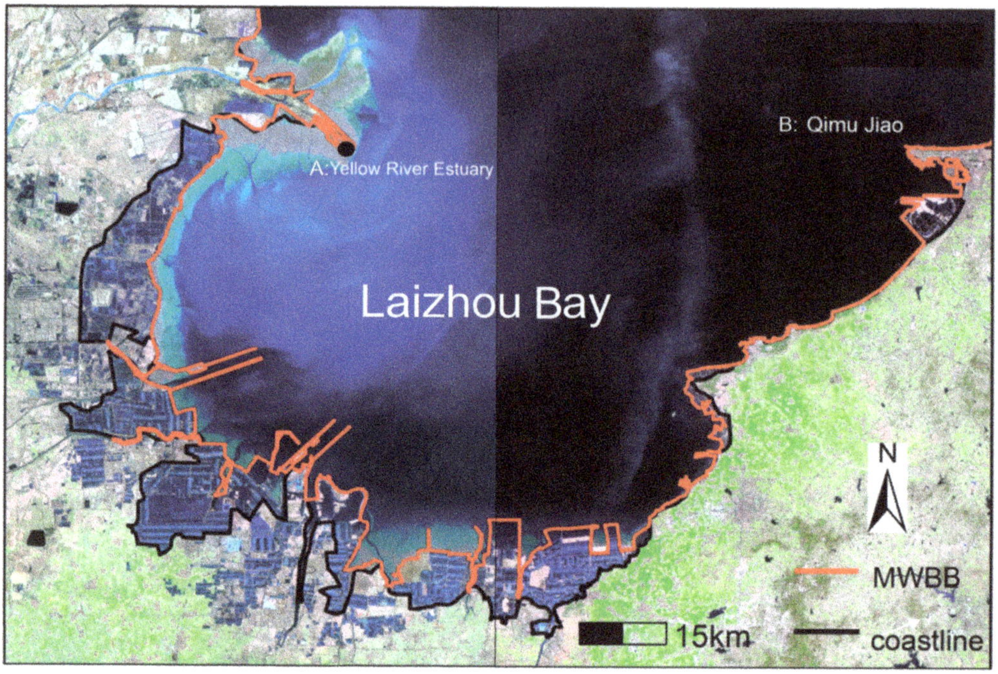

Figure 3. Comparison of positions of the MWBB and coastline in Laizhou Bay (from the Yellow River Mouth to Qimujiao).

Table 2. Boundary defining methods of MWBB with different utilization types.

<table>
<tr><th colspan="2">Boundary Types of Movable Water Bodies</th><th>Spatial Location</th><th>Remote Sensing Information (Fusion of Bands 3, 4, and 5)</th></tr>
<tr><td colspan="2">Natural boundary (including bedrock boundary and sandy boundary)</td><td>Trace line of direct meeting of land and water</td><td>Bedrock boundary: The remote sensing images have a high reflectivity white color tone and are surrounded by vegetation [21].
Sandy boundary: The reflection on remote sensing images of sandy gravel bank and beach, which are formed over time, is higher than that of other features, and their textures are white and uniform [22].</td></tr>
<tr><td rowspan="2">Reclamation boundary</td><td>Boundary of the pond dam</td><td>Boundary of the pond dam</td><td>If there are engineering facilities that protect salt pans, aquaculture ponds, and reservoirs, the aquaculture pond is dark blue on the remote sensing image, with uniform color, clear texture, flaky or striped distribution [22]; the salt field is uneven bright blue, with a clear texture, flaky distribution, and vegetation coverage; the boundary of reservoir is clear, with a gray dam.</td></tr>
<tr><td>Boundary of protective dam</td><td>Contour line of the dam on the sea side</td><td>If there are engineering facilities and linear dams that protect the reclaimed land area, the land reclamation has uneven color and clear internal textures; the linear dams have clear and gray textures, with generally small widths, and are distributed in places where water and land meet directly.</td></tr>
</table>

Table 2. *Cont.*

Boundary Types of Movable Water Bodies		Spatial Location	Remote Sensing Information (Fusion of Bands 3, 4, and 5)
	Industrial and urban boundary	Contour line on the sea side	The boundaries are clear, including urban residential areas, industrial areas, and roads, with clear internal texture and uneven color, which is easy to identify on remote sensing images.
	Marine construction boundary	Extracted contour line on the sea side if there is a port wharf or an impervious offshore structure	The boundaries are clear and easy to identify, with ports and wharves, impervious offshore structures (breakwaters and submarine tunnels), and permeable marine structures (cross-sea bridges) included, which are distributed mainly near urban and industrial areas; offshore structures have blurred, gray textures that are mostly in strips and perpendicular to the land. On the basis of the projected images of the offshore buildings from Google Maps and the continuity of the water waves on both sides, the water permeability is identified.

2.4. Ecological Cumulative Impact Assessment Model of Reclamation

2.4.1. Assessing System

Local extent and effects of coastal reclamation projects are increasingly reported, which supports our judgment of the strength of the impact intensity and looks more holistically at impacts of coastal reclamation on spatio-temporal scales through large-scale survey work to interpret these impacts (Table 3). Based on the supporting data of Table 3, the reclamation can change the original seabed topography, affect the tidal capacity of the bay, and alter sediment dynamics [23]. If habitat is destroyed, biodiversity decreases. The same holds true for pollution. For example, a port area would collect a large amount of oily wastewater discharge from ships docking in it. These pollutants will aggravate the seawater eutrophication, reducing biodiversity. In this study, according to the 6 aspects of reclamation effects on the ecosystem of the near coastal water (sedimentation environment, hydrodynamic conditions, biodiversity, habitat, pollution, and noise), a systematical assessment system was established (Figure 4).

In this system, the impact intensity and attenuation distance were used to reflect the difference in the impact of different reclamation types. If the ecological impact was great, the impact intensity was set to 1. If the ecological impact was moderate, the impact intensity was set to 0.5. The impact intensity with limited ecological impact was set to 0 [24]. The ecological impact assessment grade is divided using the equal interval breakpoint method. The values '0', '0.5', and '1' represent high, medium, and low impact, respectively.

The reclamation through infilling, with an engineering process similar to that of impervious marine construction, will change the natural properties of the ocean significantly, such as changing the sedimentary environment and hydrodynamic conditions, encroaching on the biological habitats, destroying the marine biodiversity, and generating noise and pollutants during the construction and development. However, the area of impervious marine construction is usually very small, and it has a high impact only on the sedimentary environment through noise pollution [15,25]. The enclosing reclamation and pervious marine constructions (dams or buildings in the reclaimed area) can partially change the natural properties of the ocean, and the enclosing reclamation has a high impact on the hydrodynamic conditions, marine biodiversity, and pollutant emission. Most pervious marine constructions are cross-sea bridges. During their construction and usage, pollutants can be produced which threaten the biodiversity [5]. Additionally, noise pollution caused by shipping activity in ports and marinas showed the greatest extent of modification. Compared with ports, the noise pollution caused mainly by vehicles on the road, bridges, industries, and urban areas, is relatively weak compared with ports and marinas [26].

Therefore, in this study, the cumulative ecological impact values of infilling reclamation, enclosing reclamation, impervious marine construction, and pervious marine construction were set as 6.0, 4.5, 4.0, and 1.4, respectively (Figure 4).

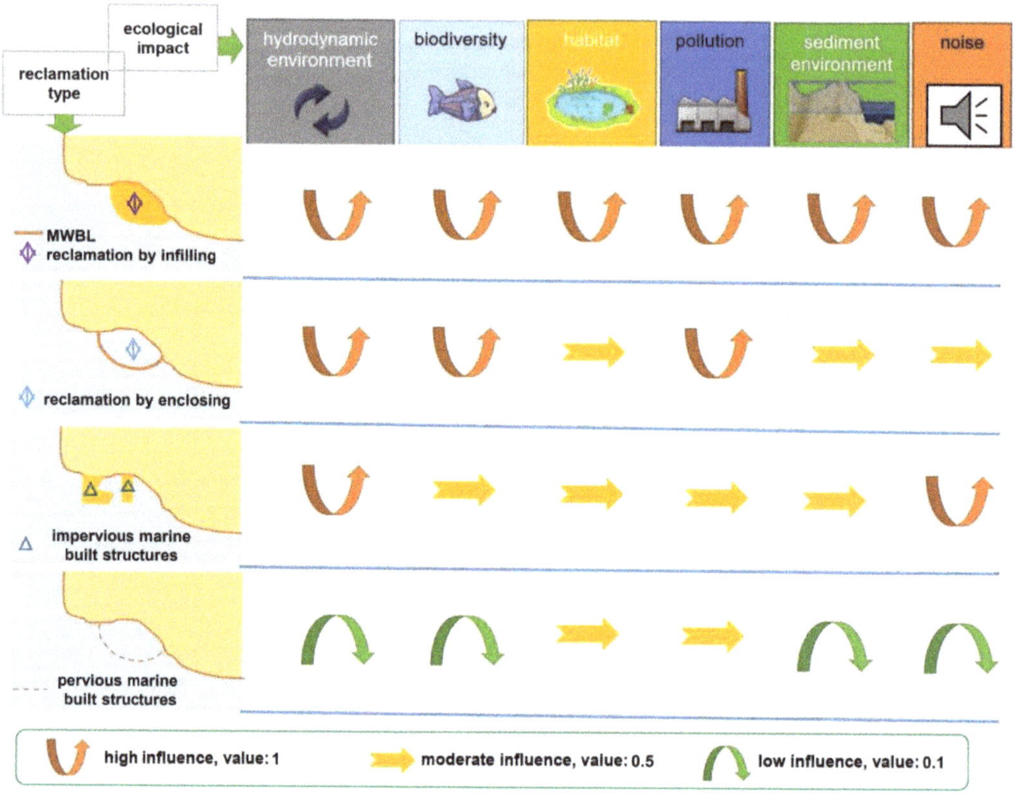

Figure 4. Assessment system of impacts of different reclamation types on the coastal ecosystem.

Table 3. Supporting data for the reclamation impact intensity on the ecosystem.

Reclamation Types	Evaluation and Survey Work of the Researcher
Infilling reclamation (e.g., industrial and urban area)	1. Shifts in the spatial extent, configuration, and dynamics of natural habitats by altering sediment dynamics and geomorphic connectivity [26]. 2. Permanent change of the geomorphology of the coastal line and the physical processes of the coastal system [27]. 3. Production of pollutants from ports and factories as well as nutrients and pesticides from croplands, resulting in deterioration of inshore and oceanic environments and severely reduced biodiversity [11]. 4. Habitat loss and sediment burial [28]. 5. Sediment dynamics and hydrodynamics were weakened [15]. 6. Destruction of habitats for fish; reduced water purification ability from narrowing and even disappearance of gulf and bay area, increased water pollution, and frequent harmful algal blooms [29]. 7. Shrinking of the tidal mudfats; loss of marine habitat; reduction of the adaptive capacity of shoreline ecological communities to sea level rise [10].

Table 3. *Cont.*

Reclamation Types	Evaluation and Survey Work of the Researcher
Enclosing reclamation (e.g., pond)	8. Same as article 6. 9. Same as article 7. 10. Production of pollutants from ponds and chemicals, such as antibiotics, resulting in deterioration of inshore and oceanic environments [29]. 11. Modification of flows of energy and materials [30]. 12. In mariculture ponds, large number of residual bait, dead fish, and shrimp would sink to the bottom of the ponds, and high-concentration organic wastes would be discharged into the surrounding water bodies [31].
Marine construction (e.g., harbor, seawall, bridge)	13. Same as article 4. 14. Port area would collect discharge large amount of oily wastewater from ships docking in [32]. 15. Noise pollution caused by shipping activity in ports and marinas showed the greatest extent of modification, affecting seascapes at regional scales [32]. 16. Physical barriers to the movement of organisms, materials, and/or energy within and among habitats [31]. 17. The hydrodynamic is significantly modified by a bridge piling [32].

2.4.2. Spatial Mapping

Previously, it was reported that coastal reclamation and development affected the occupied and the adjacent areas, with a gradual attenuation of the RCEI when the distance from the reclamation area was increased. In this study, an inverted "S"-shaped attenuation function was used, and the attenuation effect was applied to the spatial mapping [33].

Typically, steps in the mapping were: with human–computer interaction recognition as the method and the interpretation signs as reference, the different utilizations of coastal reclamation was interpreted to obtain the vector data and to determine the size of the unit according to the pixel size and the area of the evaluation area. In this study, the remote sensing data in pixel was 30 m $\times$ 30 m, the area of the Bohai Sea in 2018 was 82,404 km^2, and the unit size should meet the demand of spatial heterogeneity with sufficient area and pixel number. With the "Fishnet" tool in ArcGIS, a 3 km $\times$ 3 km cell grid was generated; thus, the Bohai Sea had 9156 cell points, and the cell size met the demand for spatial heterogeneity [34].

With the "Euclidean Distance", the distances from the 4 types of reclamation utilization to all unit points on the adjacent sea area were determined. The influence of the attenuation distance was determined according to the established attenuation formula (Formulas (1) and (2))—a theoretical model in which the ecological impact exhibits an inverted "S"-shaped attenuation as the distance increases. The attenuation effect of the RCEI on coastal waters is expressed by (x, y), and the spatial mapping is carried out. Additionally, the (x, y) was calculated by simple averaging of each of the 4 reclamation utilization's RCEI and calculated following the formula below [35].

$$f_k(x, y) = \frac{P_k}{1 + D_k(x, y)/\mathrm{w}} \tag{1}$$

$$I(x, y) = 0.25 \sum_{k=1}^{4} f_k(x, y) \tag{2}$$

where P_k is the upper limit reference value of ecological impact of the k-th type of reclamation. $D_k(x, y)$ represents the Euclidean distance from any point (x, y) in the sea to the area where the k-th reclamation type is located, w is the half-attenuation coefficient that is set to 20 km uniformly after repeated attempts, and $I(x, y)$ is the cumulative ecological impact value.

In order to realize the spatial heterogeneity expression of the RCEI, "Kriging" was used to spatially interpolate the cumulative ecological impact values of all unit points, and the cumulative ecological impact grade (I) was divided according to the equal interval breakpoint method, including low impact ($0 \leq I < 2$), moderate impact ($2 \leq I < 4$), and high impact ($I \geq 4$).

2.5. Analysis of Driving Factors of Reclamation

It is well known that the Structural Equation Modeling (SEM) is a widely used statistical analysis model. In this model, through the analyses on direct and indirect relationships between/among variables and on variables that cannot be directly measured (latent variables), the interaction strength of multiple variables and the influence transmission path can be better determined, which can effectively overcome the shortcomings of traditional statistical methods. Therefore, this model has significant application importance in biology, ecology, social sciences, education, economics, management, and other fields [36]. Meanwhile, Path Analysis (PA) is one of the main SEM models that is still currently in use. As an application model of SEM without latent variables, it tests the significance of the relationship between/among variables, and the positive and negative signs of Pearson's coefficient (r) represent the positive and negative correlations, respectively.

In the past four decades, the need for economic development and agricultural requirements have been the primary drivers cited for coastal reclamation projects, such as urban construction, port construction, oil extraction, fishery production, and saltern reclamation, in many of the coastal cities in the Bohai Sea [3,22]. Additionally, the rising human population in the coastal cities is also driving coastal expansion [37]. On the basis of the preceding description, the PA was used to construct a complex relationship between the cumulative reclamation area and the driving factors, including economy, population, marine industry, and urban and industrialization degree. The expected model of causal factors upfront is shown in Figure 5. The economic development is composed of two variables: GDP of primary industry and GDP of secondary industry. Population growth is composed of two variables: population growth rate and urbanization degree. Economic development and population growth on accelerating coastal reclamation were regarded as the most relevant indicators of accelerating coastal reclamation [37]. Given the more local land use of reclamation in the present stage, some of the reclaimed areas are used for mariculture pond, but most of them are used for ports and industrial areas, so the cargo throughput of major ports and the mariculture production were elected as access to provide for the relationship between marine industry development and reclamation [33].

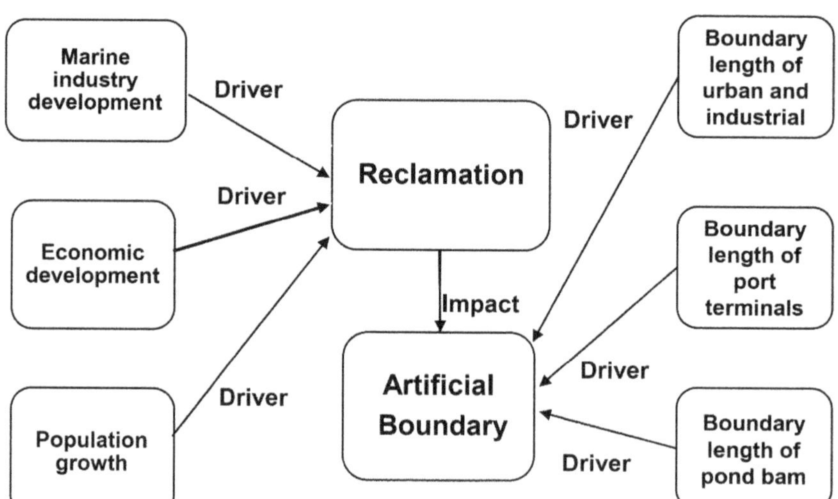

Figure 5. The expected model of causal factors upfront.

3. Results

3.1. Spatio-Temporal Characteristics of Reclamation

3.1.1. Changes in the MWBB in Near Coastal Waters

During the past 40 years, the reclaiming of the Bohai Sea has already resulted in significant spatio-temporal changes of MWBB. Marine constructions, such as ports and wharves, are bulging along the coast and extending rapidly toward the ocean, and the boundaries tend to be complex. This phenomenon is particularly obvious in Laizhou Bay, Bohai Bay, and Liaodong Bay (Figure 6).

As shown in Figure 7, in the 1980s, the MWBB was mainly natural landforms, such as bedrock, beaches, and estuaries, accounting for 74.5% of the total length of the Bohai Sea (1985). Marine aquaculture was the main development activity near the Bohai Sea, and its pond dams accounted for 21% of its boundary. In the 1990s, the types of Bohai Sea boundaries began to diversify. The length and proportion of the urban and industrial boundary and protective dams increased significantly, and the increase in the pond dams was the most significant, which jumped from 21% to 44%. Since the 21st century, the urbanization and industrialization in the coastal areas of the Bohai Sea has been accelerated, leading to a decline in the ratio of natural boundaries of the Bohai Sea to 27.4% in 2018. The proportions of urban and industrial, marine construction, and protective dam boundaries were increased, respectively, from 4.1%, 5.2%, and 2.8% in 2003 to 10.4%, 14.7%, and 12.6% in 2018.

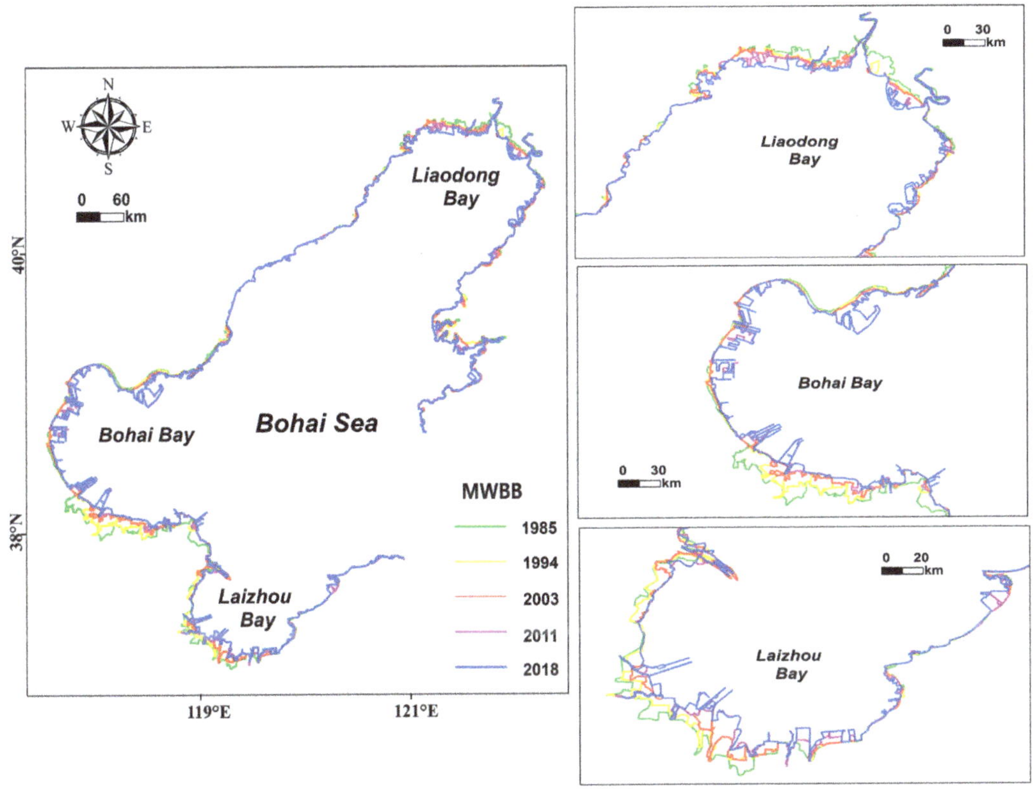

Figure 6. Spatio-temporal changes of the MWBB in the Bohai Sea.

3.1.2. Changes in Reclamation

Four development periods of Bohai Sea reclamation are shown in Figure 8a. The first period (1985–1994) was the rise of large-scale reclamation activities in the aquaculture and salt industries, which was most significant in Bohai Bay and Laizhou Bay. In the second period (1994–2003), reclamation activities were converged to some degree, because the reclamation in the first period occupied many large and small bays, and its "cutting and straightening" of the MWBB drove the limited development space close to its saturation. The third period (2003–2011) was the peak of reclamation activities. Coastal projects such as ports, wharves, and cross-sea bridges bulged along the coast, and large-scale industrial areas from reclamation settled down and were scattered in the Bohai Sea. The fourth period (2011–2018) was one of strict management and control of reclamation, and the area of newly built offshore construction was reduced. For example, according to statistics, from 1985 to 2018, the cumulative increase of reclamation area in the Bohai Sea was 5839.5 km^2, which was equivalent to half the area of Tianjin, the continuous growth rate was 3.3% per year, and 55% of the Bohai Sea reclamation could be attributed to aquaculture field and salt field. From 2003 to 2018, the area of impervious marine constructions jumped from 0.9% to 9.2%, especially for the port and wharf, which accounted for 5.8% (Figure 8b).

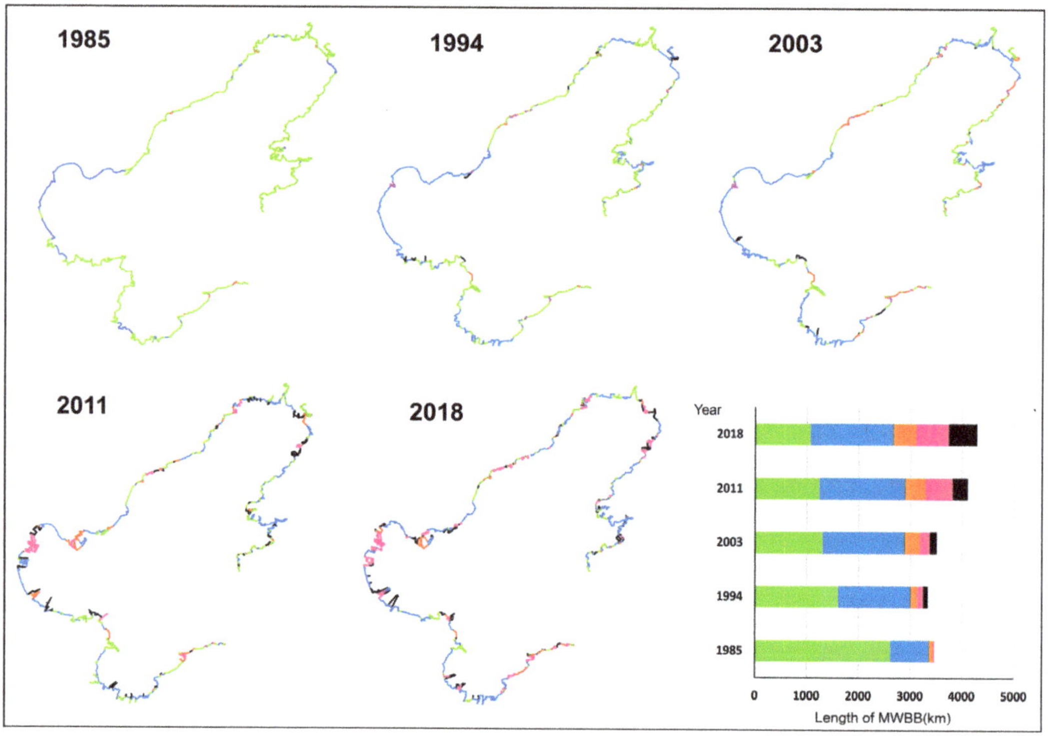

Figure 7. Spatio-temporal changes of the boundary types of the Bohai Sea in Phase 5. Green, blue, orange, red, and black denote the natural boundary, pond dam, industrial and urban area, marine construction, and protective dam, respectively.

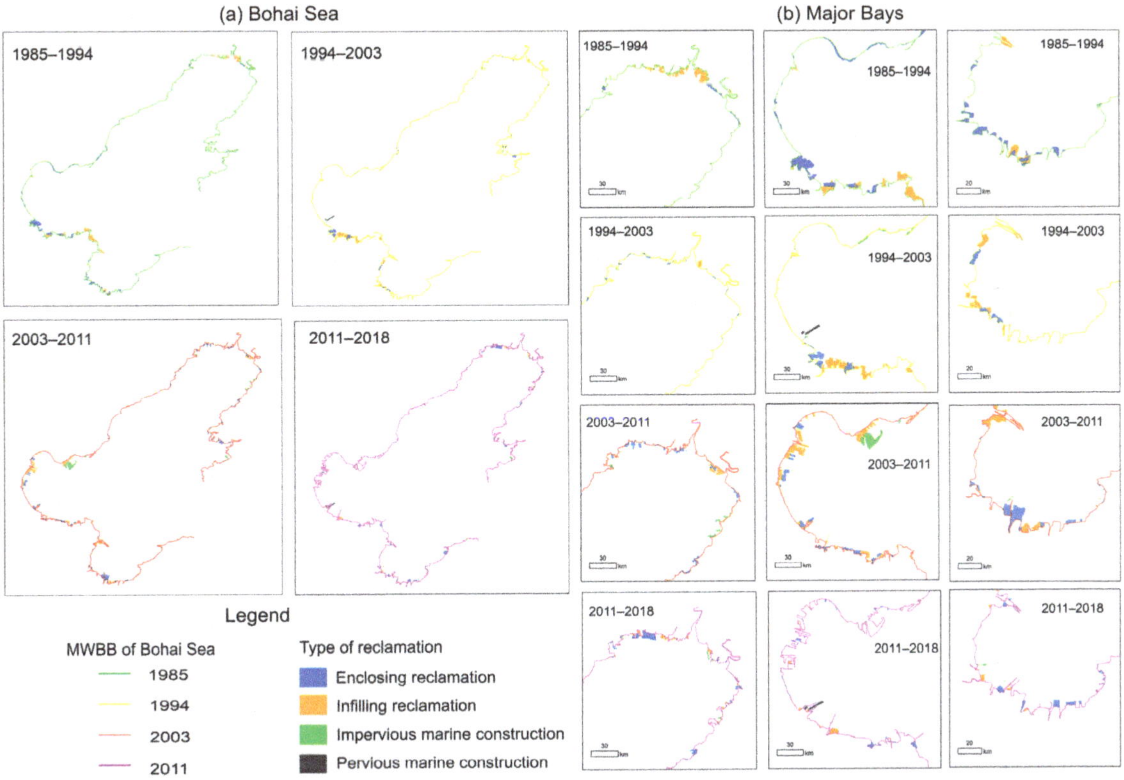

Figure 8. The development process of reclamation in the Bohai Sea.

3.2. Assessment on the RCEI

3.2.1. Evaluation Results

The RCEI scores in the Bohai Sea in 1994, 2003, 2011, and 2018 were 1.3, 1.6, 2.0, and 2.2, respectively, which exhibited a continual increase. The ecological impact level changed from low impact in 1994 and 2003 to moderate impact in 2011 and 2018. The spatial distributions of RCEI changed greatly from 1985 to 2018. In 1985, the proportions of the moderate and high impact zones were all lower than 6% and were distributed mainly in the coastal water of the Laizhou Bay and Bohai Bay. In 2018, the moderate impact zone surrounded the high impact zone, with a proportion of 22.3%. According to RCEI's contribution and the proportion of reclaimed area caused by various utilization methods in the Bohai Sea from 1985 to 2018 (Figure 9b), it was found that the completed impervious marine constructions (mainly ports and seawalls) and pervious marine constructions (cross-sea bridges) accounted for only 9.1% and 0.9% of the total reclaimed area within the Bohai Sea, whereas they had contributions of 24.4% and 4.3% of their RCEI; therefore, they were the main factors that caused the Bohai Sea's RCEI to jump from a low impact (1.3) to a moderate impact (2.2) in 2018.

3.2.2. Evaluation and Verification Results

In this study, the RCEI assessment results of the studied area were compared with the distribution diagram of the water quality status of the seas under China's jurisdiction from the China Ecological and Environmental Status Bulletin and the global marine health score published by Halpern et al. [38] over the same period. According to the 2018 China Marine Ecological Environment Bulletin, the area in the Bohai Sea that did not meet the first-class seawater quality standard in 2019 was 21.6×10^3 km^2, with an equivalent reclamation area of 17.4×10^3 km^2 having a high ecological cumulative impact, indicating that reclamation and pollution were the two most important ecological pressures in the Bohai Sea. According to the comparison of spatial distribution, the significantly affected areas in the Bohai Sea in 2018 and the areas with inferior Grade IV water quality distribution under China's jurisdiction in the summer of 2018 showed that both of them were concentrated in the main bays (Laizhou Bay, Bohai Bay, and Liaodong Bay). However, their spatial locations were heterogeneous. The significantly affected areas were concentrated in reclamation areas and ports, and the areas with inferior Grade IV water quality were concentrated in the sea entrance [12,19]. According to the global ocean health distribution map from Halpern in 2012, the scores of the Bohai Sea were relatively low at 30–40. Because the lower score means worse ocean health, the ocean health in the studied area was of low quality. With the method established in this study, the calculated cumulative ecological impact values of reclamation in the Bohai Sea from 1985 to 2011 was 47.9. Because the higher score indicates a larger marine ecological impact, these results showed that the ecological health of the Bohai Sea was of low quality.

3.3. *Analysis on the Driving Mechanism of Reclamation*

According to the PC-based analysis model of driving factors of reclamation (Figure 10), in Bohai Bay (the coastal water of Tianjin, Tangshan, and Cangzhou), the GDP of secondary industry, its cargo throughput of major ports, the boundary length of urban and industrial, and the boundary length of pond all showed a positive influence relationship with the cumulative reclamation area, and their path coefficients were 0.616, 0.282, 0.742, and 0.324, respectively. In the coastal waters of Laizhou Bay, the GDP of secondary industry, its urbanization degree, and the boundary length of ports terminals all showed a positive relationship with the cumulative reclamation area as well. Their path coefficients were 0.237, 0.279, and 0.673, respectively. These findings indicated that the spatio-temporal changes in Bohai Bay reclamation were caused by the combined effects of economic development, and marine industry development, whereas the spatio-temporal changes in Laizhou Bay reclamation were caused by the combined effects of economic development, urbanization, and marine industry development.

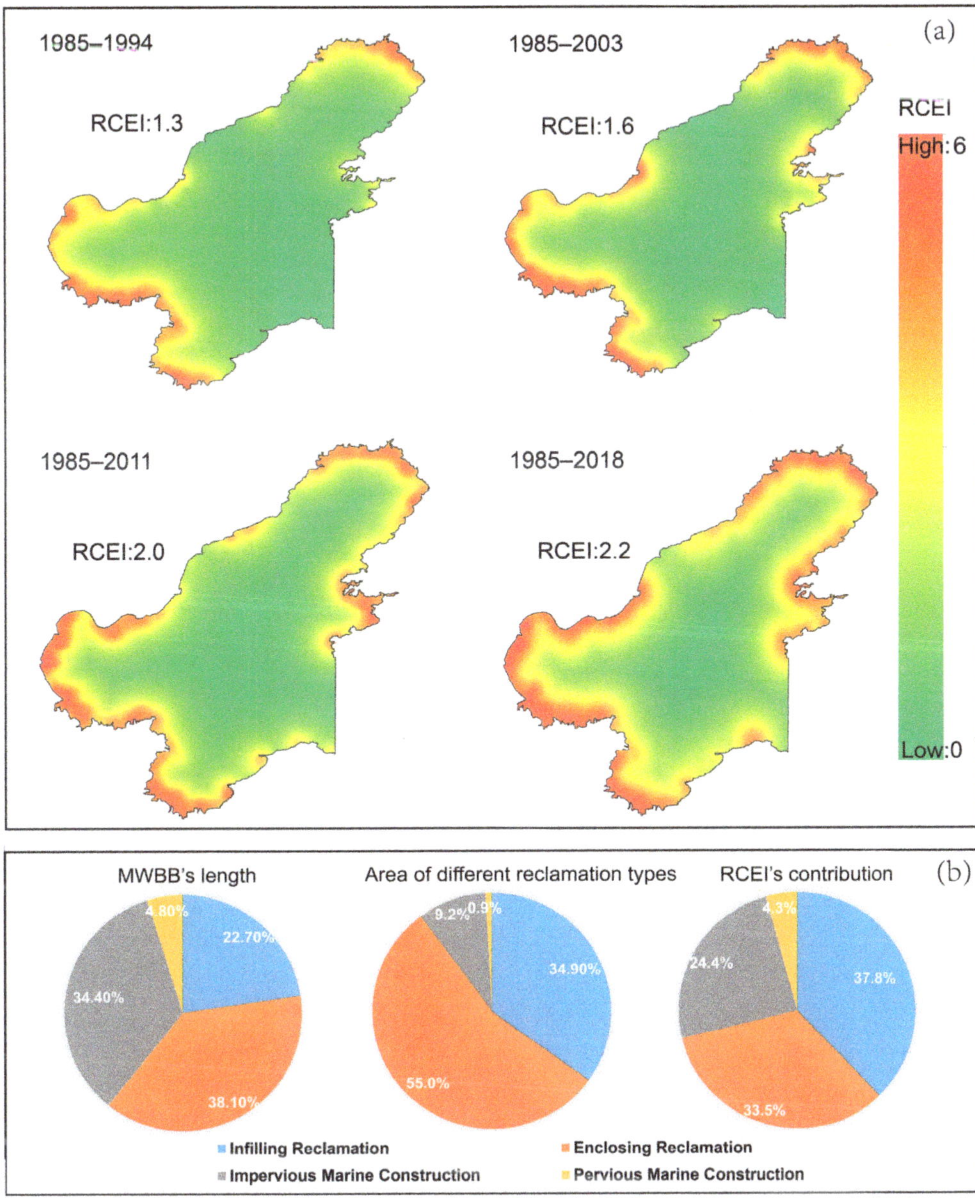

Figure 9. Spatio-temporal changes of the RCEI in the Bohai Sea (**a**) and the ratio of MWBB's length, RCEI' s contribution, and the area of different reclamation types (**b**).

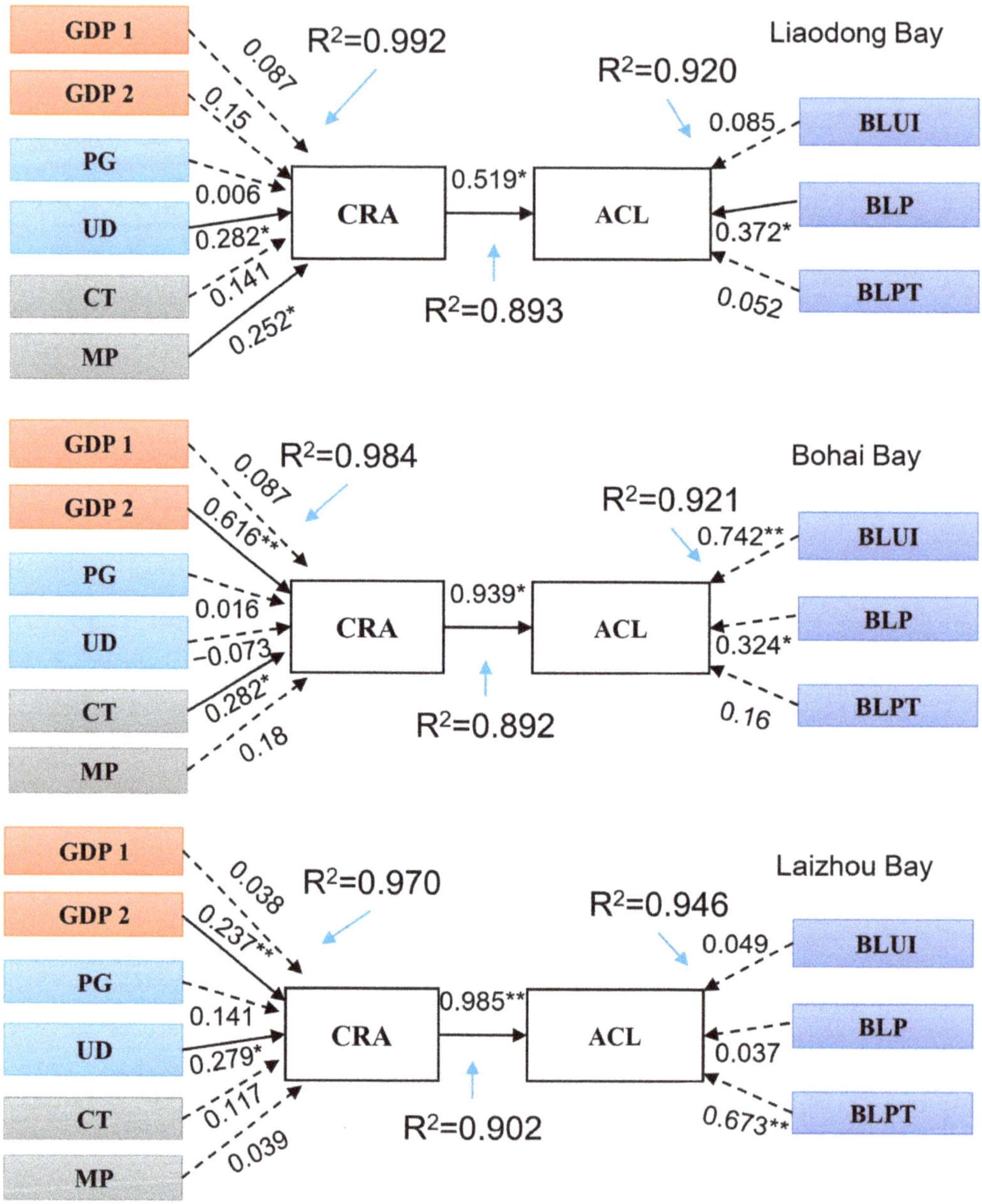

Figure 10. PC-based analysis model of driving factors of reclamation. Numbers show the standardized coefficients. Solid line indicate that the effect is significant, dotted line indicate that the effect is insignificant. The R^2 values for all the equations were high and ranged between 0.892 and 0.992, which indicate good performance. (Note: GDP1—GDP of primary industry, GDP2—GDP of secondary industry, PG—population growth rate, UD—urbanization degree, CT—cargo throughput of major ports, MP—mariculture production, BLUI—the boundary length of urban and industrial, BLPT—the boundary length of port terminals, and BLP—the boundary length of pond). * $p < 0.05$, ** $p < 0.01$.

4. Discussion

4.1. Uncertainty of RCEI

4.1.1. The Applicability and Limitation of Model

With the established model, the RCEI can be simulated with the consideration of the historical cumulative effects of reclamation on the coastal ecosystem and the influence degree of different reclamation utilization types. Through the attenuation function, the spatial heterogeneity was revealed to show the ecological impact of human activities on different locations in the coastal water. As the distance increases, the cumulative impact of reclamation development on the marine ecology gradually decreased, which was the marginal effect of the cumulative impact.

RCEI was established in our study. The features of RCEI are as follows: fixed values of different reclamation types were determined by their ecological influence characteristics (Figure 4); an inverted "S"-shaped attenuation effect was considered to represent the ecological impact of reclamation area on adjacent areas. Because the connectivity of marine ecosystems is much better than that of land—whereas roads, residential areas and cities on land often interfere with terrestrial ecological impact assessment—the assumption that the cumulative impact decayed with the distance in the RCEI study here is reasonable [39]. The reclamation types and their RCEI in the Bohai Sea indicated that the distribution of reclamation exhibited specific characteristics in different periods. Between 1985 to 2003, the reclaimed types were mainly enclosing reclamation (e.g., pond and salt pan), and the ecological impact of reclamation was concentrated as low and moderate impact. After 2003, as the proportion of harbor and seawall increased, the area designated as high and medium impact grew at the cost of low impact (Figure 8).

The RCEI model may have some limitations that need to be improved in future studies. First, because of the hydrodynamic environment of the sea, the sediment environment, noise, pollution, biological habitat, and diversity were focused on, and the ecological impact on other aspects, such as storm surges and floods, ecosystem services, and values, was not considered. Due to a lack of pertinent data, it is challenging to accurately gauge the spatial-temporal distribution and frequency of ecological disasters. Second, the attenuation of RCEI is affected not only by the distance, but also is related to the hydrodynamic conditions of the sea area. The authors used a two-dimensional tidal model to calculate the amplitude, phase lag, and tidal current in M_2 spatial simulations in the Bohai Sea [22]. However, the simulated object was a single hydrodynamic factor, and little attention was paid to the spatial simulation of the ecologically integrated condition. Taking all these factors into consideration, this study used a fixed value of the ecological impact intensity parameter for every reclamation type to evaluate the RCEI in coastal waters. The ecological impact assessment based on the six indexes has a close relationship with the reclamation type in the study areas.

4.1.2. Model Sensitivity

In the simulation of the RCEI spatial distribution (Figure 9), it was found that RCEI was highly sensitive to marine constructions. For example, if a port with a width of 1 km and a length of 3 km was built in a sea area with a good ecological environment, which was perpendicular to the MWBB, the surrounding sea area with 9.6 km^2 could become significantly affected because the ocean has good ecological connectivity [31]. Therefore, the impact of marine constructions on ecological connectivity could be reduced through minimizing the length of buildings and reducing their distances from land. Additionally, even though remote sensing data have produced a high-precision image of the RCEI and marine structures on the spatio-temporal scales, long-term studies are still necessary to determine the precise effects of artificial structures on ocean connection.

4.1.3. Sources of Model Uncertainty

The uncertainty of ecological cumulative impact assessment comes mainly from the assessment system and data [40]. On the one hand, the establishment of this evaluation system was based on widely reported results, which were described in the Sections 1 and 2.4.1. On the other hand, the data uncertainty refers to errors caused by observation and processing: first, remote sensing images cannot capture the benthic structure of the optical deep water area, and the area covered by the structures submerged by the sea cannot be tracked in the observation of the coastal reclamation and its marine constructions; and second, the application of the attenuation function to the numerical simulation will inevitably cause differences from the actual ecological cumulative impact. For example, the half attenuation coefficient may be different in different sea areas, and the use of a unified coefficient in this article will inevitably cause errors.

4.2. Economy and Urbanization Drivers

Up to present, most of the studies were based on descriptive work simply to infer the driving factors of reclamation and compare the relationships among humanistic and social factors and reclamation activities [41], whereby further experiments were needed to quantify the relative importance of different driving factors and reveal their quantitative response relationships [42]. In a survey on the area of reclamation, it was found that only indicators related to the marine industry were positively correlated with the area of reclamation, such as marine industry employees, marine industry GDP, the added value of marine oil and gas, the added value of marine chemicals, etc. [38]. Wang et al. [13] showed that reclamation expansion is strongly associated with socioeconomic variables, such as GDP, population, and urbanization level. As a result, a variety of socioeconomic factors influence coastal reclamation. The path analysis is capable of exploring the driving forces of reclamation projects using varying regression equations, and the findings showed that cargo throughput and mariculture production in the Bohai Sea is also driving coastal expansion.

4.3. Management Implications

4.3.1. Industrial Development

The Circum-Bohai-Sea is not only one of the most important economic zones in China, but also an early developed area. However, its limited land supply has led to insufficient space for its urban expansion, especially in the Bohai Bay and Laizhou Bay with silty, sandy shoals and low-level terrain. In order to obtain expanded space for further urban and industry development, intensive reclamation projects have taken place there (Figure 7). Since 1985, two large-scale reclamation activities have also occurred and led to the industrial development of cities around the Bohai Sea. In the 1980s–1990s, the reclamation area was focused on the development of agriculture and fishing. After the 21st century, the reclamation activities were shifted to aquaculture, urbanization, and industrialization, and various port terminals and coastal industries were continually settled in the coastal water of the Bohai Sea. According to statistics, from 1998 to 2018, the cargo throughput of the main ports of the Bohai Sea was increased by 11.6 times, and the aquaculture production was increased by 3.3 times. Although the high-intensity development and utilization of coastal resources has promoted the development of industrial economy significantly, the coastal cities of the Bohai Sea are facing serious homogeneity problems in the current development stage, as ports and docks, port industries, ponds, and salt fields have been settled and dispersed in 13 coastal cities; the development and utilization methods of Bohai Bay and Laizhou Bay are very similar (Figure 11). In order to avoid the vicious competition in regional industries, idle resources, and resource waste, a mechanism for efficient and intensive use of reclaimed land resources should be established because only scientific and reasonable reclamation activities could meet the efficient development of regional port terminals and port industries, while providing enough space for agriculture breeding and salt production.

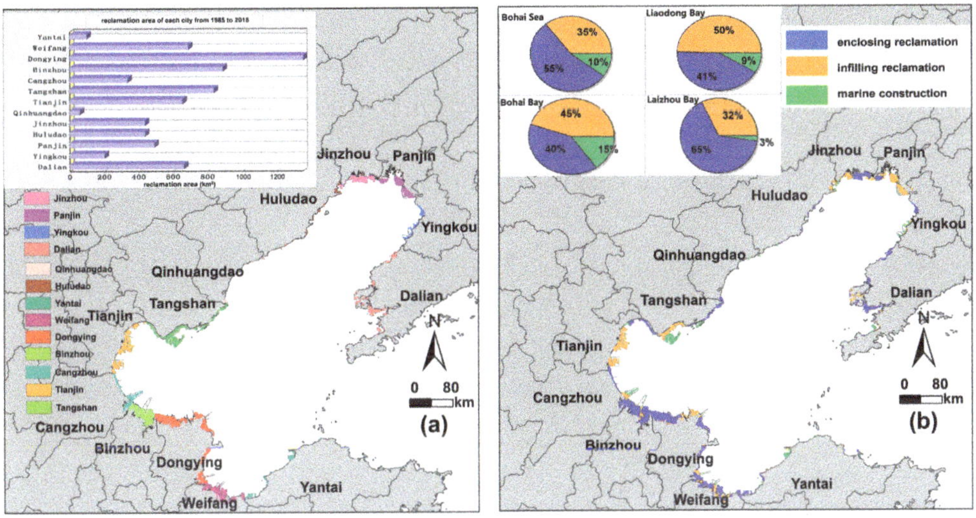

Figure 11. Reclamation area (**a**) and its utilization types (**b**) of 13 coastal cities from 1985 to 2018.

Currently, the economy of coastal waters of the Bohai Sea is developing rapidly with continual gathering of population, finance, and logistics to the sea, and the coastal ecosystem is facing severe challenges (Figure 9). Therefore, we suggest focusing on economizing the sea utilization, scientifically planning the layout of marine industries and the use of marine space, learning from the development experience of other coastal areas in China, giving full play to late-comer advantages, and promoting leaping development.

4.3.2. Ecological Protection Suggestions

Since the 21st century, the RCEI in the Bohai Sea has risen from low to moderate. In fact, the ecological impact of reclamation was concentrated generally in the main bays of the Bohai Sea. The RCEI varied greatly depending on the stage of reclamation and was exacerbated by the cumulative effect of the four stages of reclamation (Figure 9). In particular, in Bohai Bay and Laizhou Bay, with severely damaged biological communities, comprehensive measures should be used for both the restoration and protection of coastal wetland and marine biodiversity [15]. In addition, the implementation of the most stringent management of coastal reclamation, the implementation of the Bohai Sea environmental protection project with land and sea coordination, river and sea considerations, and the restoration of damaged habitats, should be carried out to gradually restore and improve the resource and environmental carrying capacity of Bohai Sea.

In order to mitigate the threat of storms and sea level rise caused by global climate change [43,44], humans have implemented synthetic engineering structures, such as sea dikes, breakwaters, and spur dams, leading to an increase in the proportion of the boundary length of the protective dams in the Bohai Sea of 7.8% from 2003 to 2018. However, in the process of constructing these marine constructions, the natural ecosystems, including grasses, trees and biological communities of the ocean and land, were destroyed, and the potential loss of ecological benefits was difficult to estimate. In order to reduce this conflict, engineering measures based on marine ecology could be used. For example, the planting and restoration of salt marshes, shellfish or coral reefs can be considered as an adaptation strategy to cope with sea level rise and storm surges and to support the restoration of multiple ecological functions, including carbon sequestration, provision of habitat, and maintenance of clean air and water [45].

5. Conclusions

Reclamation activities could directly change the hydrodynamic conditions and seawater exchange capacity of the coastal waters, resulting in natural movement of the water body instead of simple changes in the coastline. We proposed a method for identifying the extent of existing coastal reclamation, and the RCEI were established on the basis of the ecological influence characteristics of different reclamation types and their attenuation effect on adjacent areas. Results showed that reclamation had taken place in approximately 5800 km^2 of natural sea areas from 1985 to 2018. The RCEI score of the Bohai Sea in 2018 was 2.2 out of 6.0, indicating a moderate impact. The expansion of coastal projects, such as ports and wharves, was the main contributor to the increase in the RCEI. Additionally, the spatial distribution characteristics revealed that the RCEI in the offshore areas was much higher, with denser and more highly impacted areas than that in the open seas, and the intensity of the impact decreased from the offshore area to the open sea. Moreover, some local areas were significantly affected by the RCEI, especially in major bays such as the Laizhou Bay and the Bohai Bay.

Author Contributions: J.L., conceptualization, data curation, formal analysis, investigation, methodology, software, validation, visualization, writing—original draft, and writing—review and editing; H.S., conceptualization, funding acquisition, methodology, project administration, and resources; X.L., formal analysis, methodology, project administration, and resources. All authors have read and agreed to the published version of the manuscript.

Funding: This paper is supported by the National Natural Science Foundation of China (U1806214 and 42076011) and the National key research and development program (2019YFC1408405).

Data Availability Statement: The satellite altimeter data used in our study can be downloaded from the Radar Altimeter Database System (http://rads.tudelft.nl/rads/rads.shtml (last accessed on: 1 May 2021)).

Acknowledgments: This study was provided valuable suggestions on the identification and extraction methods of the coastline. Wang Yuzhe, Guo Junting, and Cao Ruichen, participated in the discussion on the study method and manuscript preparation with useful suggestions.

Conflicts of Interest: The authors declare no conflict of interest.

References

1. National Bureau of Statistics. China Statistical Yearbook (2020). Available online: http://www.stats.gov.cn/ (accessed on 2 February 2022).
2. Liu, Y.S.; Li, J.T. The pattern sand driving mechanisms of reclaimed land use in China's coastal areas in recent 30 years. *Scientia. Sinica. Terrae.* **2020**, *50*, 761–774. (In Chinese)
3. Wang, K.F. Evolution of Yellow River Delta Coastline Based on Remote Sensing from 1976 to 2014, China. *Chin. Geogr. Sci.* **2019**, *29*, 181–191. [CrossRef]
4. Yan, H.K.; Wang, N.; Yu, T.L.; Fu, Q.; Liang, C. Comparing effects of land reclamation techniques on water pollution and fishery loss for a large-scale offshore airport island in Jinzhou Bay, Bohai Sea, China. *Mar. Pollut. Bull.* **2013**, *71*, 29–40. [CrossRef]
5. Cao, R.; Chen, H.; Rong, Z.; Lv, X. Impact of ocean waves on transport of underwater spilled oil in the Bohai Sea. *Mar. Pollut. Bull.* **2021**, *171*, 112702. [CrossRef]
6. Chen, W.; Wang, D.C.; Huang, Y.; Chen, L.; Zhang, L.; Wei, X.; Sang, M.; Wang, F.; Liu, J.; Hu, B. Monitoring and analysis of coastal reclamation from 1995-2015 in Tianjin Binhai New Area, China. *Sci. Rep.* **2017**, *7*, 3850. [CrossRef]
7. Ministry of Ecology and Environment of the People's Republic of China. Bulletin on the State of Marine Ecology and Environment of China (2018). Available online: https://www.mee.gov.cn/ (accessed on 3 February 2022).
8. He, Q.; Bertness, M.D.; Bruno, J.F.; Li, B.; Chen, G.; Coverdale, T.C.; Altieri, A.H.; Bai, J.; Sun, T.; Pennings, S.C.; et al. Economic development and coastal ecosystem change in China. *Sci. Rep.* **2014**, *4*, srep05995. [CrossRef]
9. Chu, N.; Yao, P.; Ou, S.; Wang, H.; Yang, H.; Yang, Q. Response of tidal dynamics to successive land reclamation in the lingding bay over the last century. *Coast. Eng.* **2022**, *173*, 104095. [CrossRef]
10. Bi, X.L.; Liu, F.Q.; Pan, X.B. Coastal projects in China: From reclamation to restoration. *Environ. Sci. Technol.* **2012**, *46*, 4691–4692. [CrossRef] [PubMed]
11. Zhu, G.; Xie, Z.; Xu, X.; Ma, Z.; Wu, Y. The landscape change and theory of orderly reclamation sea based on coastal management in rapid industrialization area in Bohai Bay, China. *Ocean Coast. Manag.* **2016**, *133*, 128–137. [CrossRef]

12. Liang, S.; Li, X.; Wang, J. A systematic view of remote sensing. In *Advanced Remote Sensing*, 1st ed.; Academic Press: Cambridge, MA, USA, 2012; pp. 1–31.
13. Wang, W.; Liu, H.; Li, Y.; Su, J. Development and management of land reclamation in China. *Ocean Coast. Manag.* **2014**, *102*, 415–425. [CrossRef]
14. Mai, S.; Bartholomä, A. The missing mud flats of the Wadden Sea: A reconstruction of sediments and accommodation space lost in the wake of land reclamation. *Proc. Mar. Sci.* **2000**, *2*, 257–272.
15. Shen, C.C.; Shi, H.H.; Zheng, W.; Li, F.; Peng, S.; Ding, D. Study on the cumulative impact of reclamation activities on ecosystem health in coastal waters. *Mar. Pollut. Bull.* **2016**, *103*, 144–150. [CrossRef]
16. Hebei Provincial Bureau of Statistics. Hebei Statistical Yearbook (1995–2018). Available online: http://tjj.gxzf.gov.cn (accessed on 5 February 2022).
17. Liaoning Provincial Bureau of Statistics. Liaoning Statistical Yearbook (1995–2018). Available online: http://tjj.ln.gov.cn (accessed on 5 February 2022).
18. Shandong Provincial Bureau of Statistics. Shandong Statistical Yearbook (1995–2018). Available online: http://tjj.shandong.gov.cn (accessed on 5 February 2022).
19. Tianjin Bureau of Statistics. Tianjin Statistical Yearbook (1995–2018). Available online: http://stats.tj.gov.cn (accessed on 5 February 2022).
20. Crowell, M.; Leatherman, S.P.; Buckley, M.K. Shoreline change rate analysis: Long term versus short term data. *Shore Beach* **1993**, *61*, 13–20.
21. Suo, A.N.; Zhang, M.H. Sea areas reclamation and coastline change monitoring by remote sensing in coastal zone of Liaoning in China. *J. Coast. Res.* **2015**, *73*, 725–729. [CrossRef]
22. Lu, J.F.; Zhang, Y.B.; Lv, X.Q.; Shi, H.H. The Temporal Evolution of Coastlines in the Bohai Sea and Its Impact on Hydrodynamics. *Remote Sens.* **2022**, *14*, 5549. [CrossRef]
23. Pan, H.D.; Jiao, S.Y.; Xu, T.; Lv, X.Q.; Wei, Z.X. Investigation of tidal evolution in the Bohai Sea using the combination of satellite altimeter records and numerical models. *Estuar. Coast. Shelf Sci.* **2022**, *279*, 108140. [CrossRef]
24. Zhang, Z.F.; Suo, A.N.; Yang, Z.X. *Evaluation and Early-Warning Technical Methouds for Marine Resources & Environmental Carrying Capacity and Its Application*; China Ocean Press: Beijing, China, 2019.
25. Heery, E.; Bishop, M.J.; Critchley, L.P.; Bugnot, A.B.; Airoldi, L.; Mayer-Pinto, M.; Sheehan, E.V.; Coleman, R.A.; Loke, L.H.; Johnston, E.L.; et al. Identifying the consequences of ocean sprawl for sedimentary habitats. *J. Exp. Mar. Biol. Ecol.* **2017**, *492*, 31–48. [CrossRef]
26. Lee, S.Y.; Dunn, R.J.K.; Young, R.A.; Connolly, R.M.; Dale, P.E.R.; Dehayr, R.; Lemckert, C.J.; McKinnon, S.; Powell, B.; Teasdale, P.R.; et al. Impact of urbanization on coastal wetland structure and function. *Austral Ecol.* **2006**, *31*, 149–163. [CrossRef]
27. Bugnot, A.B.; Mayer-Pinto, M.; Airoldi, L.; Heery, E.C.; Johnston, E.L.; Critchley, L.P.; Strain, E.M.A.; Morris, R.L.; Loke, L.H.L.; Bishop, M.J.; et al. Current and projected global extent of marine built structures. *Nat. Sustain.* **2020**, *4*, 33–41. [CrossRef]
28. Naser, H.A. Effects of reclamation on macrobenthic assemblages in the coastline of the Arabian Gulf: A microcosm experimental approach. *Mar. Pollut. Bull.* **2011**, *62*, 520–524. [CrossRef]
29. Chen, C.; Ji, R.; Zheng, L. Influences of physical processes on the ecosystem in Jiaozhou Bay: A coupled physical and biological model experiment. *J. Geophys. Res. Oceans* **1999**, *104*, 29925–29949. [CrossRef]
30. Park, S.R.; Kim, J.H.; Kang, C.K.; An, S.; Chung, I.K.; Kim, J.H.; Lee, K.-S. Current status and ecological roles of Zostera marina after recovery from large-scale reclamation in the Nakdong River estuary, Korea. Estuar. *Coast. Shelf. Sci.* **2009**, *81*, 38–48. [CrossRef]
31. Bishop, M.J.; Mayer-Pinto, M.; Airoldi, L.; Firth, L.B.; Morris, R.L.; Loke, L.H.L.; Hawkins, S.J.; Naylor, L.; Coleman, R.A.; Chee, S.Y.; et al. Effects of ocean sprawl on ecological connectivity: Impacts and solutions. *J. Exp. Mar. Biol. Ecol.* **2017**, *492*, 7–30. [CrossRef]
32. Zarzuelo, C.; López-Ruiz, A.; Valle-Levinson, M.; Díez-Minguito, M. Ortega-Sánchez. Bridge-piling modifications on tidal flows in an estuary. *Coast. Eng.* **2022**, *173*, 104093. [CrossRef]
33. Wu, Z.X.; Yu, Z.M.; Song, X.X.; Li, Y.F.; Cao, X.H.; Yuan, Y.Q. A methodology for assessing and mapping pressure of human activities on coastal region based on stepwise logic decision process and GIS technology. *Ocean Coast. Manag.* **2016**, *120*, 80–87. [CrossRef]
34. Chi, Y.; Shi, H.H.; Zheng, W.; Sun, J.; Fu, Z. Spatiotemporal characteristics and ecological effects of the human interference index of the yellow river delta in the last 30 years. *Ecol. Indic.* **2018**, *89*, 880–892. [CrossRef]
35. Shi, H.H.; Zheng, W.; Wang, X.L. *Investigation and Carrying Capacity Evaluation of Carbon Sequestering Biological Resources of Typical Island Ecosystems in Northern China*; China Ocean Press: Beijing, China, 2017; 305p.
36. Murray, C.C.; Agbayan, S.; Alidina, H.M.; Ban, N.C. Advancing marine cumulative effects mapping: An update in Canada's Pacific waters. *Mar. Policy* **2015**, *58*, 71–77. [CrossRef]
37. Meng, W.Q.; Hu, B.B.; He, M.X.; Liu, B.Q.; Mo, X.Q.; Li, H.Y.; Wang, Z.L.; Zhang, Y. Temporal-spatial variations and driving factors analysis of coastal reclamation in China. *Estuar. Coast. Shelf Sci.* **2017**, *191*, 39–49. [CrossRef]
38. Halpern, B.S.; Longo, C.; Hardy, D.; McLeod, K.L.; Samhouri, J.F.; Katona, S.K.; Kleisner, K.; Lester, S.E.; O'Leary, J.; Ranelletti, M.; et al. An index to assess the health and benefits of the global ocean. *Nature* **2012**, *488*, 615–620. [CrossRef]

39. Geary, W.L.; Bode, M.; Doherty, T.S.; Fulton, E.A.; Nimmo, D.G.; Tulloch, A.I.T.; Tulloch, V.J.D.; Ritchie, E.G. A guide to ecosystem models and their environmental applications. *Nat. Ecol. Evol.* **2020**, *4*, 1459–1471. [CrossRef]
40. Sengupta, D.; Chen, R.S.; Meadows, M.E. Building beyond land: An overview of coastal land reclamation in 16 global megacities. *Appl. Geogr.* **2018**, *90*, 229–238. [CrossRef]
41. Lu, J.F.; Zhang, Y.B.; Shi, H.H.; Lv, X.Q. Coastal vulnerability modelling and social vulnerability assessment under anthropogenic impacts. *Front. Mar. Sci.* **2022**, *9*, 1015781. [CrossRef]
42. Lu, J.F.; Zhang, Y.B.; Shi, H.H.; Lv, X.Q. Spatio-temporal changes and driving forces of reclamation based on remote sensing: A case study of Guangxi Beibu gulf. *Front. Mar. Sci.* **2023**, *10*, 1112487. [CrossRef]
43. Nicholls, R.J.; Cazenave, A. Sea-level rise and its impact on coastal zones. *Science* **2010**, *328*, 1517–1520. [CrossRef] [PubMed]
44. Webb, E.L.; Friess, D.A.; Krauss, K.W.; Cahoon, D.R.; Guntenspergen, G.R.; Phelps, J. A global standard for monitoring coastal wetland vulnerability to accelerated sea-level rise. *Nat. Clim. Chang.* **2013**, *3*, 458–465. [CrossRef]
45. Cheong, S.; Silliman, B.; Wong, P.P.; Van Wesenbeeck, B.; Kim, C.K.; Guannel, G. Coastal adaptation with ecological engineering. *Nat. Clim. Chang.* **2013**, *3*, 787–791. [CrossRef]

remote sensing

MDPI

Article

Monitoring of Coastal Boulder Movements by Storms and Calculating Volumetric Parameters Using the Volume Differential Method Based on Point Cloud Difference

Yao Yao [1,*], Helene Burningham [1], Jasper Knight [2] and David Griffiths [3]

1 Department of Geography, University College London Gower Street, London WC1E 6BT, UK
2 School of Geography, Archaeology and Environmental Studies, University of the Witwatersrand, Johannesburg 2050, South Africa
3 Department of Civil, Environmental and Geomatic Engineering, University College London, Gower Street, London WC1E 6BT, UK
* Correspondence: yao.yao.19@alumni.ucl.ac.uk

Abstract: The measurements of boulder volume and axial length play significant roles in exploring the evolution of coastal boulder deposition, which provides a theoretical framework to examine the hydrodynamics of extreme wave events. At present, the application of structure-from-motion (SfM) to unmanned aerial system (UAS) imagery is one of the most used boulder surveying techniques. However, the monitoring of boulder movement and the accurate measurement of boulder morphometrics are rarely investigated in combination. In this study, UAS surveys were used to monitor moving boulders and measure boulder volumes using the volume differential method based on the differences of dense point clouds. This was undertaken at a site on the rocky shoreline of northwest Ireland in three repeated UAS surveys conducted in 2017, 2018, and 2019. The results from UAS monitoring and mapping of the distribution of 832 moving boulders in the study area over the 3-year period showed that boulders located in different zones of the coast vary significantly in their mobility. The main findings reveal that the theoretical error of the volume, obtained using the volume differential method, was estimated as 1–3.9%, which is much smaller than that of the conventional method of estimating volume using a tape measure.

Keywords: structure-from-motion; boulder morphometry; storm wave forcing; rocky shoreline; UAS; UAV; drone mapping

Citation: Yao, Y.; Burningham, H.; Knight, J.; Griffiths, D. Monitoring of Coastal Boulder Movements by Storms and Calculating Volumetric Parameters Using the Volume Differential Method Based on Point Cloud Difference. *Remote Sens.* **2023**, *15*, 1526. https://doi.org/10.3390/rs15061526

Academic Editors: Ramón Blanco Chao, Germán Flor-Blanco and José Juan de Sanjosé Blasco

Received: 1 February 2023
Revised: 7 March 2023
Accepted: 8 March 2023
Published: 10 March 2023

1. Introduction

In recent decades, various extreme sea wave events, triggered by storms and tsunamis, have caused significant damage and loss of life in many coastal areas around the world. For instance, Typhoon Haiyan, in the Philippines in 2013, resulted in an economic loss of up to 2 billion USD and at least 6300 deaths [1]. To more accurately reconstruct the intensity of the extreme wave events affecting a region, there has been extensive research into the movement and evolution of coastal boulders [2–12]. Since coastal boulders typically weigh several tons to tens of tons, they are resistant to average wave or tide conditions, and thus, changes in boulder distribution patterns across rocky shorelines can be considered a record of extreme coastal climate events [13–15]. Although several existing mathematical models have achieved the restoration of storm intensity and wave height [6,7,9,13,16–19], they remain heavily reliant on the accurate acquisition of certain parameters, such as the volume and axial length of boulders [20] in addition to regional wave climate buoy or wave model data.

The volume and axial length measurements of coastal boulders have been investigated in various regions around the world [3–6]. Before the emergence of new measurement techniques, the volume of a boulder was estimated by the length of a-, b- and c-axes (corresponding to the long, intermediate, and short axes of the boulder, respectively, measured with a tape measure in the field [4,6]), and it was assumed that the boulder was an idealised geometry. However, there is a large potential difference between the estimated volume and the real volume of a detached boulder. To solve this problem, Liu et al. [6] and Gennady et al. [4] used terrestrial laser scanner (TLS) and structure-from-motion (SfM) technologies on the ground to determine the real volume of the boulders more accurately. However, these two ground-based measurement technologies are restricted to measuring a single large boulder each time (low efficiency), and the cost of the TLS system is quite high. Therefore, studies have also applied unmanned aerial system- (UAS) SfM technology to collect datasets of study sites at a kilometre scale, using nadir or oblique photography and 3D photogrammetry software (Agisoft, Pix4DMapper), to perform modelling for volume calculations [3,5] or the monitoring of boulder movements [2]. Since multi-angle images can be captured using SfM technology based on oblique photography, the reconstruction of boulder size, shape, volume, and distribution is usually better than with SfM technology based on nadir photography. However, the cost of doing so means that the size of the survey area can be compromised. Although consideration is given to battery life and oblique photography, at the same time, when a 5-lens camera is used, the cost of UAS with 5-lenses is comparable to that of TLS technology. In addition, the above-mentioned techniques are not able to collect information from the bottom of a boulder, which adds to the uncertainty about volume calculations. The current approach is to simply close the bottom of the boulder model by using an artificial plain surface [4–6], but this can lead to more significant errors for the boulders on uneven surfaces.

At present, the calculation of boulder volume is not integrated with the monitoring of boulder movement from UAS surveys, and it is challenging in coastal environments comprising a large number of small boulders. This is because it is very difficult to detect the movement of small boulders (1–2 m in length) using even high-resolution satellite imagery. In this context, a method that combines the monitoring of boulder movement with accurate calculations of boulder volume is needed so that boulder movement and directions can be detected promptly and the volumes can be obtained with fewer UAS surveys. Further, the availability of different types of coastal boulder parameters facilitates the analysis of how extreme wave events caused by storms affect the movement of boulders.

The Multiscale Model-to-Model Cloud Comparison (M3C2) algorithm has been widely used to monitor environmental change, including coastal boulder movements [2], landslides [21], and riverbank erosion [22]. The M3C2 distance has a physical meaning, and in most cases, it can represent the thickness of a certain part of the boulder, which makes it possible to calculate the volume of the boulder. This study has built from this basis to calculate boulder volume using the M3C2 algorithm, and it has been applied to boulders mapped from repeated UAS surveys conducted along a rocky shoreline in northwest Ireland that has been previously examined for boulder dynamics [23,24]. The purpose of this application was to obtain boulder volumes and axial lengths in tandem with boulder movement over time and space from the UAS surveys. To support this analysis, a box group methodology (for simulating boulders) was used to verify the theoretical accuracy of this method.

2. Study Area

The study area is located on the Atlantic-facing northwest coast of Ireland at Falchorrib, County Donegal (Figure 1). This coastline has a northwest–southeast trend, and the rock shoreline examined in this study has a total length of about 1 km. Bedrock geology is rudaceous (coarse-grained) biotite granite that has developed distinctive sets of contraction joints that act as lines of weakness to weathering and erosion. The joint sets have acted as a first-order control in shore platform geometry, giving rise to a stepped profile in some places and deeply eroded, joint-aligned gullies in others. Hydraulic wave action contributes to the removal of individual bedrock boulders, as well as joint-defined slabs, from the shoreface; this provides the boulder supply for subsequent transport alongshore and into the upper intertidal zone, where distinctive boulder ridges are found [24]. Boulders are contained in the bedrock shore platform or within 20 m of it and are sourced entirely from the bedrock shoreline; the adjacent hinterland comprises a gentle slope covered almost entirely in peat bog. Northwest Ireland has a temperate maritime climate with onshore southwesterly winds, and this gives rise to its susceptibility to storms. The role of the storm wave forcing of boulder dynamics along the western Ireland coast has been examined in several studies [11,24–26].

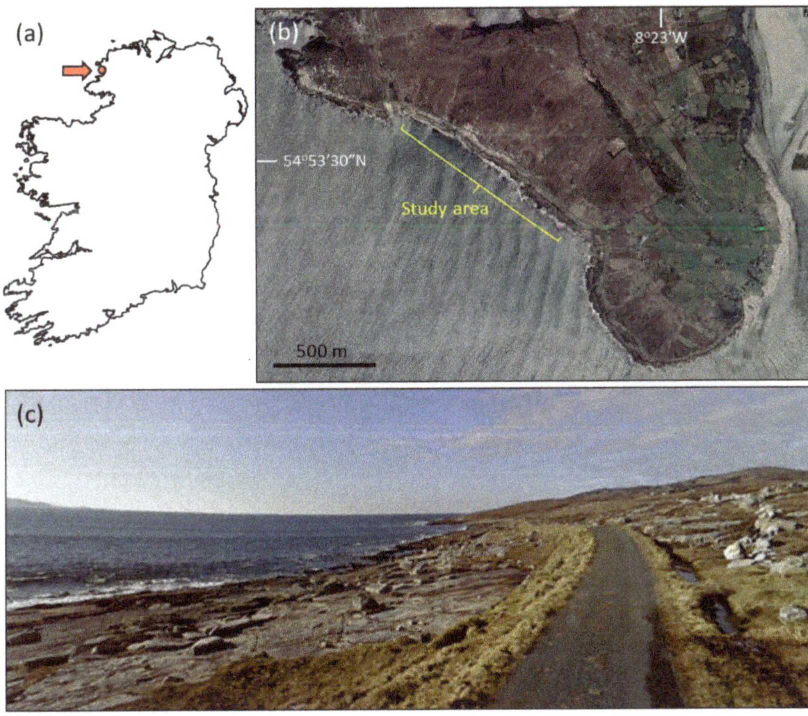

Figure 1. (**a**) The location of the study area in northwest Ireland, indicated by the red dot. (**b**) The study area is shown using satellite imagery. (**c**) A photo of the study area looking northwest. The surveyed area is located between the road and the sea and consists of a bedrock shore platform and boulders distributed on its surface.

3. Materials and Methods

The approach followed here involved four stages: (1) UAS surveys to collect nadir and oblique images of the boulders in the study area in 2017, 2018, and 2019, supported by dGPS surveying for ground control; (2) the application of Pix4DMapper 4.5.6 to generate high-density point clouds, orthomosaics, and digital surface models (DSMs) of the study area; (3) the application of CloudCompare to process the point clouds and calculate the M3C2 distance; (4) the use of Python and QGIS to calculate the M3C2 volume of the boulder and output the lengths of the a-, b-, and c-axes.

Practically, it is difficult to accurately measure the volume of boulders without moving them, and some are extremely large (several metres in diameter). To verify the feasibility and accuracy of the above method, multiple large shipping boxes with regular shapes were used to simulate boulders. The same approach outlined in stages 1–4 was followed to calculate the volume of the boxes and the length of the axes; these measurements were then compared with the actual dimensions of the boxes. This was used as an analogue to the boulders observed in the field in order to test the M3C2 method against boxes of known sizes.

3.1. Materials and Equipment

The initial data required for this study were divided into two parts: UAS images of the study area and ground control points (GCPs) obtained in the field. The UAS images were captured using the DJI Phantom 4 Pro drone, which is a relatively low-cost UAS that is applicable for the production of relatively high-quality image data. Having a built-in RGB camera with a 1″ CMOS sensor, the Phantom 4 Pro is capable of capturing images with an effective pixel count of 20 MP. In addition, the camera has a nominal 8 mm focal length. It uses a global shutter instead of a rolling shutter, which is essential for the reconstruction of coastal boulders, as distortion can be caused by the rolling shutter. The built-in camera of the Phantom 4 Pro is also fitted with a stabilization system (gimbal) intended to prevent the shooting angle of the camera from being affected by drone movement. In flight, the camera system is in constant communication with the flight control system of the drone, which enables the camera to collect real-time data on the geographic position and attitude of the drone before writing it to each captured image file for later reconstruction.

The GCPs data were acquired using a Leica 1200 base station and a GS15 rover differential GPS system; surveyed ground control points and checkpoints were acquired with centimetre accuracy (average 3D coordinate quality of 0.014 m).

3.2. Acquisition of UAS Imagery and GCP Data

The ground sampling distance (GSD) was set to 2 cm, and the image overlap for each flight was set to at least 80% front overlap and 60% side overlap. High overlap has become a standardized operating procedure, ensuring that every point on the ground of the flight route is shown in a minimum of 5 images [27], and the benefits have been demonstrated in previous studies [28–32]. A single flight was used in the 2017 survey with a route direction aligned west–east to reduce cross-winds in light westerly winds. To obtain denser point clouds and more details from the reconstruction results, two flights were used in the 2018 and 2019 surveys, resulting in a total of 933 and 504 images, respectively.

Although each UAS image records the geographic coordinates and altitude at the time when the image is taken, the accuracy of the UAS GNSS receiver remains at the metre-scale, which is insufficient for the reconstruction of boulders on the spatial scale of less than 1 m. Therefore, it is necessary to process the reconstruction with accurate 3D GCPs. In total, this survey acquired 30 GCPs that are common to each survey by using natural discrete features across the bedrock platform, road, and jetty. The GCPs were surveyed into an Ordnance Survey of Ireland GPS benchmark located near the jerry at the northwestern end of the study area. Horizontal (x- and y-direction) and vertical (z-direction) coordinate quality (CQ) averages were 0.74 cm and 1.13 cm, respectively; these measures were used as the accuracy of GCPs in the Pix4DMapper.

3.3. Reconstruction of the Study Area Using Pix4DMapper

Separate projects were created for each year in Pix4D Mapper, and all UAS images for each year were imported. Pix4D uses the positioning, elevation, and shooting angle data contained in the image file for initial processing. Before adding GCPs, the full key point image scale is used, and the initial process (sparse reconstruction) is conducted. Once the sparse reconstruction is complete, a dense number of key points are generated across the study area—each of which can be found within multiple images. GCPs in the images for each year were marked manually and were first imported to identify a key point close to the GCP in the Pix4D software (Figure 2). Second, a new 3D control point was created, with the values of the 3D coordinates and the horizontal and vertical accuracy of this control point manually defined. With the position of the control point manually marked in at least two images, Pix4D can find all images where the control point is captured and mark them automatically. After the completion of automatic marking, each marked image is checked for fine adjustments, which ensures that they are close enough to the actual measurement point. If the control point features of an image are not captured clearly, the image is scaled down before marking, which results in the image carrying less weight in the subsequent reconstruction. Finally, the sparse reconstruction is re-optimized after all control points have been marked. Out of the 30 control points, 25 were used for sparse reconstruction, and 5 were selected to evaluate the accuracy of the final results.

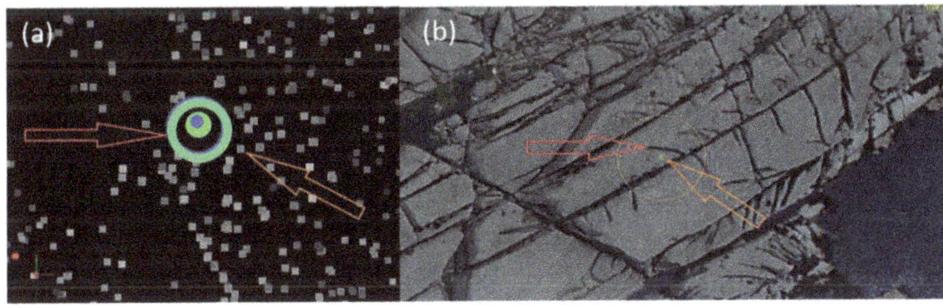

Figure 2. (**a**) A key point close to the GCP is chosen for manually marking the GCP on images. The orange arrow represents the selected key point, and the red arrow represents the measurement point of the GCP, which is the endpoint of a bedrock fracture (**b**).

The final constructed models include point clouds, DSMs, and orthomosaics. The following tests were conducted on multiple sets of parameter combinations:

- For point cloud generation (dense reconstruction): The image scale in the point cloud densification was set to 1/2 and the multiscale was turned off to reduce noise. The point density was set to high, with each point requiring at least 2 matches, which can increase the detail and density of point cloud reconstruction.
- DSM and orthomosaicing: Triangulation was used instead of inverse distance weighting, as it can improve clarity for the edges of the boulders. Since the resolution setting is consistent with the GSD (2 cm), each boulder can be better identified.

3.4. Calculation of M3C2 Distance

The Multiscale Model to Model Cloud Comparison (M3C2) algorithm was applied to calculate the distance distribution between two point clouds [21], which is a key step in determining change, boulder volume, and c-axis length. Typically, the two clouds should be aligned with each other before the M3C2 process, which causes a complete overlap of most areas (i.e., unchanged areas) of the point cloud model. The open-source software, CloudCompare, provides the tools used for point cloud alignment as well as access to the M3C2 plugin. The intuition provided by the tools and the plugin for monitoring the movement of coastal boulders has been tested and used in a previous study [2].

The core workflow using M3C2 is detailed as follows:

1. Importing the point cloud. Calculating the M3C2 distance requires two clouds, with one for reference and the other for comparison. In this project, the point cloud of the previous year was uniformly treated as the reference cloud; for example, the point cloud of 2017 was used as the reference cloud in the comparison between 2017 and 2018.
2. Setting the core point. The distance distribution between two point clouds was calculated based on several cell areas, which are usually divided from the original point cloud. The centres of these cell areas are core points, which are usually the reference point cloud itself or a subsample set. Here, the point cloud of the middle year (2018) was taken in this project as the core point to calculate the M3C2 distances of 2017–2018 and 2018–2019, thus reducing the systematic error of the volume calculations.
3. Defining normal vectors. Since the M3C2 algorithm calculates the distance between the reference cloud and the comparison cloud in terms of each core point, the normal vector of each core point is critical and defines the direction in which the distance from one cloud to another is calculated. Usually, a given value, D, is required to confirm

the normal direction of the core point. Then, the M3C2 algorithm is applied to create a sphere with the core point as the sphere centre and $D/2$ as the radius. The points contained within the sphere are fitted to a plane, and the normal vector of this plane is treated as the normal of this core point. However, any normal in a non-vertical direction was meaningless here (see Section 3.5 for an explanation), and the vertical direction was used directly.

4. Calculating the distance between two clouds. After the core point (i) and the normal vector (N) were determined, another parameter d was set for the M3C2 algorithm to make a circle with core point i as the centre and $d/2$ as the radius. Subsequently, a cylinder was created along the axis of the normal vector that passed through the core point i. The parts of the two point clouds located inside the cylinder were defined as the n_1 and n_2 point cloud subsets. All the points in n_1 and n_2 were projected onto the cylindrical axis, with core point i as the origin, thus determining their distance distributions. The mean of these two distributions defined the average cloud positions, i_1 and i_2, along the normal direction, while the distance between the two point clouds (L_{M3C2}) was defined by the distance between i_1 and i_2. In addition, it was necessary to input the maximum length of the cylinder to speed up the calculation process. In the cases of no corresponding point cloud data within the set length of the cylinder in the comparison cloud, the distance was not calculated.
5. Outputting M3C2 distance. The M3C2 distance value, as calculated based on each core point, was temporarily saved in a new point cloud composed of core points (the 2018 point cloud with RGB information removed herein). For the convenience of editing and reading, the results of this project were exported as a CSV file.

3.5. Calculation of the M3C2 Volume of Moving Boulders and Determination of the Length of the a-, b-, and c-Axes

The principle of differentiation was used to calculate the volume of the moving boulders, each of which was divided into several polygonal columns parallel to the z-axis of the coordinate system, as shown in Figure 3. The volume of the boulder was obtained by summing the volumes of all polygonal columns. It was expressed as:

$$\sum ds \cdot dh \tag{1}$$

where ds and dh represent the cross-sectional area and height of the polygonal column, respectively. This method involves the following steps:

1. Identifying moving boulders. In the M3C2 results, areas where values were significantly above or below 0 indicated boulder movement, keeping in mind that the surrounding bedrock does not change. QGIS was used to visualize the M3C2 results, with two orthomosaics, corresponding to the year, combined to identify the moving boulders.
2. Determining the edges of the boulder. Since accurate boulder edges are required for volume calculation, this study generated boulder edges manually to eliminate errors introduced by edge detection tools. The boulders whose volumes could not be calculated were excluded. Based on the visualization in step 1, the edges of the moving boulders were outlined using the orthomosaic of the study area and saved as a polygonal vector file.
3. Boulder outline gridding. This step was a process of differentials, using GIS tools to grid the boulder edges at set distances and calculate the area of each grid (which is the cross-section of the polygonal column, ds). In this study, the grid was set to 2 cm, which was the GSD of the UAS surveys.
4. Determining the height of the polygonal column. Since the two point clouds had been finely aligned before the M3C2 algorithm was operated, the unchanged areas in the study area tended to be zero in the M3C2 results. Therefore, when a single boulder was moving, the M3C2 value (absolute value) of each core point in the area where the boulder was located could represent the height of that boulder in the vicinity of

the core point (Figure 4). In the grid generated in step 3, there may have been one or more core points (M3C2 values) in some grids. In this situation, the maximum value of M3C2 was taken as the height (*dh*) of the polygonal column. In the case of no core point in the grid, the M3C2 value of the core point which was closest to the grid centre was treated as the height (*dh*) of the polygonal column. The advantage of using M3C2 core points to find the depth of boulders is reflected here. If two point clouds are directly used to calculate the elevation difference, the closest points found in the two point clouds may not be at the same position, which does not represent the depth of the boulder at a specific position very well. It is worth noting that the Z values of these core points are meaningless, as the grids only select core points in the horizontal direction and read the M3C2 value. This explains why only the vertical normal is selected in the M3C2 algorithm, as only the vertical *dh* is meaningful.

5. Calculating the boulder volume. The volume of each polygonal column was expressed as $ds \cdot dh$ (Equation (1)), and the sum of the volumes of all polygonal columns represents the theoretical volume of the boulder. According to the principle of differentiation, the smaller the grid scale and the higher the density of the core point, the closer the calculated theoretical volume is to the real volume.

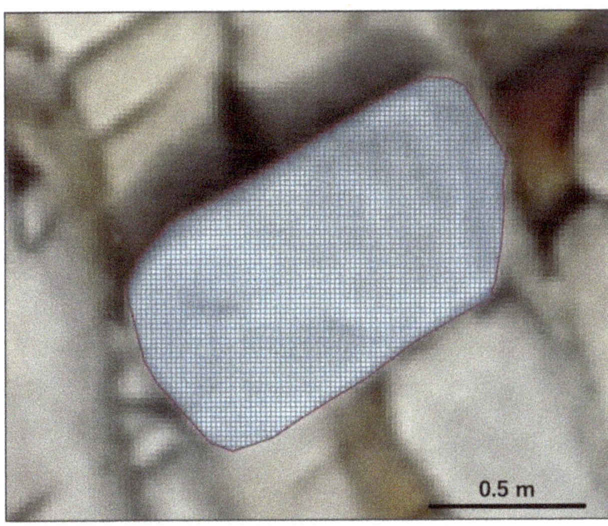

Figure 3. Illustration of the boulder gridding procedure.

It was relatively simple to obtain the length of the a-, b-, and c-axes of the boulder. The a- and b-axes could be directly measured using QGIS (Figure 5), while the c-axis was the maximum *dh* of all polygonal columns. In this section, the python code was used to go through steps 3–5 and output the length of the c-axis.

3.6. Accuracy Verification of M3C2 Volumes and Boulder Axis Lengths

In order to verify the feasibility and theoretical accuracy of the above method (Sections 3.1–3.5), the most straightforward solution is to compare the calculation results with the real values by referencing some boulders with known real volume and axes length. However, it is difficult to obtain the true volume and axes length of boulders in the field. As an alternative, large, regular-shaped shipping boxes were used to simulate coastal boulders. As shown in Figure 6, several boxes of the same specification were combined into three box groups. Group 1 and Group 3 each consisted of 6 boxes arranged in different ways, while Group 2 consisted of 2 boxes. The boxes in each group were strictly aligned without gaps and placed on flat ground. The length of the a-, b-, and c-axes of the box were 1.054 m,

0.248 m, and 0.742 m, respectively, as obtained through accurate measurement with a tape measure. The volume of each box group is expressed as:

$$V_t = a \cdot b \cdot c \cdot n \tag{2}$$

where a, b, and c represent the length of the a-axis, b-axis, and c-axis (m), respectively, of the original box; and n refers to the number of boxes in combination.

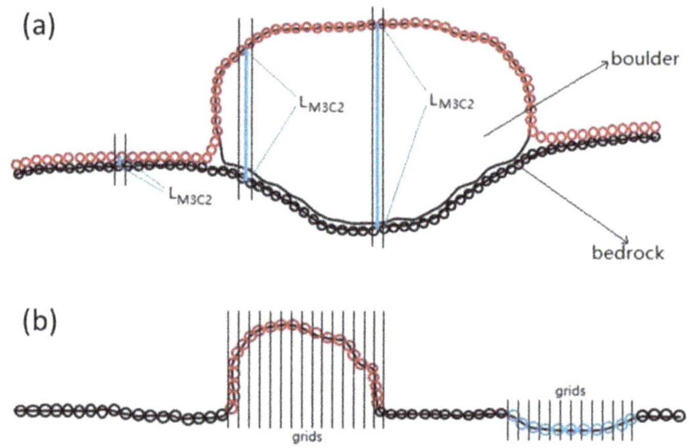

Figure 4. (**a**) M3C2 distance at different locations. Red and black circles represent the point clouds of the same region for two different years. The M3C2 values in the boulder-moving areas were significantly larger than those in unchanged areas. (**b**) The method used to determine the height (*dh*) of the polygon column. This is a side view of the M3C2 output, which is a point cloud comprised of core points (2018 point cloud with RGB information removed and M3C2 results added). The red, blue, and black circles indicate the core points where the M3C2 values were significantly above 0 (boulder appears), far below 0 (boulder disappears), and approaching 0 (no change), respectively.

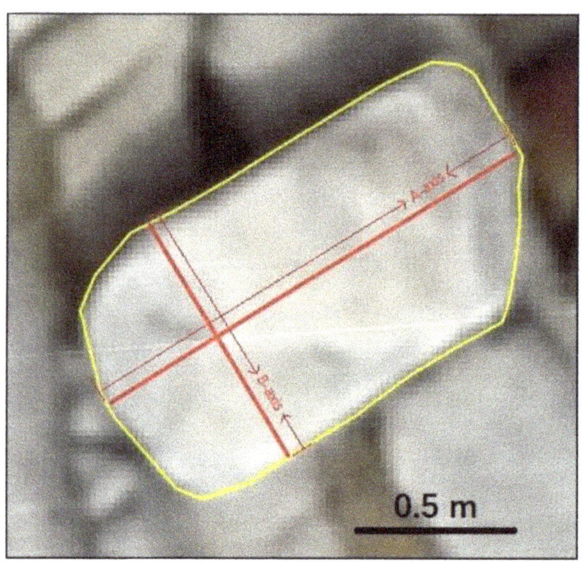

Figure 5. Measuring the length of the a-axis and b-axis in orthomosaics.

Figure 6. Three box groups used to train the M3C2 model.

By using the same equipment, parameters, and analysis process followed for the coastal boulder analysis, UAS surveys were conducted on a planar site with (simulating the appearance of the boulder) and without (simulating the disappearance of a boulder) the boxes. The accuracy of the method employed was determined by comparing the calculation results with the measured dimensions.

4. Results

4.1. Results of Model Reconstruction

Figure 7 shows the 3-year point cloud models of the study area, as constructed by Pix4DMapper. The point clouds extended landward and seaward of the shore platform, so they were clipped to eliminate the vegetated hinterland, the sea, and areas not covered by all three surveys. The point clouds were clipped using a common boundary: after clipping, the number of points in 2017, 2018, and 2019 was 63,209,768, 107,638,302, and 138,540,087, respectively.

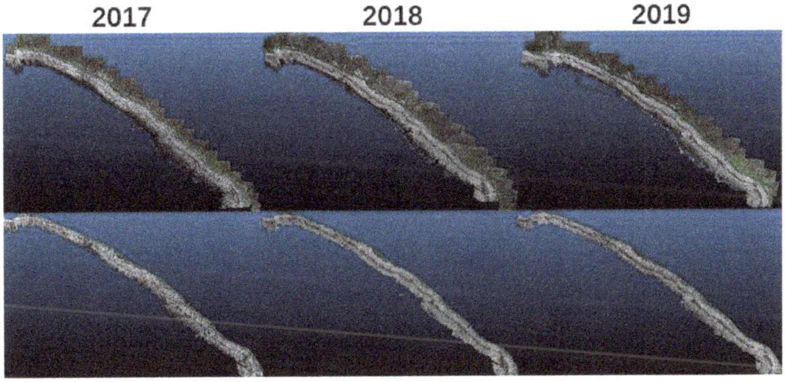

Figure 7. Point cloud models for the three-year span, where the top row shows the original clouds outputted by Pix4DMapper, and the bottom row indicates the clipped clouds.

Orthomosaics and DSMs are generated using Pix4DMapper based on the point clouds, the accuracy of which can directly represent the accuracy of the cloud. Given the difficulty of visualizing the details of the boulder after the point cloud was enlarged and the computer power required for loading the point cloud, most of the follow-up work was based on orthomosaics, such as finding moving boulders, determining the boundaries of boulders, measuring the length of the a- and b-axes, and determining the horizontal accuracy of the point cloud. Although the DSMs were not used in the calculation of M3C2 volumes and boulder axis lengths, they could be loaded into QGIS along with the orthomosaics to determine the elevation accuracy of the point clouds.

The accuracy of the models in 2017, 2018, and 2019 was determined using five uniformly distributed GCPs (Figure 8, Table 1) not used in the Pix4DMapper reconstruction. The error near the northwest pier (id = 19, 46) was within 2 cm, which was clearly insignificant. In comparison, the error in the middle part (id = 52, 64) was relatively significant, reaching up to 36 cm. In addition, there were no GCPs in the southeastern corner of the study area. Through the misalignment of the orthomosaics, it could be found that the orthomosaic in 2019 had an overall horizontal misalignment in the southeastern corner, with an error of about 20 cm. This is because the southeastern corner of the study area in 2019 was located at the edge of the UAS survey area; there are fewer UAS images in this area, and the GCP correction is negligible.

Figure 8. Distribution of GCPs. The blue dots represent the 25 GCPs included in the Pix4DMapper calculations. The yellow dots represent the 5 GCPs used to confirm the accuracy achieved by the model, and the red range represents the research area.

Table 1. Error values for the 2017, 2018, and 2019 models at the positions of the 5 GCPs.

ID	2017		2018		2019	
	Vertical Error (m)	**Horizontal Error (m)**	**Vertical Error (m)**	**Horizontal Error (m)**	**Vertical Error (m)**	**Horizontal Error (m)**
19	0.004	0.026	0.010	0.026	0.002	0.015
46	0.039	0.034	0.006	0.021	0.010	0.048
52	0.361	0.061	0.111	0.077	0.325	0.038
64	0.193	0.041	0.286	0.061	0.209	0.083
73	0.009	0.045	0.041	0.068	0.023	0.051

4.2. M3C2 Results

The errors and misalignments, as described in Section 4.1, tend to have a substantial impact on the accuracy of M3C2 results, to which the only solution is to align the point clouds. According to the test results, the alignment of the full point cloud was ineffective because of the long and narrow extent of the survey area, with unevenness shown in the distribution of misalignment and error (Table 1). The scale of the unevenness was about 300 m (about 1.8×10^4 m^2), which was the largest theoretical scale for point cloud alignment in this study area. However, QGIS is more prone to collapse when loading point clouds (M3C2 results) at that size (approximately 200 million points), which is a necessary step to determine the edges of the moving boulder later. To deal with this issue, the area was divided into 18 sub-areas (the length range was 60–100 m, and the average area was 4800 m^2, Figure 9), within which the alignment and M3C2 algorithms were run separately for each sub-area, thereby also significantly reducing the runtime. The alignment algorithm relies on fine registration (iterative closest point (ICP)). To improve the outcome of alignment, the root mean square (RMS) difference in the alignment algorithm was set to 10^{-6} m, the final overlap was set to 100%, and the random sampling limit was set to 100,000 points.

Figure 9. Sub-areas of the site used to improve point cloud alignment and computational efficiency.

The cylinder diameter, d, was set to 0.1 m in the M3C2 algorithm, which is conducive to avoiding invalid M3C2 distances (i.e., no point in the comparison cloud can be found to calculate the distance). The maximum length of the cylinder was set to 5 m, and only the vertical normal vectors were used. After calculation, the core point with the added M3C2 result was maintained in its original 3D position and outputted as a temporary file. The visualisation of the resulting M3C2 distances (Figure 10) clearly highlighted unchanged regions, thus indicating successful alignment and boulder movement.

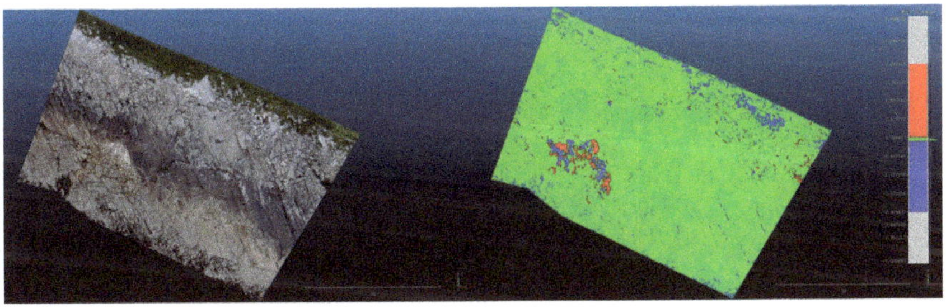

Figure 10. Visualization of M3C2 results in CloudCompare, where red (positive value) indicates the appearance of the boulder, blue (negative value) indicates the disappearance of the boulder, and green (value close to 0) indicates no change. To highlight boulder movements, the colour bar was adjusted to display the core points with M3C2 values between −0.15 m and 0.15 m in green, and the core points below −0.15 m and above 0.15 m in blue and red, respectively.

4.3. The Volume and Axial Lengths of the Boulder

The differential volume method based on the M3C2 distance is not suitable for volumetric calculations in all cases of moving boulders. If the moving distance is less than the scale of the boulder itself, the M3C2 algorithm cannot output all the elevation changes of the whole boulder unit. As shown in Figure 11, the M3C2 value of a part of this boulder (the blank area between red and blue) is close to 0, which makes it impossible to calculate the volume. Furthermore, if the area where the boulder contacts the surface changes, as shown in Figure 12, the M3C2 value is unfit to represent the thickness of the boulder. The stacking of boulders is common on this platform; again, this causes problems, as part of a stacked boulder can be suspended with a void between it and the underlying surface (Figure 13). Since the volume of the suspended part cannot be removed by the M3C2 distance, a considerable error arises.

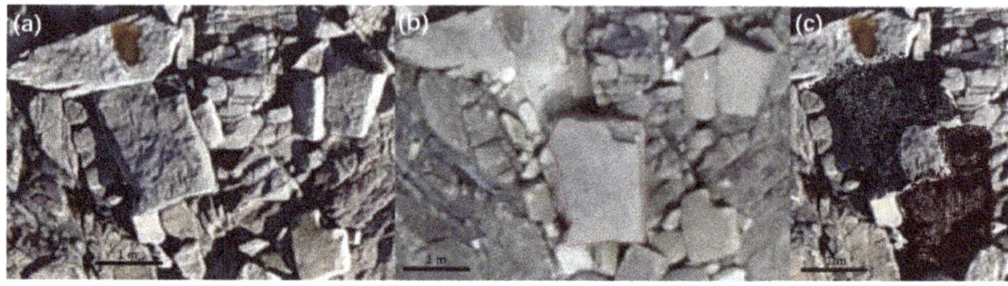

Figure 11. Example of a boulder moving less than the scale itself. (**a**,**b**) show the location of the boulder in 2017 and 2018, respectively; (**c**) shows the M3C2 result of this boulder.

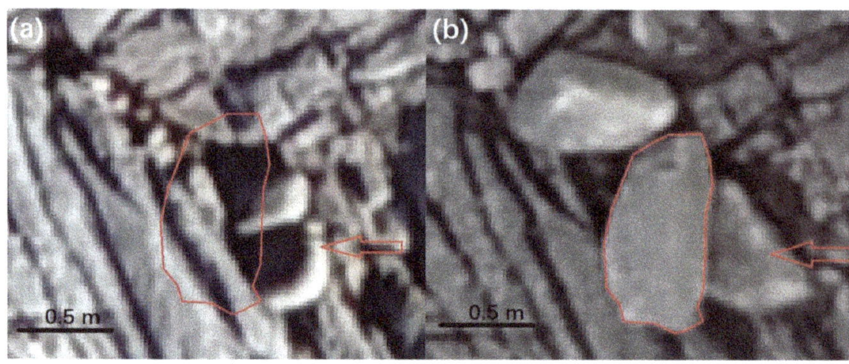

Figure 12. An example of surface changes, where boulders in 2017 (**a**) moved away and new boulders moved in by 2018 (**b**), meaning that the surface beneath the boulders was not acquired in either survey.

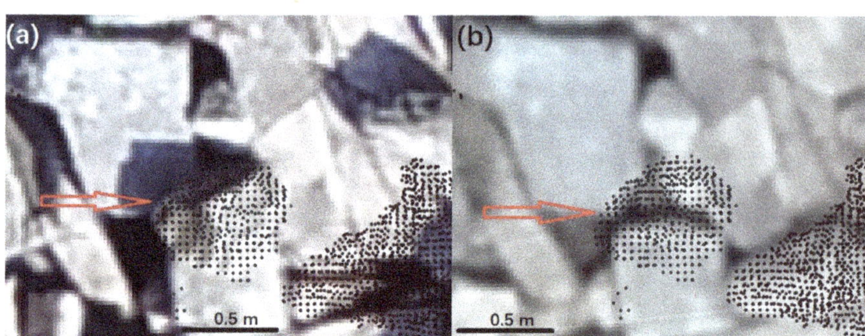

Figure 13. Partially suspended boulder, where (**a**,**b**) indicate the location of the boulder in 2017 and 2018 (disappeared), respectively.

Further issues arise when a boulder boundary is unrecognizable. Such cases include boulders with blurry borders, being shadowed, or being obscured by other boulders. Finally, orthomosaic misalignment (which occurs in the southeastern corner of the 2019 image) leads to issues in capturing the appropriate surface relative to the boulder. Boulder boundaries in sub-areas 16–18 in 2019 were not used for volume calculations. These three sub-areas can only be used to calculate the volume of the boulders whose boundaries were determined by the images from 2017 and 2018, and there is no way to calculate the volume of the boulders photographed in 2019. Finally, if it is determined that the same boulder appears twice in the M3C2 results (a boulder disappeared from one place in 2017 but appeared in another place in 2018, as shown in Figure 14), only the better one, such as the one with a clearer boundary, is used to calculate its volume. In this study, the same boulders corresponded with the same surface textures and edge shapes, which could identify sliding or flipping boulders. Most of the larger boulders in the study area are flattened, which helps them avoid complex rotational movements.

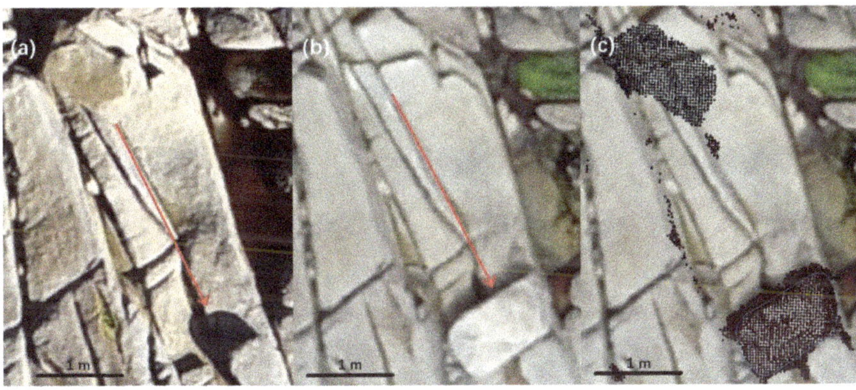

Figure 14. The movement path of a boulder. (**a**,**b**) indicate the location of the boulder in 2017 and 2018, respectively, while (**c**) shows the M3C2 results. The location in 2018 (red) was used to calculate the boulder volume.

Comparing the change in shore platform point clouds between 2017 and 2018 (Figure 15a), 517 moving boulders, whose volume can be calculated, were identified in the study area. Among them, 202 boulders were located in the intertidal zone and 315 in the supratidal zone. According to the comparison between 2018 and 2019, the number of boulders whose volume could be calculated decreased to 378 (Figure 15b), of which, 256 boulders were in the intertidal and 122 in the supratidal zone.

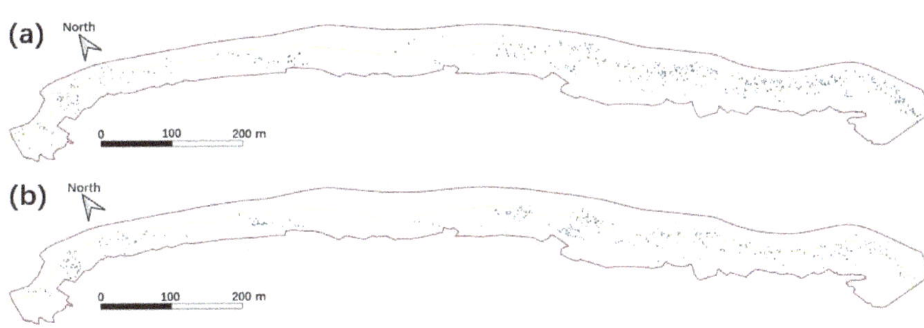

Figure 15. The distribution of volume-computable boulders, where (**a**) shows the comparison of 2017–2018, and (**b**) shows the comparison of 2018–2019. The red boundary represents the study area, the blue boundaries represent the boulders, and the yellow line represents the boundary between the intertidal and supratidal zones.

All volume-computable boulders used the code mentioned in Section 3.5 to determine volume and c-axis length; then, the a- and b-axis lengths of the 20 largest boulders in both comparisons were measured in QGIS, and the estimated volumes of these boulders were calculated (Table 2). In conventional boulder volume estimates, the lengths of the a-, b-, and c-axes are used, with the boulders assumed to be idealised geometries, such as cuboids or ellipsoids. Some researchers use the correction factor V_{dGPS}/V_{abc} to make the estimated volumes (product of a, b, and c) more accurate, where the V_{dGPS} is the boulder volume, calculated using the coordinates of the boulder vertices and edges that were measured by dGPS [10,33,34]. According to previous studies [10,33,34], the value of the correction factor is in the range of 0.5–0.8, and the rather cubic boulders have higher correction factors. The standard ellipsoid volume was used to calculate the estimated volume in this study due to the large number of boulders. The volume of the ellipsoid is expressed as $V = \pi abc/6$, which corresponded to a correction factor of $\pi/6$ (0.524).

Table 2. The 20 largest boulders in the two comparisons, with the axial length, estimated volume (V_e), M3C2 volume (V_M), and error percentage (e_p) calculated.

Years	Sequence	a-Axis (m)	b-Axis (m)	c-Axis (m)	$V_e(m^3)$	$V_M(m^3)$	e_p
2017–2018	1	2.488	1.476	0.835	1.605	1.854	−13.4%
	2	1.952	1.454	1.008	1.498	1.512	−1.0%
	3	1.314	1.178	0.885	0.717	0.935	−23.3%
	4	1.624	0.917	0.854	0.666	0.837	−20.5%
	5	1.839	1.245	0.609	0.730	0.685	6.7%
	6	1.548	1.077	0.852	0.744	0.640	16.2%
	7	1.670	0.925	0.560	0.453	0.594	−23.8%
	8	1.204	1.118	0.872	0.614	0.549	11.9%
	9	1.598	0.830	0.713	0.495	0.498	−0.4%
	10	1.554	0.786	0.663	0.424	0.474	−10.5%
	11	1.405	0.814	0.776	0.465	0.471	−1.3%
	12	1.389	0.962	0.721	0.504	0.468	7.7%
	13	1.348	1.013	0.483	0345	0.429	−19.5%
	14	1.644	1.207	0.476	0.495	0.427	16.0%
	15	1.642	0.725	0.619	0.386	0.424	−9.1%
	16	1.191	0.852	0.758	0.403	0.418	−3.7%
	17	1.119	1.100	0.796	0.513	0.405	26.8%
	18	1.151	0.827	0.637	0.318	0.404	21.4%
	19	1.614	1.071	0.314	0.284	0.385	26.3%
	20	1.559	1.094	0.365	0.326	0.378	13.9%
2018–2019	1	1.311	1.119	0.982	0.754	0.954	21.0%
	2	1.305	1.088	0.663	0.493	0.582	15.3%
	3	1.269	0.844	0.858	0.481	0.409	17.6%
	4	1.869	0.867	0.490	0.416	0.390	6.6%
	5	1.410	0.849	0.624	0.391	0.363	7.7%
	6	1.219	0.815	0.643	0.335	0.356	−6.0%
	7	1.316	0.892	0.561	0.345	0.340	1.5%
	8	1.407	0.888	0.698	0.457	0.323	41.2%
	9	1.281	0.548	0.665	0.244	0.309	−21.0%
	10	0.895	0.892	0.904	0.378	0.303	24.8%
	11	1.349	0.850	0.570	0.342	0.292	17.2%
	12	1.080	0.780	0.602	0.266	0.291	−8.6%
	13	1.286	0.773	0.441	0.230	0.286	−19.7%
	14	1.429	0.791	0.503	0.298	0.278	7.2%
	15	1.060	0.744	0.565	0.233	0.255	−8.4
	16	1.329	0.697	0.513	0.249	0.249	0.0%
	17	0.958	0.850	0.590	0.252	0.248	1.4%
	18	1.302	0.843	0.483	0.278	0.238	16.7%
	19	0.972	0.679	0.552	0.191	0.226	−15.7%
	20	1.215	0.645	0.521	0.214	0.219	−2.2%

As revealed by the comparison between 2017 and 2018 (Table 2), 17 out of the 20 largest boulders were in the intertidal zone. The M3C2 volume of these boulders ranged between 0.38 m^3 and 1.85 m^3, while the a-axis length varied between 1.12 m and 2.49 m. Comparatively, the conventional estimates of volume range from 0.28 m^3 to 1.61 m^3. The estimated volumes are usually smaller than the M3C2 volumes, with their error between 0.43% and 26.75%. Of all the 517 boulders, 86.8% (449) had an M3C2 volume of less than 0.2 m^3 (Figure 16).

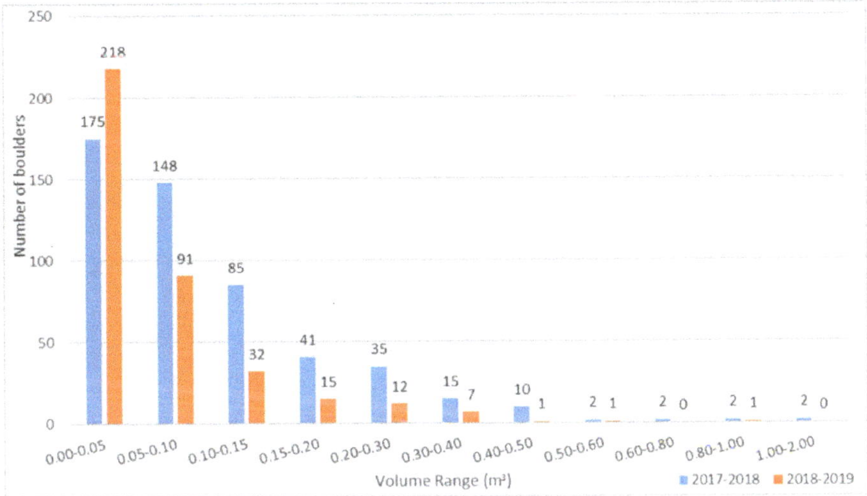

Figure 16. Volume distribution of moving boulders in different time periods. More than half of the movable boulders were smaller than 0.1 m^3, especially in the 2018–2019 period. The 2017–2018 period saw a greater number of larger boulders being moved.

According to the comparison between 2018 and 2019 (Table 2), 18 out of the 20 largest boulders were in the intertidal zone. Their M3C2 volume and a-axis length were lower compared to the 2017–2018 period and were 0.22–0.95 m^3 and 0.90–1.87 m, respectively. The conventional estimate of volume ranges between 0.19 m^3 and 0.75 m^3, and their error ranges from 0.03% to 41.19%. The M3C2 volume of all 378 boulders was less than 1 m^3 and was mostly below 0.2 m^3 (94.2%, 356 boulders, Figure 16).

4.4. Accuracy Verification Results

The volumes and axial lengths of the three test box groups were calculated using the same method as for the real boulders (Figure 17); the results of which are shown in Table 3. For the volume of M3C2, the error of the three box groups was around 0.01 m^3, which was more favorable to the box group with a larger real volume. The error of the larger box groups (groups 1 and 3) was around 1%, while it was 3.9% for the smaller box group (group 2). For the a-axis and b-axis, the error of measurement result was about 1 cm (the error was less than 2%), while that of the c-axis ranged from 1.5 cm to 3 cm (the error percentage was up to 4.4%).

Table 3. Theoretical accuracy of the M3C2 method.

ID	a-Axis			b-Axis		
	Measured Value (m)	True Value (m)	e_p	Measured Value (m)	True Value (m)	e_p
1	1.504	1.488	1.1%	1.057	1.054	0.3%
2	1.053	1.054	−0.1%	0.506	0.496	2.0%
3	1.056	1.054	0.2%	0.754	0.742	1.6%
ID	c-Axis			Volume		
	Computed Value (m)	True Value (m)	e_p	Computed Value (m^3)	True Value (m^3)	e_p
1	0.775	0.742	4.4%	1.151	1.164	−1.1%
2	0.773	0.742	4.2%	0.373	0.388	−3.9%
3	1.502	1.486	1.1%	1.176	1.164	1.0%

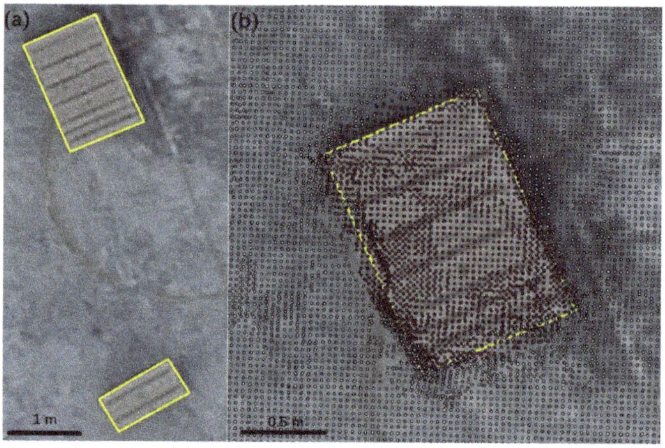

Figure 17. Accuracy was verified using box groups. (**a**) The orthomosaic and the boundaries of box groups 1 and 2. (**b**) The M3C2 results of box group 1, where the red dots indicate the values in excess of 0.15 m (box height), and the white dots indicate the values close to 0 (unchanged area).

5. Discussion

5.1. Benefits of Differential Volume Calculation Based on M3C2 Distance

The theoretical accuracy that can be achieved using differential volume calculation based on the M3C2 distance has been demonstrated here. The error of the volume calculation result was only 1–3.9%, which is much smaller than the error of the boulder volume estimated using the conventional method of tape measuring (up to 36.9% [5]). In this study, there remained a significant error in the boulder volume estimated using the conventional method (up to 41.19%, Table 2) compared with the M3C2 volume, even though the accuracy of axial length calculation in this study has been improved compared to the manual tape measure [5,6]. The reduction of error in the M3C2 volume was attributed to the use of the real boundary of the boulder and the gridded boulder thickness statistics, which reduce the error caused by the conventional method of boulder volume estimation where the boulder is assumed to have an ideal geometry. In addition, the M3C2 distance was treated as the thickness of the boulder when the elevation change of the ground was taken into account, which solved the error caused by simplifying the bottom of the boulder to a plain surface. This is because other volume measurement methods are not applicable to obtaining the bottom information of the boulder. In addition to reducing volume errors, there are other advantages shown by the equipment and UAS survey parameters required for differential volume calculations based on M3C2 distances, such as low cost, high efficiency, and excellent safety performance. The cost of the UAS and dGPS systems are significantly lower than that of the TLS system, with higher measurement accuracy. Further, it is needless for researchers to perform dangerous measurement work on-site (such as climbing boulders, handling equipment, etc.). In terms of efficiency, with only nadir photography required, there will be more time saved for coastal boulder surveys in larger sites (e.g., a range of several kilometres) compared with oblique photography. In future work, UAS survey parameters can be adjusted according to actual needs to change the balance between accuracy and efficiency. For instance, given a confined research area, it is worthwhile to consider reducing GSD (e.g., 1 cm) or adding oblique images, which can help improve the accuracy of the 3D reconstruction model and volume calculation, despite the longer survey time.

5.2. Analysis of the Error between the M3C2 Volume and Real Volume of the Boulders

There remain errors between the M3C2 volumes and the real volumes of boulders, which to a large extent, result from the measurement of boulders that are not located on the rock platform surface but are suspended above it by resting on other boulders. Although all boulders were screened before the calculation of M3C2 volumes, and the evidently suspended boulders have been removed, it is unavoidable to have hidden suspensions that are not obvious from the orthomosaic. As indicated by the blue area in Figure 18, this inconspicuous suspension is caused by the shape of the boulder itself rather than other boulders. Located under the boulder, it can neither be displayed in the orthomosaic nor inferred from the surrounding objects. Since the image taken, based on the nadir photography, fails to provide sufficient detail required for the 3D reconstruction of the suspension, the actual outcome of the point cloud model reconstruction is marked by the red circles in Figure 18. The M3C2 value is ineffective at removing the height of the suspended part, which makes the M3C2 volume slightly larger than the real volume.

Figure 18. Errors caused by differential volume calculation based on M3C2 distance. The blue areas indicate the error area resulting from the suspension of the boulder. Red circles represent the point cloud model.

5.3. The Relationship between Coastal Boulder Movement and Storm Intensity

Previous studies and measurements [6,35] have shown that the initiating wave height of coastal boulders is proportional to the boulder size (axial length and volume). According to the results in Figure 16 and Table 2, the total number and maximum volume of boulders moved in 2017–2018 (517, 1.854 m^3) were significantly larger than those in 2018–2019 (378, 0.954 m^3). Storm intensity and wave height in 2017–2018 were greater than in 2018–2019, as confirmed by a review of storm observation data collected within the time frame of this study (Table 4). The reduction in the intensity of the storms in 2018–2019 resulted in lower boulder mobility, which allowed more boulders to remain in place without moving or moving out of the study area.

Table 4. Storms affecting the northeast Atlantic between 18 September 2017 (survey in 2017) and 24 May 2019 (survey in 2019), with references to the maximum wave height and wind speed recorded in each [36–38].

Years	Name	Appearance Time	Maximum Wind Speed (mph)	Maximum Wave Height (m)
2017–2018	Ophelia	16–20 October 2017	119	13.828
	Brian	19–23 October 2017	85	15.938
	Caroline	6–11 December 2017	93	11.25
	Dylan	30 December 2017–3 January 2018	77	20.156
	Eleanor	2–5 January 2018	140	20.156
	Fionn	14–21 January 2018	147	20.625
	David	17–21 January 2018	126	20.625
	Georgina	23–27 January 2018	140	13.125
	Emma	26 February–7 March 2018	142	No Data

Table 4. *Cont.*

Years	Name	Appearance Time	Maximum Wind Speed (mph)	Maximum Wave Height (m)
2018–2019	Hector	13–17 June 2018	70	10.078
	Helene	16–21 September 2018	78	9.375
	Ali	17–22 September 2018	102	9.375
	Bronagh	20–25 September 2018	96	8.438
	Callum	10–16 October 2018	76	12.188
	Diana	27–30 November 2018	110	15.703
	Enk	7–14 February 2019	86	10.547
	Gareth	11–14 March 2019	81	10.078
	Hannah	25–28 April 2019	82	10.078

6. Conclusions

Through differential volume calculations based on M3C2 distances, the details of moving boulder bottoms were identified through comparisons, at least between different datasets obtained at different times, and the simplified treatment of boulder bottoms used in previous studies was optimized. By using accurate boulder boundaries and cm-scale gridded thickness statistics, this method improved the accuracy of boulder volume calculations (as low as 1.0–3.9%) compared to assuming the boulder has an ideal geometry. When only nadir photography was used, there was a considerable improvement in the endurance time of the UAS, making it possible to survey greater lengths of the coast. In addition, compared with the TLS system and the 5-lens oblique camera with higher precision, the differential method based on M3C2 distance can significantly reduce the cost of conducting a coastal survey while ensuring a relatively high level of accuracy, which is conducive to the development of higher-accuracy boulder (as well as ecological) surveys in the future.

In this study, volumes were calculated for 517 moving boulders along a rocky shoreline in northwest Ireland, for the 2017–2018 period. However, the number of moving boulders decreased to 378 in the 2018–2019 period.

In future research, the accuracy of boulder volume calculation can be further improved by reducing the GSD or adding oblique images. UAS surveys of coastal boulders could also be conducted before and after storms, and the model can be used to track the movement of single boulders in response to this forcing.

Author Contributions: Conceptualization, H.B. and D.G.; methodology, H.B. and Y.Y.; software, H.B. and Y.Y.; formal analysis, Y.Y.; investigation, Y.Y., H.B. and J.K.; data curation, H.B.; writing—original draft preparation, Y.Y.; writing—review and editing, H.B. and J.K.; supervision, H.B. and D.G. All authors have read and agreed to the published version of the manuscript.

Funding: This research received no external funding.

Data Availability Statement: Publicly available datasets were analysed in this study. These data can be found here: [https://1drv.ms/u/s!AkrDDM8CislFhWLPtnwcs7bwW7_E?e=1C0IUz] (accessed on 30 January 2023).

Conflicts of Interest: The authors declare no conflict of interest.

References

1. Li, J.; Xue, J.; Wang, W.; Zhang, J.; Xu, Y. Analysis and enlightenment of the overseas important typhoon disaster in 2013. *Meteorol. Disaster Reduct. Res.* **2014**, *37*, 50–54.
2. Nagle-McNaughton, T.; Cox, R. Measuring Change Using Quantitative Differencing of Repeat Structure-From-Motion Photogrammetry: The Effect of Storms on Coastal Boulder Deposits. *Remote Sens.* **2020**, *12*, 42. [CrossRef]
3. Sedrati, M.; Morales, J.A.; El M'rini, A.; Anthony, E.J.; Bulot, G.; Le Gall, R.; Tadibaght, A. Using UAV and Structure-From-Motion Photogrammetry for the Detection of Boulder Movement by Storms on a Rocky Shore Platform in Laghdira, Northwest Morocco. *Remote Sens.* **2022**, *14*, 4102. [CrossRef]
4. Gienko, G.A.; Terry, J.P. Three-dimensional modeling of coastal boulders using multi-view image measurements. *Earth Surf. Process. Landf.* **2014**, *39*, 853–864. [CrossRef]

5. Boesl, F.; Engel, M.; Eco, R.C.; Galang, J.B.; Gonzalo, L.A.; Llanes, F.; Quix, E.; Brückner, H. Digital mapping of coastal boulders—High-resolution data acquisition to infer past and recent transport dynamics. *Sedimentology* **2020**, *67*, 1393–1410. [CrossRef]
6. Liu, Z.; Zhou, L.; Gao, S. Application of the terrestrial laser scanner to the coastal boulders on the southern coast of Hainan Island. *Haiyang Xuebao* **2019**, *41*, 127–141.
7. Terry, J.P.; Dunne, K.; Jankaew, K. Prehistorical frequency of high-energy marine inundation events driven by typhoons in the Bay of Bangkok (Thailand), interpreted from coastal carbonate boulders. *Earth Surf. Process. Landf.* **2016**, *41*, 553–562. [CrossRef]
8. Kennedy, A.B.; Mori, N.; Yasuda, T.; Shimozono, T.; Tomiczek, T.; Donahue, A.; Shimura, T.; Imai, Y. Extreme block and boulder transport along a cliffed coastline (Calicoan Island, Philippines) during Super Typhoon Haiyan. *Mar. Geol.* **2017**, *383*, 65–77. [CrossRef]
9. Goto, K.; Okada, K.; Imamura, F. Characteristics and hydrodynamics of boulders transported by storm waves at Kudaka Island, Japan. *Mar. Geol.* **2009**, *262*, 14–24. [CrossRef]
10. May, S.M.; Engel, M.; Brill, D.; Cuadra, C.; Lagmay, A.M.F.; Santiago, J.; Suarez, J.K.; Reyes, M.; Brückner, H. Block and boulder transport in eastern Samar (Philippines) during Supertyphoon Haiyan. *Earth Surf. Dyn.* **2015**, *3*, 543–558. [CrossRef]
11. Erdmann, W.; Kelletat, D.; Scheffers, A. Boulder transport by storms—Extreme-waves in the coastal zone of the Irish west coast. *Mar. Geol.* **2018**, *399*, 1–13. [CrossRef]
12. Terry, J.P.; Jankaew, K.; Dunne, K. Coastal vulnerability to typhoon inundation in the Bay of Bangkok, Thailand? Evidence from carbonate boulder deposits on Ko Larn Island. *Estuar. Coast. Shelf Sci.* **2015**, *165*, 261–269. [CrossRef]
13. Frohlich, C.; Hornbach, M.J.; Taylor, F.W.; Shen, C.-C.; Moala, A.; Morton, A.E.; Kruger, J. Huge erratic boulders in Tonga deposited by a prehistoric tsunami. *Geology* **2009**, *37*, 131–134. [CrossRef]
14. Terry, J.P.; Lau, A.Y.A.; Etienne, S. *Outlook for Boulder Studies within Tropical Geomorphology and Coastal Hazard Research*; Springer: Berlin/Heidelberg, Germany, 2013; pp. 97–102.
15. Terry, J.P.; Lau, A.Y.A.; Etienne, S. *Reef-Platform Coral Boulders*; Springer: Berlin/Heidelberg, Germany, 2013; pp. 97–102.
16. Morton, R.A.; Gelfenbaum, G.; Jaffe, B.E. Physical criteria for distinguishing sandy tsunami and storm deposits using modern examples. *Sediment. Geol.* **2007**, *200*, 184–207. [CrossRef]
17. Hall, A.M.; Hansom, J.D.; Williams, D.M.; Jarvis, J. Distribution, geomorphology and lithofacies of cliff-top storm deposits: Examples from the high-energy coasts of Scotland and Ireland. *Mar. Geol.* **2006**, *232*, 131–155. [CrossRef]
18. Spiske, M.; Böröcz, Z.; Bahlburg, H. The role of porosity in discriminating between tsunami and hurricane emplacement of boulders—A case study from the Lesser Antilles, southern Caribbean. *Earth Planet. Sci. Lett.* **2008**, *268*, 384–396. [CrossRef]
19. Etienne, S.; Paris, R. Boulder accumulations related to storms on the south coast of the Reykjanes Peninsula (Iceland). *Geomorphology* **2010**, *114*, 55–70. [CrossRef]
20. Nandasena, N.A.K.; Paris, R.; Tanaka, N. Reassessment of hydrodynamic equations: Minimum flow velocity to initiate boulder transport by high energy events (storms, tsunamis). *Mar. Geol.* **2011**, *281*, 70–84. [CrossRef]
21. Lague, D.; Brodu, N.; Leroux, J. Accurate 3D comparison of complex topography with terrestrial laser scanner: Application to the Rangitikei canyon (N-Z). *ISPRS J. Photogramm. Remote Sens.* **2013**, *82*, 10–26. [CrossRef]
22. James, M.R.; Robson, S.; Smith, M.W. 3-D uncertainty-based topographic change detection with structure-from-motion photogrammetry: Precision maps for ground control and directly georeferenced surveys. *Earth Surf. Process. Landf.* **2017**, *42*, 1769–1788. [CrossRef]
23. Knight, J.; Burningham, H.; Barrett-Mold, C. The geomorphology and controls on development of a boulder-strewn rock platform, NW Ireland. *J. Coast. Res.* **2009**, *56*, 1646–1650.
24. Knight, J.; Burningham, H. Boulder dynamics on an Atlantic-facing coastline, northwest Ireland. *Mar. Geol.* **2011**, *283*, 56–65. [CrossRef]
25. Scheffers, A.; Kelletat, D.; Haslett, S.; Scheffers, S.; Browne, T. Coastal boulder deposits in Galway Bay and the Aran Islands, western Ireland. *Z. Geomorphol.* **2010**, *54*, 247–279. [CrossRef]
26. Cox, R.; Jahn, K.L.; Watkins, O.G.; Cox, P. Extraordinary boulder transport by storm waves (west of Ireland, winter 2013–2014), and criteria for analysing coastal boulder deposits. *Earth Sci. Rev.* **2018**, *177*, 623–636. [CrossRef]
27. Furukawa, Y.; Curless, B.; Seitz, S. Towards internet-scale multi-view stereo. In Proceedings of the 2010 IEEE Conference on Computer Vision and Pattern Recognition, San Francisco, CA, USA, 13–18 June 2010; pp. 1434–1441.
28. Grenzdörffer, G.J.; Engel, A.; Teichert, B. The photogrammetric potential of low-cost UAVs in forestry and agriculture. *Int. Arch. Photogramm. Remote Sens. Spat. Inf. Sci.* **2008**, *31*, 1207–1214.
29. Mosbrucker, A.; Major, J.; Spicer, K.; Pitlick, J. Camera system considerations for geomorphic applications of SfM photogrammetry. *Earth Surf. Process. Landf.* **2017**, *42*, 969–986. [CrossRef]
30. Henry, J.; Malet, J.; Maquaire, O.; Grussenmeyer, P. The use of small-format and low-altitude aerial photos for the realization of high resolution DEMs in mountainous areas: Application to the Super-Sauze earthflow (Alpes-de-Haute-Provence, France). *Earth Surf. Process. Landf.* **2002**, *27*, 1339–1350. [CrossRef]
31. Torres-Sánchez, J.; López-Granados, F.; Borra-Serrano, I.; Peña, J. Assessing UAV-collected image overlap influence on computation time and digital surface model accuracy in olive orchards. *Precis. Agric.* **2017**, *19*, 115–133. [CrossRef]
32. Mesas-Carrascosa, F.; Notario García, M.; Meroño de Larriva, J.; García-Ferrer, A. An Analysis of the Influence of Flight Parameters in the Generation of Unmanned Aerial Vehicle (UAV) Orthomosaicks to Survey Archaeological Areas. *Sensors* **2016**, *16*, 1838. [CrossRef]

33. Engel, M.; May, S.M. Bonaire's boulder fields revisited: Evidence for Holocene tsunami impact on the Leeward Antilles. *Quat. Sci. Rev.* **2012**, *54*, 126–141. [CrossRef]
34. Hoffmeister, D.; Ntageretzis, K.; Aasen, H.; Curdt, C.; Hadler, H.; Willershäuser, T.; Bareth, G.; Brückner, H.; Vött, A. 3D model-based estimations of volume and mass of high-energy dislocated boulders in coastal areas of Greece by terrestrial laser scanning. *Z. Geomorphol.* **2014**, *58*, 115–135. [CrossRef]
35. Nott, J. Waves, coastal boulder deposits and the importance of the pre-transport setting. *Earth Planet. Sci. Lett.* **2003**, *210*, 269–276. [CrossRef]
36. UK Storm Season 2017/18. Available online: https://www.metoffice.gov.uk/weather/warnings-and-advice/uk-storm-centre/uk-storm-season-2017-18 (accessed on 24 April 2019).
37. UK Storm Season 2018/19. Available online: https://www.metoffice.gov.uk/weather/warnings-and-advice/uk-storm-centre/uk-storm-season-2018-19 (accessed on 17 August 2020).
38. Irish Weather Buoy Network Observations. Available online: http://www.marine.ie/site-area/data-services/real-time-observations/irish-weather-buoy-network-observations (accessed on 21 December 2022).

remote sensing

Article

Mapping Environmental Impacts on Coastal Tourist Areas of Oceanic Islands (Gran Canaria, Canary Islands): A Current and Future Scenarios Assessment

Leví García-Romero [1,2,*], Teresa Carreira-Galbán [1,3], José Ángel Rodríguez-Báez [4], Pablo Máyer-Suárez [1], Luis Hernández-Calvento [1] and Amalia Yánes-Luque [4]

1 Grupo de Geografía Física y Medio Ambiente, Instituto de Oceanografía y Cambio Global, IOCAG, Universidad de Las Palmas de Gran Canaria, ULPGC, 35214 Telde, Spain
2 Grupo de Geopatrimonio y Geoturismo en Espacios Volcánicos, GeoTurVol-Departamento de Geografía e Historia, Universidad de La Laguna, 38200 San Cristobal de La Laguna, Spain
3 ECOS, Estudios Ambientales y Oceanografía S.L., 35013 Las Palmas de Gran Canaria, Spain
4 Territorial Management and Risks Group (GEORIESGOS), University of La Laguna, 38200 San Cristobal de La Laguna, Spain
* Correspondence: levi.garcia@ulpgc.es

Citation: García-Romero, L.; Carreira-Galbán, T.; Rodríguez-Báez, J.Á.; Máyer-Suárez, P.; Hernández-Calvento, L.; Yánes-Luque, A. Mapping Environmental Impacts on Coastal Tourist Areas of Oceanic Islands (Gran Canaria, Canary Islands): A Current and Future Scenarios Assessment. *Remote Sens.* **2023**, *15*, 1586. https://doi.org/10.3390/rs15061586

Academic Editors: José Juan de Sanjosé Blasco, Germán Flor-Blanco and Ramón Blanco Chao

Received: 5 February 2023
Revised: 6 March 2023
Accepted: 12 March 2023
Published: 14 March 2023

Abstract: The combination of the current sea level rise (SLR) and an increase in the frequency and severity of sea storm events and surges, will have important environmental impacts on coastal tourist areas located in oceanic islands that are often heavily dependent economically on tourism. Importantly, coastal tourist resorts and their associated beaches are commonly located in the S-SW of these islands where they are protected from the effects of the NE trade winds but exposed to these storms. This study analyses the current and future conditions of SW sea storm events. Then, through mapping using orthophotos, LiDAR data, and bathymetry with high spatial resolution, it studies the environmental effects that these events are having now and will have in the future on tourist zones of oceanic islands. This is of particular importance, as in the case of the Macaronesia islands, SW sea storm events are occurring with ever greater frequency. The case study considers the SW sector of the island of Gran Canaria (Canary Islands, Spain). Regarding the analysis of SW storms, the results obtained show an increase in the frequency of events, especially since the 1990s, with 104 events detected (67.5% between 1958 and 2018). The most affected areas during these storms are usually beaches (severe erosive processes). Breakwaters and seafronts usually resist the impact of waves well, however, as is discussed that in the future scenario, their implementation will mean a high economic cost for local administrations.

Keywords: climate change; tourist beach; socio-environmental effects; SW sea storms; Macaronesia; mapping

1. Introduction

Tsunamis and sea storm events around the world have highlighted the vulnerability of many coastal populations, including those serving tourists (e.g., the 26 December 2004 tsunami around the Indian Ocean). In Kellman et al. (2008), four main themes were considered with relevance to disaster risk reduction: information and awareness, warning systems, personal preparation, and local livelihoods [1]. The authors suggested three important areas for further investigation: the connections between sustainable tourism and disaster vulnerability, the role of tourists in disasters, and disaster risk reduction education.

As well as the potential impact of sea storm events, the vulnerability of coastal areas increases when other stress-inducing factors come into play, including, for example, the rise in human population, urbanisation processes, the sea level rise (SLR), and what has been termed 'coastal rigidity' [2,3]. These stress-inducing factors produce the phenomenon known as coastal squeeze [4], defined as "one form of coastal habitat loss, where intertidal habitat is lost due to the highwater mark being fixed by a defence or structure (i.e., the high water mark residing against a hard structure such as a sea wall) and the low water mark migrating landwards in response to SLR" [5]. This process is aggravated "in areas where land claim or coastal defence has created a static, artificial margin between land and sea, or where the land rises relative to the coastal plain, and habitats become squeezed into a narrowing zone" between the ocean (with a rising level) and an inland obstruction [4].

Coastal scientists are currently predicting a higher incidence of storm surges and a general SLR [6–8] based on the available climate data record. These data also reveal an effect known as coastal narrowing, which has been defined as "decreases in coastal zone width", with one example being "saltmarsh erosion caused by increases in the wind/wave climate or the migration of intertidal channels" [5]. Coastal narrowing, however, is not related to SLR.

The importance of coastal areas for the tourism industry (the largest industry in the world according to the World Tourism Organisation (2001) [9] is clear, as indeed is the importance of the industry for coastal areas. The need to protect coastal resources in view of the anticipated SLR is fundamental for the economy of many countries whose main source of income is tourism [10]. This can present certain dilemmas at various spatial scales for the coastal areas themselves [11–13], and especially island coastal areas [14–16]. Coastal managers need to adopt tactics that function in a complementary way to natural processes rather than simply implementing hard engineering solutions [11].

About three-quarters of small countries are islands of fewer than one million inhabitants, and small island regions tend to suffer from economic limitations. At the same time, these regions commonly specialise in tourism due to the availability of accessible natural resources such as beaches, landscapes, and sunny weather. In many cases, this type of region presents a comparative advantage for tourism, and therefore its economy is often strongly dependent on this industry [15].

An evaluation of risks due to sea storm events on the coast requires knowledge about and an understanding of their behaviour, evolution, and consequences (topics that will be considered in this work). In this way, it is also possible to analyse the specific risk for the population and their reaction capacity to adapt and, at the very least, reduce their effects, especially when it comes to small, highly exposed spaces with tourism-based economies [16,17]. For this, the most commonly applied method in the study of risk at a local scale, which aims to explore both the risk and its relationship with exposure and vulnerability, and on this basis estimate the possible economic and other costs [18].

A rise in the mean sea level is considered one of the main effects of climate change. Its impact on coastal areas includes the danger of coastal floods, alterations to erosion rates, and changes and/or losses of coastal ecosystems, among many others. Impacts are also expected on, for example, land use, coastal population settlements, and ecosystem services [19]. The consequences of an SLR in a specific coastal territory can be evaluated by determining the flooding sea level in wave- and swell-exposed areas, and the mapping of these floods is essential for land use planning, supporting engineering decisions regarding flood control measures and the coordination of emergency responses [20]. The anticipated SLR used for the purposes of the present work is taken from the Fifth Assessment Report of the Intergovernmental Panel on Climate Change (IPCC) which, on the basis of the Representative Concentration Pathway 8.5 (RCP 8.5), estimated a mean SLR of 0.74 m (within a possible range of 0.53 to 0.98 m) for the time horizon of 2100 [21].

According to different groups of experts, in Spain, the coasts of the Canary Islands and Galicia will be the most affected by processes linked to climate change [22]. Ocean-related climate changes are especially worrying for the Canary coasts in general, and the Canary beach-dune systems in particular. On the basis of a sea level rise of 0.35 m in the area surrounding the islands, a coastline retreat of between 15 m and 35 m has been predicted, although, due to changes in wave direction, this value could be as high as 70 m in southern coastal stretches [22,23].

The analysis carried out by Yánes-Luque (2021) [24] highlights the high degree of exposure of the coastal tourist areas of the SW Canary Islands, especially Tenerife and S-SW Gran Canaria, to the impacts of waves, due to their coastline having experienced incessant urban growth, which favours the disturbance of these enclaves by destructive physical processes (e.g., marine storms). The marine storms recorded in the study area occur with defined atmospheric patterns, with up to 298 sea storms detected. These storms are the result of fluctuations in the pressure gradient between the Icelandic depression cores and the Azores anticyclone, especially during the winter. According to our research, which assessed the effects of these storm events based on reports in the local press, sea storms ultimately result in drownings, damage to street furniture and harbour infrastructure, and loss of sand on beaches.

Gran Canaria is the second-largest island in the Canary archipelago (1560 km^2). As a result of its location in the Atlantic Ocean and an urbanisation process that has been ongoing over the last few decades (since the 1960s), the coastal tourist resorts of the S-SW of the island have a high degree of exposure to the impacts of waves [24]. Tourism is the mainstay of the Canarian economy, responsible for more than 40% of employment in the archipelago and over 35% of GDP [25,26]. In this area of the island, the density of buildings parallel to the shore is high, further facilitating potential damage through destructive processes such as sea storm events [24].

The study presented in this work is structured following the approach of Pontee (2013) [5], considering and analysing coastal narrowing and coastal squeeze as separate concepts and processes. In this context, the study also has the following specific aims: (i) to analyse sea storm events in the S-SW of Gran Canaria (normally from a SW direction) since 1958 to 2018; (ii) to spatially quantify through mapping the impact of sea storm events under present climate conditions on the coastal narrowing zone and to predict, on the basis of the RCP 8.5 climate change scenario, their impact on the coastal squeeze zone, in both cases on the basis of maps using flooding sea levels obtained from bidimensional numerical models; and (iii) to spatially analyse an approximation of the effects on the study area, both at the present time and in the context of the mean SLR according to RCP 8.5. Furthermore, this work fulfils one of the main tasks to be carried out by the Member States of the European Union in Directive (2007/60/EC) [25], adopted at the end of 2007, where it is proposed that in order to comply with this Directive, the flood hazards and risks in the territory of member states must be mapped.

2. Study Area

The Canary Islands are a typical example of a "sun, sand and sea" tourism destination. The large number of visitors to the islands has resulted in the tourist industry playing a vital role in the economic development of the archipelago. Tourism accounts for between 50% and 80% of its GDP (according to various authors) [26] and it was the second most-visited Spanish regional tourist destination between 2016 and 2019, representing 17.6% of all tourism in Spain [27]. Around 28% of the visitors to the archipelago come to Gran Canaria (the island where the object of the case study presented in this paper is situated), a percentage only surpassed by the island of Tenerife. More specifically, the municipality of Mogán, the focus of this research), represents more than 6% of the tourism in the archipelago, and more than 23% of the tourist visitors to Gran Canaria. It is surpassed only by the municipality of San Bartolomé de Tirajana (south of Gran Canaria), where the tourist activity that began in the 1960s is mainly associated with the natural resources of the beaches and dunes of Maspalomas [28] (Figure 1).

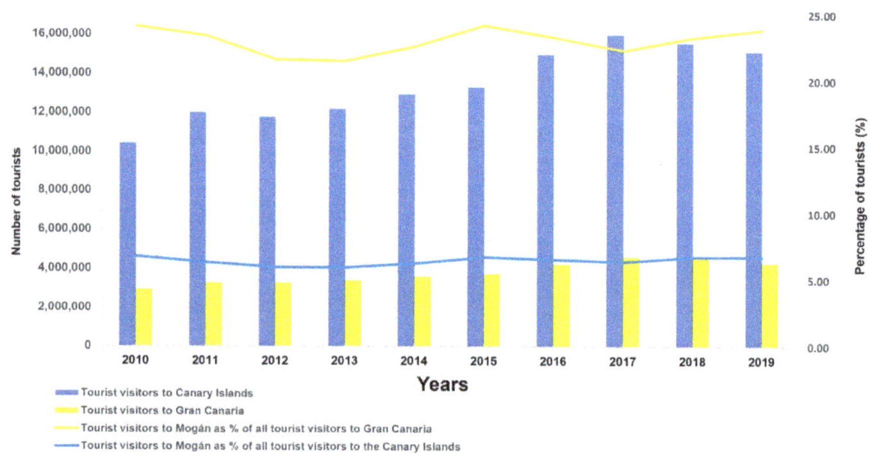

Figure 1. Tourist visits to the Canary Islands, Gran Canaria, and Mogán over a 10-year period (2010–2019).

As a result of the emergence and rapid rise of the tourist industry, urban expansion in Gran Canaria has taken place at a fast pace [24], with a more or less constant rate of construction since the 1960s–1970s [29]. The urbanisation process progressed spatially from the southeast and south (the case of San Bartolomé de Tirajana as explained above) to the northwest and inland from the coast [24]. The area considered in the present study is situated in the southwest of the island and is relatively sheltered from the most important winds that affect the island, the so-called trade winds (known locally as Alisios) which reach the islands through the Azores anticyclone with a predominantly N-NE direction. This protection from these winds is due to mountainous regions of the island blocking their path. The area is additionally protected against the effects of polar front disturbances [30]. These conditions of climatic comfort are an important tourist attraction [31] and have led to considerable urban-tourist development in this area. This pattern is commonly observed on islands in the Atlantic Ocean where the main tourist resorts are located in the shadow that the island itself produces towards the south, southwest, and west in other Macaronesia Islands, including São Miguel, Terceira, São Jorge, Faial and Flores in the Azores archipelago, Madeira, Porto Santo, the islands of Maio, Santiago, Fogo, Boa Vista and Sal (Cape Verde), and other Canary Islands such as Tenerife, La Gomera, Lanzarote, and Fuerteventura.

In Gran Canaria, the area between the beaches of La Carrera and Amadores (13.63 km of coastline) (Figure 2) is characterised by small embedded beaches with short coastal drift and located around the mouths of dry ravines, therefore, with scarce or practically nonexistent sediment supplies. This coastal stretch with seven beaches and five ports is highly vulnerable when storms from the southwest occur, which are especially frequent in winter [24]. The sea storm events recorded in the study area originate in an area with clearly defined atmospheric patterns which are the result of fluctuations in the pressure gradient between the Icelandic depressive nuclei and the Azores anticyclone, particularly during winter. The greater or lesser intensification of this gradient in the North Atlantic translates into a variable coastal impact as the result of increases in swell energy. This does so to the point that the maximum significant wave height (Hs) can be 4 or 5 times higher than its usual value [24]. The same behaviour has also been observed in the different Macaronesian archipelagos, including the Azores [32] and Cape Verde [33], as well as in Madeira [34].

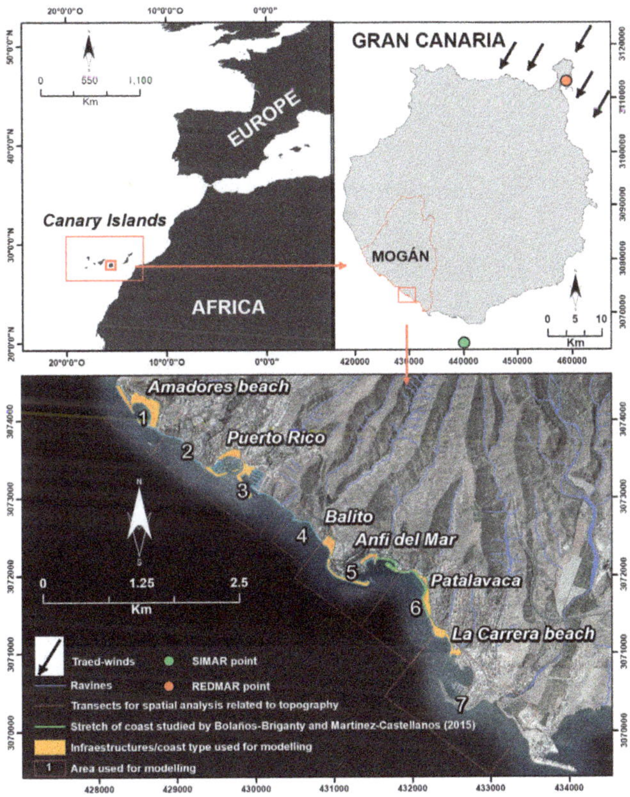

Figure 2. Study area and infrastructure/coastal type used for modelling in the present research study (see legend in the figure; left, down).

3. Materials and Methods

3.1. SW Storms in the Study Area. Estimation of Storm Frequency

A number of sources were used to obtain the data for the analysis of sea storm events in the study area: (i) wind and wave data from the SIMAR-44 wave generation and propagation mathematical models (Spanish State Ports Oceanographic Database). Of these, the "4035005" for the S-SW of Gran Canaria (27.67°N; 15.58°W) was selected (Figure 2). The data provided include 24 daily registers of H_s (m), peak period (T_p, s), wind direction (degrees), and wind direction and speed (degrees; m/s). On the basis of the first two parameters, the maximum wave height (H_{max}, m) and wave length (L, m) can be calculated. Using these data, it is possible to define the usual wave regime and extreme events. The criterion for the extreme events identification is based on the application of the cutoff threshold corresponding to the P_{99} of the mean daily height of the H_s (1.7 m calculated for Mogán). Therefore, a marine storm event is considered to have occurred when the established cutoff wave height threshold is equalled or exceeded for twelve consecutive hours. In addition, the Peak Over Threshold method (Puertos del Estado, Spain) of extreme values is applied, whereby storms are considered to be independent if five days elapse between them; (ii) atmospheric conditions data on the basis of Wetterzentrale and AeMET (the Spanish State Meteorological Agency) weather reports. These were used to determine the synoptic situation that concurs in the origin and evolution of the storms; (iii) newspaper articles which, given their role as a mass media communicator, tend to rapidly report on natural disasters [35–37]. Information on the evolution of the sea storm events, the areas affected by them, and the extent of any damage caused were obtained from the following regional newspapers: Diario de Avisos, El Día, Diario de Las Palmas, La Provincia, and Canarias 7. References to damage reported in previous studies were also included [38].

3.2. Flood Level Modelling Data

To determine the coastal flood sea level, the swell from deep waters was simulated with the aid of bidimensional numerical models. The flood sea level is the sum of the tidal level (meteorological and astronomical) and the runup, whose value—in the case of projection for beaches—depends on the significant wave height at the shoreline, the incident wave peak period, and the slope of the beach [39]. To determine the breakwater overflow level, the wave height at the foot of the structure has to be determined. The criterion used to extract the wave height at the foot of the structure is its depth calculated using the equation $H_{s12} \times 1.57$ [40]. For the purposes of the present study, the Coastal Modelling System was used together with the MOPLA tool, which is a Spanish-developed beach wave propagation, current, and morphodynamic evolution model [41]. The spectral wave propagation was performed in this tool using the OLUCA-SP module [42,43].

Propagation from deep waters requires a statistical characterisation of the swell in order to select the sea state, defined by a wave height and peak period, with respect to which it is intended to establish the flood sea level. The selected sea state for the study area in question is that with a SW component, an $H_{s99\%}$ of 3.84 m, and $T_{p99\%}$ of 8.80 s. The H_{s12} value for this directional component is 4.22 m. The study area is sheltered from the N and NNE winds and swells, which are the most common type on the island, and is primarily affected, in terms of storm floods and damage, by SW or WSW swells. On the basis of the wave characterisation described above, wave propagation was studied in various sea conditions: for the present-day scenario, half-tide and high tide values of 0 and 1.53 m, respectively, were used and, in order to evaluate the effects under an RCP 8.5 scenario change, a half-tide value of 0.74 m and a high tide value of 2.27 m were used. These ranges were taken on the basis of the Las Palmas 2 tide gauge data series (the closest tide gauge to the study area) and with a data series dating back to 1992. The Las Palmas 2 tide gauge is located in Las Palmas Port (28.14°N; 15.40°W, Figure 2) and forms part of the Spanish State Ports Tide Gauge Network [44].

The Coastal Modelling System employed requires the bathymetric and topographic data of the terrain over which the swell is propagated. A continuous digital land-sea model (topography and bathymetry) (Table 1, D L-S M) designed by SDI Canarias-Grafcan S.A (Canary Islands Government) was used. This model has joined a corrected digital terrain model (DTM) from a LiDAR flight (2015; RMSE $\leq$ 0.50 m) and the corrected bathymetry from the eco-cartography of the Canary Islands carried out with an echo sounder (year 2000; RMSE $\leq$ 1 m), making a special analysis of the difference between the geographical zero and the hydrological zero to fit both models (difference calculated for Gran Canaria: 1560 m) and the differences in the coastline between both dates so that both models also coincide. The propagation is performed using a grid mesh, with wave heights obtained for each node and with the specific coordinates of the coastal area of interest. To obtain more precise results, the area was divided into seven zones (Figure 1) to enable modelling with a higher mesh resolution. The resulting point clouds were processed in a GIS for rasterisation (Figure 3(1)) (the Kriging interpolation method was employed) and the wave height at the foot of the beach was subsequently determined. In the case of the Puerto Rico coastal strip, this is at elevation −6.62 m. For evaluation of the breakwater overflow level, the wave height at the base of the structure was used according to the bathymetric data of the study area. After obtaining the wave height at the foot of the beach, it is possible to determine the runup which, in the case of beaches, as explained above, also depends on the slope and incident wave peak period [41]. For the breakwaters, the formula proposed by McCartney and Ahrens (1975) [45] was used. The slope was determined using the digital elevation model (DEM) of the Spanish National Geographic Institute. For the coastal strip of 1700 m in the study area, the flood sea level was determined for a total of seven beaches and the breakwater overflow level for five ports, as well as some built-up areas along the coast protected by rockfill.

Table 1. Digital surface models (DSMs) used in this research.

Type	Year (Source)	RMSE (m)			Spatial Resolution (m)
		x	y	z	
D L-S M (from LiDAR and echosounder data) (a)	LiDAR (December 2015) Echosounder (2000)			≤0.5 ≤1 m	2.5
DSMs (2) (from LiDAR data, density: 0.5 (p/m^2))	October 2009 (b) December 2015 (b)	0.3 0.3	0.3 0.3	0.4 0.2	2.5

(a). SDI Canarias-Grafcan S.A (Canary Islands Government); (b). Instituto Geográfico Nacional (IGN).

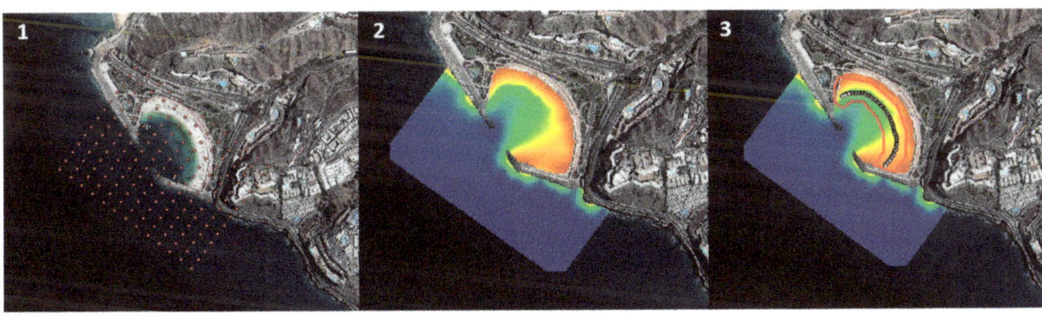

Figure 3. Example of the steps employed to calculate the flood sea level by SW sea storm events. (**1**) Mesh to interpolate Hs. (**2**) Raster obtained through Hs interpolation. (**3**) (red polygon), extraction of Hs values at elevation −2 (considered as beach foot).

3.3. Topographic Variables, Coastal Narrowing and Coastal Squeeze Zones Calculation, and Mapping

LiDAR data from flights in 2009 and 2015 were used for the spatial analyses of topographic variables such as erosion range (data from 2009 and 2015) and mean slope (data from 2015). Digital Surface models (DSMs) were obtained (ground + buildings) that allowed us to include built-up and non-built-up surfaces with flood risk (Table 1). By means of a digital elevation model of difference (DoD), it was possible to gain an approximate knowledge of the changes in coastal erosion. These two LiDAR flight datasets were the only ones freely available from Spain's National Geographic Institute through the National Aerial Orthophotography Plan. Extraction of the erosion range and mean slope data, together with the mean width of coastal narrowing and coastal squeeze (explained below), was done using perpendicular transects to the coast (every 20 m) in all the modelled areas (Figure 2).

Coastal narrowing and coastal squeeze zones were calculated from the 2015 DSM. For this, the raster reclassification tool, which can be found in any GIS software, was used with the flood sea levels. Each zone was then reclassified before the GIS mask extraction tool was used. Finally, GIS was used to vectorise the results of the reclassification and the resulting polygon vectors were merged. In accordance with the distinction made by Pontee (2013) [5], the coastal narrowing zone was determined as the difference between the flood sea levels at half-tide and high-tide shown in Table 2 (current climate). Additionally, in accordance with the distinction made by Pontee (2013), the coastal squeeze zone, which is related to SLR predictions, was determined as the difference between the result of the flood sea level at half-tide in Table 2 (current climate) and half-tide in Table 3 (RCP 8.5).

Table 2. Data used to calculate the current flood sea level (last column) in the different zones studied (beaches, grey), and with two situations (half-tide and high tide), when SW marine storms occur (1958–2018).

CURRENT CLIMATE									
Case Study (SW Sea Storm Events)		[1] *Hs*	[2] Slope	[1] Breakwater Foot Depth	[1] Crest Elevation	[1] Runup Significant	[1] Half-Tide	[1] High Tide	[1] Flood Sea Level
Zone 1. Amadores beach	Half-tide	1.96	0.0575	-	-	0.87	0	0	0.87
Zone 1. Amadores beach	High tide	2.16	0.0575	-	-	0.91	0	1.53	2.44
Zone 1. Breakwater zone	Half-tide	2.74	2.21	-	-	2.21	0	0	2.41
Zone 1. Breakwater zone	High tide	2.84	2.31	-	-	2.43	0	1.53	3.96
Zone 3. Puerto Rico beach	Half-tide	2.05	0.036	-	-	0.89	0	0	0.89
Zone 3. Puerto Rico beach	High tide	2.84	0.036	-	-	1.05	0	1.53	2.58
Zone 3. Breakwater zone 1	Half-tide	1.99	1.91	-	4	1.97	0	0	1.97
Zone 3. Breakwater zone 1	High tide	2.14	1.91	-	4	2.09	0	1.53	3.62
Zone 3. Breakwater zone 2	Half-tide	2.28	2.58	8	4	1.91	0	0	1.91
Zone 3. Breakwater zone 2	High tide	2.87	2.58	8	4	2.27	0	1.53	3.8
Zone 5. Balito beach	Half-tide	2.32	0.174	-	-	1.94	0	0	1.94
Zone 5. Balito beach	High tide	3.01	0.174	-	-	2.21	0	1.53	3.74
Zone 5. Anfi del Mar	Half-tide	0.98	0.062	-	-	0.62	0	0	0.62
Zone 5. Anfi del Mar	High tide	1.03	0.062	-	-	0.86	0	1.53	2.39
Zone 5. Breakwater zone 1	Half-tide	2.6	3.25	15.5	5	2.12	0	0	2.12
Zone 5. Breakwater zone 1	High tide	3.39	3.25	15.5	5	2.61	0	0.53	3.14
Zone 5. Breakwater zone 2	Half-tide	1.39	3.5	3.5	5	1.24	0	0	1.24
Zone 5. Breakwater zone 2	High tide	1.6	3.5	3.5	5	1.39	0	1.53	2.92
Zone 6. Patalavaca beach	Half-tide	3.07	0.034	-	-	1.09	0	0	1.09
Zone 6. Patalavaca beach	High tide	3.16	0.034	-	-	1.11	0	1.53	2.64
Zone 6. Breakwater zone	Half-tide	2.48	1.33	−2	5.5	4.03	0	0	4.03
Zone 6. Breakwater zone	High tide	2.75	1.33	−2	5.5	4.39	0	1.53	5.92
Zone 7. La Carrera beach	Half-tide	2.33	0.104	-	-	1.16	0	0	1.16
Zone 7. La Carrera beach	High tide	2.5	0.104	-	-	1.2	0	1.53	2.73
Zone 7. Resort and coast zone	Half-tide	2	2.9	−1.5	4	1.79	0	0	1.79
Zone 7. Resort and coast zone	High-tide	1.96	2.9	−1.5	4	1.76	0	1.53	3.29

[1] meters; [2] degrees.

Table 3. Data used to calculate the flood sea levels projected in the RCP 8.5 scenario (last column) in the different zones studied (beaches, grey), and with two situations (half-tide and high tide), when SW sea storm events occur.

RCP 8.5									
Case Study (SW Sea Storm Events)		[1] *Hs*	[2] Slope	[1] Breakwater Foot Depth	[1] Crest Elevation	[1] Run-Up Significant	[1] Half-Tide	[1] High Tide	[1] Flood Sea Level
Zone 1. Amadores beach	Half-tide	2.03	5.75	-	-	0.89	0	0.74	1.63
Zone 1. Amadores beach	High tide	2.24	5.75	-	-	0.93	0	2.27	3.2
Zone 1. Breakwater zone	Half-tide	2.73	2	-	-	2.36	0	0.74	3.1
Zone 1. Breakwater zone	High tide	2.8	2	-	-	2.41	0	2.27	4.68
Zone 3. Puerto Rico beach	Half-tide	2.76	0.036	-	-	1.03	0	0.74	1.77
Zone 3. Puerto Rico beach	High tide		0.036	-	-	1.07	0	2.27	3.34
Zone 3. Breakwater zone 1	Half-tide	2.4	1.91	-	4	2.29	0	0.74	3.03
Zone 3. Breakwater zone 1	High tide	2.41	1.91	-	4	2.29	0	2.27	4.56
Zone 3. Breakwater zone 2	Half-tide	2.86	2.58	8	4	2.26	0	0.74	3
Zone 3. Breakwater zone 2	High tide	2.9	2.58	8	4	2.28	0	2.27	4.55
Zone 5. Balito beach	Half-tide	2.87	0.174	-	-	2.16	0	0.74	2.9
Zone 5. Balito beach	High tide	2.93	0.174	-	-	2.18	0	2.27	4.45
Zone 5. Anfi del Mar	Half-tide	0.91	0.062	-	-	0.59	0	0.74	1.33
Zone 5. Anfi del Mar	High tide	0.92	0.062	-	-	0.6	0	2.27	2.87
Zone 5. Breakwater zone 1	Half-tide	3.36	3.25	15.5	5	2.59	0	0.74	3.33
Zone 5. Breakwater zone 1	High tide	3.4	3.25	15.5	5	2.62	0	2.27	4.89
Zone 5. Breakwater zone 2	Half-tide	1.4	3.5	3.5	5	1.25	0	0.74	1.99
Zone 5. Breakwater zone 2	High tide	1.67	3.5	3.5	5	1.44	0	2.27	3.71
Zone 6. Patalavaca beach	Half-tide	3.13	0.034	-	-	1.1	0	0.74	1.84
Zone 6. Patalavaca beach	High tide	3.17	0.034	-	-	1.11	0	2.27	3.38
Zone 6. Breakwater zone	Half-tide	2.62	1.33	−2	5.5	4.22	0	0.74	4.96
Zone 6. Breakwater zone	High tide	2.81	1.33	−2	5.5	4.47	0	2.27	6.74
Zone 7. La Carrera beach	Half-tide	2.36	0.104	-	-	1.17	0	0.74	1.91
Zone 7. La Carrera beach	High tide	2.7	0.104	-	-	1.25	0	2.27	3.52
Zone 7. Resort and coast zone	Half-tide	1.79	2.9	−1.5	4	1.63	0	0.74	2.37
Zone 7. Resort and coast zone	High tide	2.14	2.9	−1.5	4	1.89	0	2.27	4.16

[1] meters; [2] degrees.

3.4. Physical and Economic Impacts of the SW Sea Storm Events and Estimated Impacts in View of SLR

3.4.1. Fieldwork

A six-field campaign was conducted during April and May of 2021, right after winter when more frequent storms are observed [24], to identify the environmental impacts of SW storms in the study area. The work consisted of tracing the entire coastline observed in the study area (Figure 2) in order to detect in particular three types of damage that may show signs of wave effect: (i) damage to infrastructure; (ii) cliff slides; (iii) outcrops of volcanic lava flows and pebbles on sandy beaches indicating the absence of sand. Using GPS to situate all the affected sites, the dimensions of the affected areas were also characterised and measured (major axis × minor axis = m^2). The field campaign was conducted alternating between weeks of low tide and weeks of high tide. In the former, it was possible to identify all the impacts that high tide could cover, and in the latter to verify that the impact was caused by wave intensity.

3.4.2. Economic Cost Calculation

A report (in Spanish with English title: Design of the seafront from Anfi del Mar to Arguineguín) drawn up by Bolaños-Briganty and Martinez-Castellanos (2015) [46] was used as a reference document for the calculation of the economic cost of SW storms in the study area. This document was written with the construction in mind of a 1587 m long seafront promenade over an already existing pathway that was unmaintained and lacked any type of amenity due to strong surf. The report included specifications in accordance with the maritime climate of the area in relation to protection against sea damage. The document in question follows the same reasoning as is being applied throughout the study area of the present research where proposals have been made for the reconstruction of beaches and infrastructures such as seafront promenades and piers protected by breakwater walls. According to the reference document, the estimated construction, infrastructure, health and safety, and waste management costs for the proposed work amount to 19,274,540.91€. This figure was extrapolated to the rest of the coast, taking into consideration the stretch of seafront (in m) protected against storms in each of the areas used to generate the aforementioned bidimensional models (Figure 3) (Equation (1)).

$$\mathrm{Ce} = (\mathrm{SLm} \times \mathrm{SLp})/\mathrm{Cp} \tag{1}$$

where Ce is the cost incurred as the result of current SW sea storm effects, SLm is the length of the seafront in the model zone, SLp is the length of the seafront in the reference project (1587 m), and Cp is the total cost of the reference project (19,274,540.91 €).

With respect to the estimation of the SLR-related economic cost, the same extrapolation was performed but with the addition of an adjustment factor (Equation (2)) due to the anticipated flood area increase, which relates the mean width of the coastal squeeze zone in each modelled area with the mean width of the proposals included in the reference project (13.25 m). This adjustment factor was calculated through a GIS using the mean length of the transects perpendicular to the coast (understood as the mean width of the coastal squeeze zone by its orientation) traversed by the coastal squeeze zone (Table 4).

$$\mathrm{Ce} = (\mathrm{SLm} \times \mathrm{SLp})/\mathrm{Cp} \times \mathrm{a} \tag{2}$$

where a is the adjustment factor (Equation (3)).

$$\mathrm{a} = \mathrm{Wm}/\mathrm{Wp} \tag{3}$$

where Wm is the mean width in each zone for modelling and Wp is the mean width of the reference project, which behaves as a constant of 13.25 m.

Table 4. Adjustment factor used to calculate the prediction of the SLR-related economic cost.

Study Case or Zone	*a*
Zone 1. Amadores beach	2.43
Zone 1. Breakwater zone	2.19
Zone 3. Puerto Rico beach	4.40
Zone 3. Breakwater zone 1	0.39
Zone 3. Breakwater zone 2	0.31
Zone 5. Balito beach	7.43
Zone 5. Anfi del Mar	2.57
Zone 5. Breakwater zone 1	0.14
Zone 5. Breakwater zone 2	0.25
Zone 6. Patalavaca beach	1.06
Zone 6. Breakwater zone	1.02
Zone 7. La Carrera beach	0.60
Zone 7. Resort and coastal zone	0.18

4. Results and Discussion

Coastal planning and management require the use of risk assessment frameworks that consider the combined effect of a variety of both climatic and non-climatic factors, as well as associated uncertainties such as SLR, changes in the probability of occurrence of extreme events, and socioeconomic trends that are difficult to predict [47]. The presentation below of this variety of factors has the aim of helping in the management-related decision-making process in tourist areas of oceanic islands that suffer the effects of sea storm events given the anticipated SLR.

4.1. Current Sea Storm Conditions

4.1.1. Evolution of Sea Storm Events and Their Characteristics

An important issue in coastal studies is the frequency and magnitude of storm events and their impacts [32]. In this regard, Yanes-Luque et al. (2021) [24] studied records of sea storm events between 1958 and 2018 based on newspaper archives, verifying the occurrence of 217 documented events and the resulting damage to coastal areas of Gran Canaria and Tenerife. As can be seen in Figure 4, a total of 154 sea storm events were detected in the study area of the present research over the last 60 years of the study period and, of these, 67.53% (104 events) took place in the last 30 years (Figure 4, left). That is, a polynomial trend (R^2: 0.5876) is observed in relation to the increase in storm events from 1958 to 2018. A similar trend has been observed throughout the North Atlantic zone [47,48], where modelling also revealed an increase in the height of the most-damaging waves [47]. It is also important to observe how, although the linear trend is low (R^2: 0.1577) with a *p*-value of 3.4155E-07, the duration of the wave tends to increase with wave height (Figure 4, right). In view of all the above, it is concluded that, as well as a greater frequency of sea storm events, the storms are becoming more severe as the highest waves are present over longer periods of time.

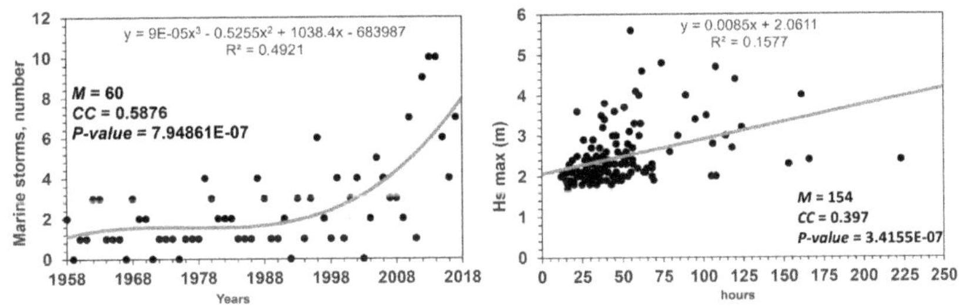

Figure 4. Evolution of the number of sea storm events between 1958 and 2018 (sixty years) (polynomial trend) (**left**) and correlation between maximum wave height (Hs max) and storm duration (linear trend) (**right**). M is the sample size and CC is the correlation coefficient.

4.1.2. Current Effects of SW Sea Storm Events

Table 2 shows the variables used to calculate the flood sea level (last column) on the basis of bidimensional numerical models. The data are ordered by the modelled study zones (1, 3, 5, 6, and 7), coastal type (beach or infrastructure), and half-tide or high tide conditions when the SW storm events occur. It can be seen that, generally, beaches (shaded areas in the table) have a lower flood sea level than infrastructures in both half-tide (range 0.62–1.94 m) and high tide (range 2.39–3.74 m) conditions. However, as can be seen in Figure 5, which shows the calculated coastal narrowing zone, the beaches, normally the main attraction of coastal tourist destinations [49], are subject to greater flooding.

In this sense, in the North Atlantic, there are indications that there is an increase in the frequency of very powerful storms, especially at higher latitudes, at the same time that coastal ecosystems (e.g., beaches) have decreased. The North Atlantic Oscillation (NAO) phenomenon is associated with the frequency and monitoring of these ecosystems or the loss of these ecosystems and their services. In addition, recent roughness in the sea seems to be related to high values of the NAO index, which are also expected to increase with global warming. Therefore, when coupled with the projected sea level rise, which is addressed in this research, this trend is likely to lead to increased sediment loss resulting in widespread coastal erosion [50–53].

4.1.3. Coastal Narrowing Calculation

Coastal narrowing is defined by Pontee (2013) [5] as "a general description for decreases in coastal zone width" and has no relationship to SLR. In this sense, in order to analyse the current conditions of sea storm events (especially SW events) that take place close to tourist areas and their socio-environmental impact, it is necessary to determine the affected physical area. The coastal narrowing concept helps to achieve this goal. In the study area, it has been estimated that SW sea storm events currently affect 38,460.04 m^2 (half-tide) and 93,072.26 m^2 (high tide). However, in the RCP 8.5 scenario, the affected area rises to 68,035.66 m^2 (half-tide) and 159,116.11 m^2 (high tide), the equivalent of a 176.9% increase under half-tide conditions and a 170.9% increase under high tide conditions. As mentioned above, beaches are the coastal areas most affected by surface flooding. Rock outcrops have been detected as the result of beach erosion, even in breakwater-protected artificial beaches like Amadores beach [54], which has only been open to the public since 1998. Old beach rocks, which were covered by imported sand 22 years ago, are currently resurfacing all year round, as determined by fieldwork and as can be seen in the dates of the orthophotos shown in Figure 6A,B (top)). Seasonal effects can thus be discarded.

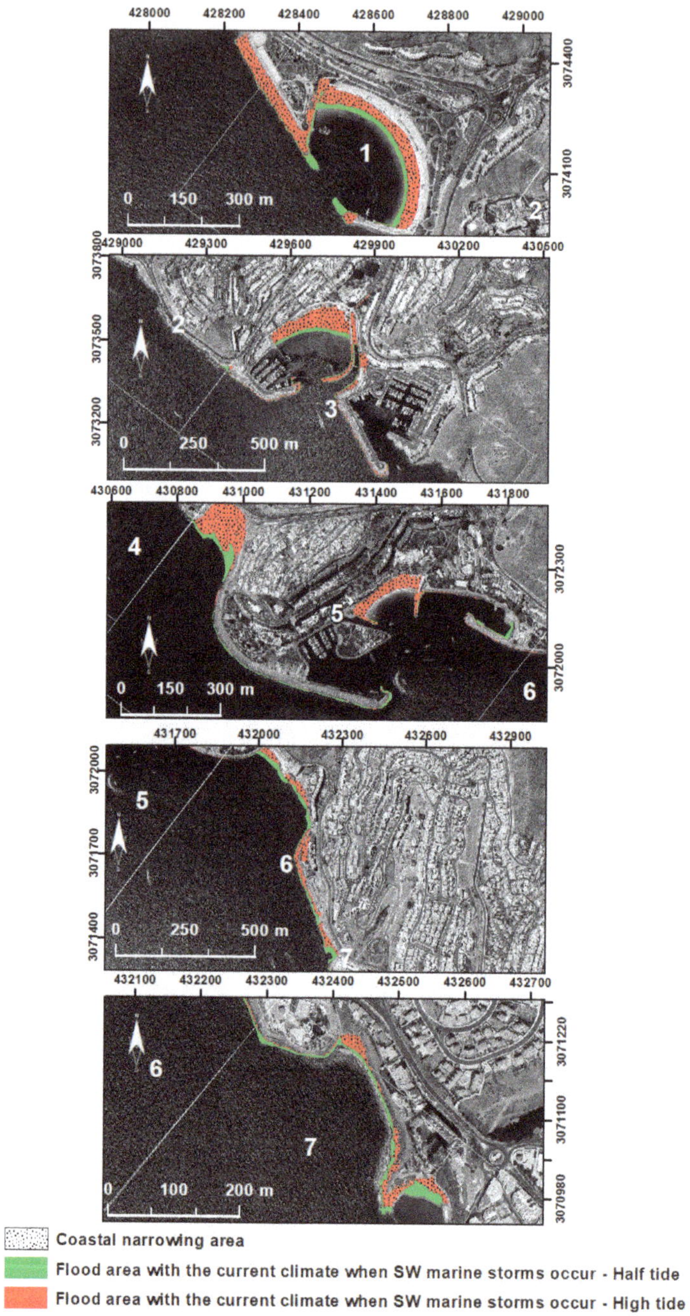

Figure 5. Study area zones affected by SW marine storms. Differences between half-tide and high tide.

In this sense, it is important to calculate not only the flood sea levels under the present climate scenario when SW sea storm events take place but also the areas that are affected (Figure 5). This type of map can be used in the management decision-making process and the establishment of new regulations on the part of the authorities with respect to coastal use and planning in island regions [55]. It additionally allows prioritisation in coastal classification by considering variables such as floodable surface area. In Zone 1 (Figure 5—Amadores beach), the half-tide flooded area when SW sea storm events occur is 9607.38 m^2 and the high tide value is 29,980.12 m^2, resulting in a coastal narrowing zone of 21,022.52 m^2 (2.12 times larger than the area flooded under half-tide conditions). In Zone 3 (Figure 5—Puerto Rico), half-tide and high-tide floods affect 6029.19 m^2 and 26,109.50 m^2, respectively, and the coastal narrowing zone is 20,080.31 m^2 (3.33 times larger than the area flooded under half-tide conditions). In Zone 5, half-tide and high-tide floods affect 10,283.07 m^2 and 18,412.13 m^2, respectively, and coastal narrowing is 8129.06 m^2 (79.05% of the area flooded under half-tide conditions). In Zone 6, half-tide and high-tide floods affect 9794.20 m^2 and 15,190.17 m^2, respectively, and coastal narrowing is 5395.97 m^2 (55.09% of the area flooded under half-tide conditions). Finally, in Zone 7, half-tide and high-tide floods affect 10,746.20 m^2 and 14,380.34 m^2, respectively, and coastal narrowing is 3634.14 m^2 (33.82% of the area flooded under half-tide conditions). The three most affected zones in terms of surface area by SW sea storm events (Zones 1, 3, and 5) are principally comprised of beaches and breakwaters (both for vessels and beach protection), whereas the least affected (Zones 6 and 7) have no form of beach protection. It can thus be affirmed that the breakwaters are no guarantee of beach stability (Figure 6). Similar results have been reported by [56–58], among others. Two possible reasons for the erosion despite the presence of breakwaters are that: (i) the breakwater itself is modifying the sand dynamics and contributions to the beach, as reported in islands like Oahu (Hawaiian Islands) [59]; (ii) these are enclosed beaches with no longshore drift sediment contribution, a condition very common in islands with limited sediment supplies [60].

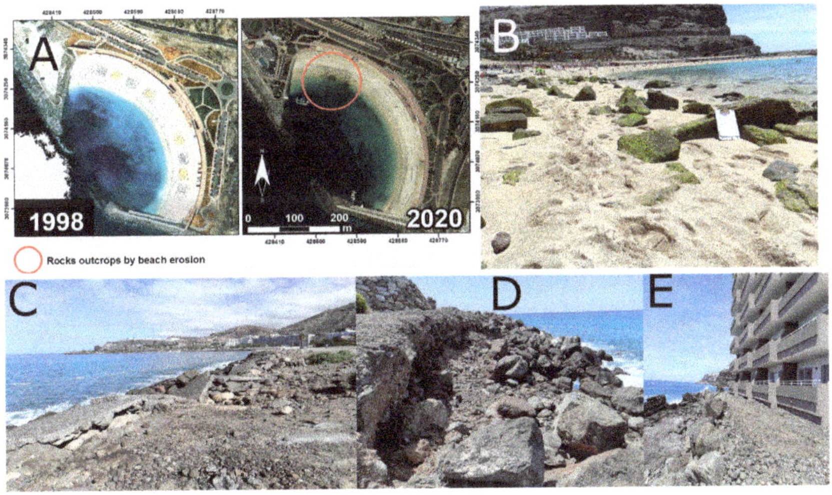

Figure 6. Environmental effects by sea storm events. (**A**,**B**): Rock outcrops as the result of beach erosion in Amadores between 1998 (flight carried out between October and December 1998) and 2020 (flight carried out between May 2020 and February 2021) (located in Figure 7; legend 4. Beach erosion). Fieldwork photograph (top right) taken on 19 April 2021. Orthophoto source: SDI Canarias (Canary Islands, GRAFCAN, S.A.). (**C**): seafront damage (located in Figure 7; legend 1. Seafront damage only). (**D**): breakwater damage (located in Figure 7; legend 2. Breakwater damage only). (**E**): affected seafront and breakwater located within 4 m of a tourist building (located in Figure 7; legend 3. Seafront and breakwater damage).

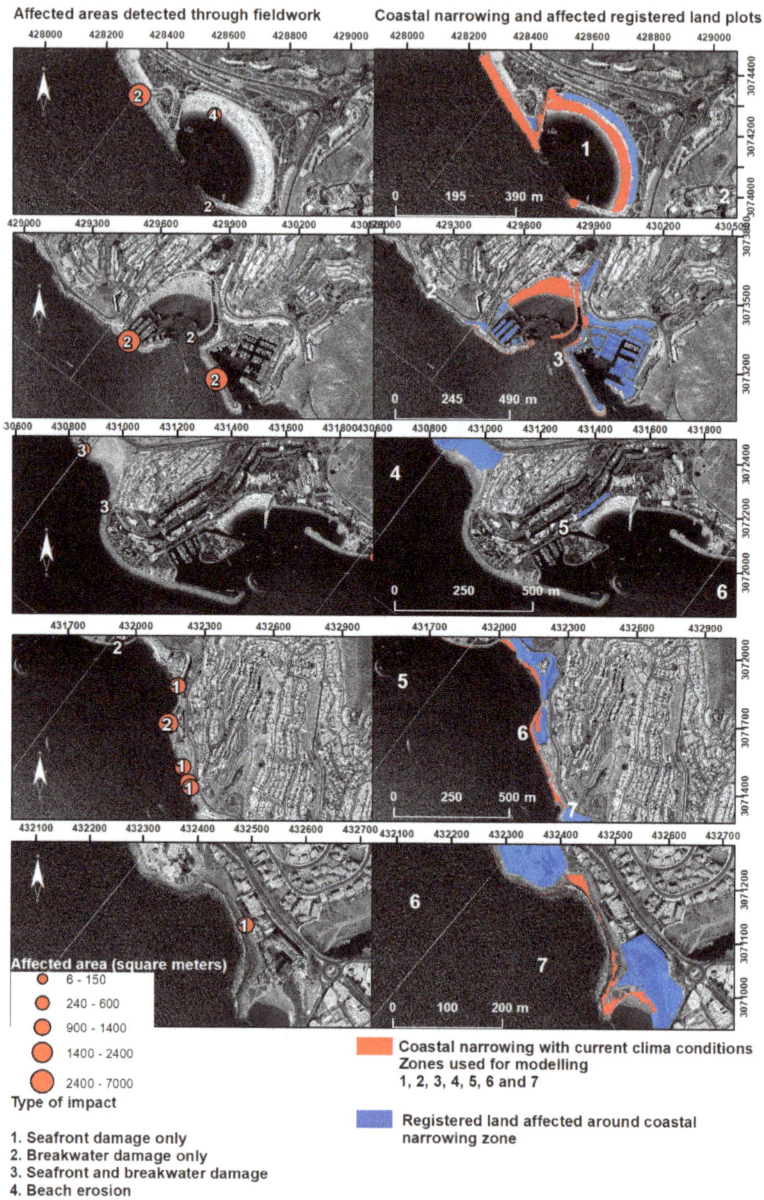

Figure 7. Distribution of the environmental effects identified through fieldwork.

4.2. Current Environmental Problems through the Coastal Narrowing Zone

It was found (Figure 4) that the frequency of SW sea storm events is increasing in the study area and, through the bi-dimensional models to calculate the flood sea level and map the floodable surfaces, that the coastal spaces are more easily flooded as the coastal slope decreases (R2 = 0.5885) (Figure 8, right). This pattern has been widely reported in the scientific literature, showing that zones of lower height and slope are the most affected [61–65]. It was also found through the DoDs obtained from the LiDAR data (2009 and 2015) that the erosion range increases with flooded area measured through the length of the transects explained in the methodology section (R2 = 0.6872) (Figure 8, left). The combination of these two topography-related variables allows the identification of coastal zones at greater risk of flooding [63].

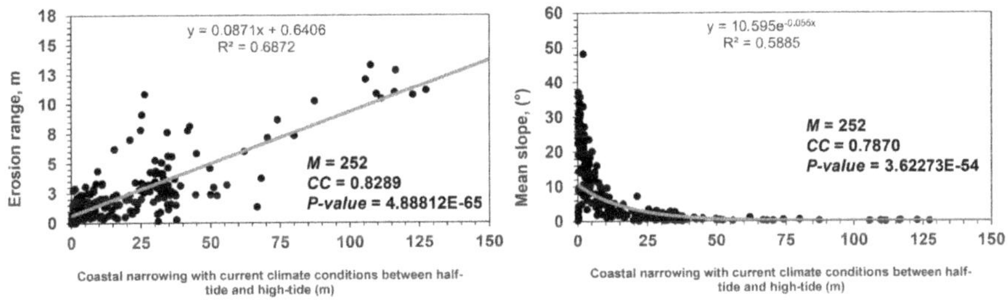

Figure 8. Correlation between coastal narrowing and erosion range (**left**, linear trend), and mean slope (**right**, exponential trend). M = sample; CC = correlation coefficient.

Coastal Impacts by Marine Storms Identified through Fieldwork

Four main problems caused by coastal erosion in the study area were identified (and measured) in situ during the field campaign (Figures 6 and 7): (i) seafront damage only; (ii) breakwater damage only; (iii) seafront and breakwater damage; and (iv) beach erosion. With respect to seafront damage only, it was calculated that 6,100.72 m^2 of the study area is affected. The highest and most significant impact was calculated with respect to breakwater damage only, affecting 20,048.11 m^2. Damage to both seafront and breakwater in the same place affected a total of 391.61 m^2, and the corresponding calculation for beach erosion was 604.43 m^2. Practically all of the designated zones of the study area were found to be affected. Beach erosion, however, was mostly in Zone 1 (Amadores), and damage to both seafront and breakwater in Zone 5 (especially Balito beach) (Figure 7, third row). In addition, a total of 49 registered plots situated in the coastal narrowing zone (Figure 7, blue) were found to be affected as they overlie the coastal narrowing zone. These plots are mostly exploited to offer tourist services (42.65%), including restaurants, marinas, and rental services. Tourist hotels and apartments are constructed on a further 34.48% of the plots and residential homes on 20.87%. The remaining 2% is presently unoccupied, but construction projects are being proposed, for example in the case of Balito beach. Plans have been drawn up to build on a surface area of 430,000 m^2, and the necessary modifications to subsidiary regulations of Mogán Council have been underway since 2019. It is expected that the forthcoming General Urban Development Plan of the Council will include this project. The results obtained in the present research strongly suggest that this type of project, whose main use will be for touristic and residential purposes, is unadvisable, especially in view of the fact that Balito beach is one of the zones most affected by SW sea storm events.

4.3. Effects of Sea Level Rise under the Current Scenario

4.3.1. Effects of Projected RCP 8.5 SW Sea Storm Events on Flood Sea Level

Table 3 shows the RCP 8.5 projected flood sea levels. Although the trend is linear in relation to the results shown in Table 2 (R^2 = 0.9648), this is not necessarily the case in the breakwater zones (especially Zones 1 and 5). The results are difficult to interpret as they will depend on the morphology of the structure, their state of conservation, and the management practices carried out, as well as the success or otherwise in meeting the challenges of ecological conservation [66–68].

4.3.2. Coastal Squeeze Calculation

The coastal squeeze phenomenon (see Section 1) is one of the main problems faced in coastal management, as it affects practically all coastal strips around the world where diverse socio-ecological systems coexist [69]. Such systems include beaches, the areas most affected according to the results of the present study. In Zone 1 (Figure 9), Amadores beach, the estimated flooded surface areas in an RCP 8.5 scenario when SW sea storm events occur during half-tide is 14,448.76 m^2, and during high tide 44,428.90 m^2. The corresponding values in Zone 3 (Figure 9), Puerto Rico, are 14,718.42 m^2 and 50,779.86 m^2; in Zone 5 (Figure 9), they are 19,943.32 m^2 and 35,079.31 m^2; in Zone 6, 15,013.77 and 20,681.99 m^2; and, finally, in Zone 7 (Figure 9) the corresponding values are 3911.39 m^2 and 8682.63 m^2. The coastal squeeze flooded surface areas in these five Zones are therefore (ordered by Zone number) 50.39%, 144.11%, 93.94%, 53.29%, and 42.42% (with half-tide conditions) and 48.19%, 94.48%, 90.52%, 36.15%, and 156.85% (with high-tide conditions) higher than the corresponding coastal narrowing calculations given in Section 4.1.3.

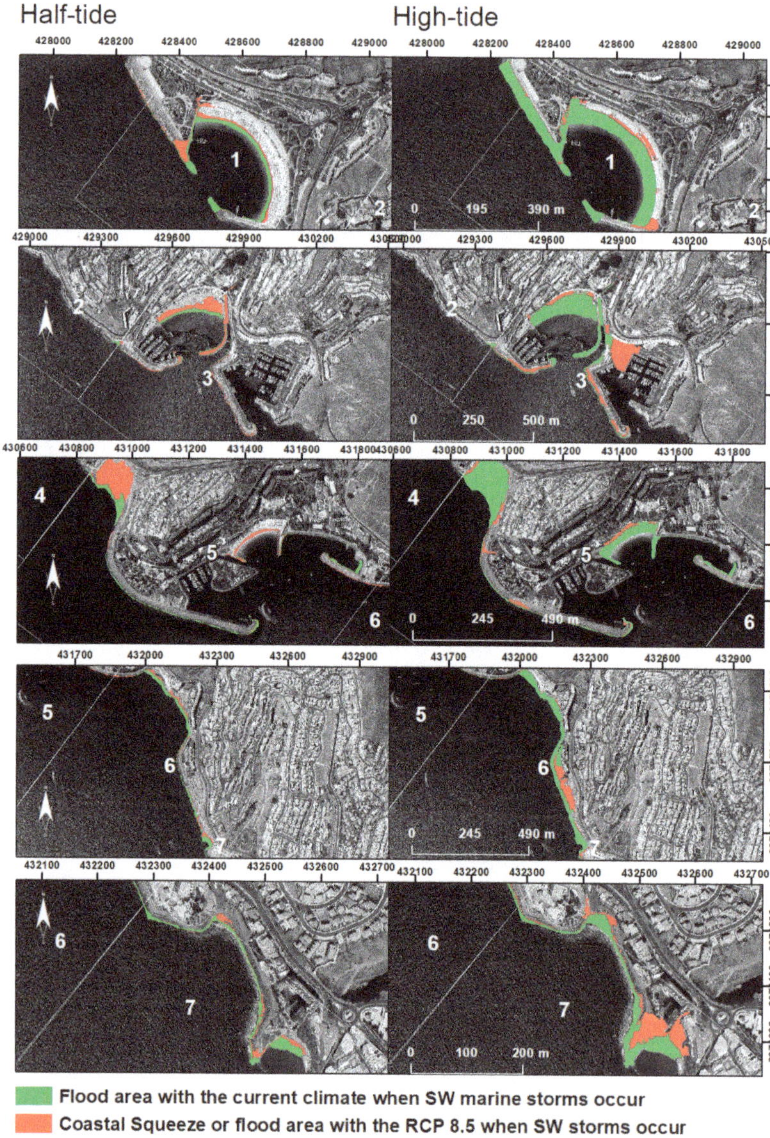

Figure 9. Coastal squeeze area. (**Left**): Coastal narrowing area (green) adding the flooded surface in an RCP 8.5 scenario (red) under half-tide conditions and when SW sea storms occur. (**Right**): Coastal narrowing area (green) adding the flooded surface in an RCP 8.5 scenario (red) under high-tide conditions and when SW sea storms occur.

There is some debate around the retreat capacity of some coastal ecosystems in the face of SLR and the need for more precise remodelling than solely the use of the flood sea level. With beach-dune systems, for example, certain limitations need to be taken into consideration in the case of application of the Bruun rule (1962) [70], as this should strictly be done in areas with environmental conditions similar to those of the original hypothesis [71]. It has been reported that erosion of these sedimentary systems can be overestimated by 10–60% [72]. Modifications to the Bruun rule to avoid such overestimations include adding overwash and wind-driven onshore transport [73]. In the case of the present research, the results obtained to date show that sections of the study area that could have retreat capacity, like the beaches (rigid structures like the breakwaters have no such capacity), are totally flooded. In addition, the studied beaches are sheltered inland by rigid structures (buildings, etc.), which impede beach recovery [74].

4.3.3. Calculation of the Cost of the Effects of Sea Storm Events

The first studies that evaluated the potential cost of SLR using a cost-benefit model weighed the cost of protecting a property with the value of the property at the time of flooding [75–77]. This approach assumes or understands that property will be protected if its value exceeds the cost of protection at the time of flooding. Certain limitations of this approach have been pointed out. Firstly, modelling is done only of the damage that is caused by the SLR, ignoring possible extreme storm events and the erosion exacerbated by the SLR [78], as is the case in the present study; secondly, it is assumed that the owners and those responsible for decision-making have undertaken perfect planning, whereas, in reality, the envisaged protection structures may be for an SLR that may not occur [78]; and, thirdly, this approach only examines the net social cost of property and ignores other economic impacts (e.g., interruptions to commerce) of SLR [79,80]. More recent studies have taken into account the economic impact of sea storm events in the context of SLR [81,82]. For beaches, the most important direct economic use tends to be related to recreational and leisure activities and there is general agreement among economists (within a reasonable range) that the most appropriate unit to use in calculations in this respect is the so-called beach day [83–85]. In this regard, Pendleton et al. (2011) [85] found that an increased beach width had important benefits, increasing the value of coastal properties. Another way to assess the value of beaches and their disappearance is through the use of hedonic models [86]. The method employed in the present study quantifies the cost on the basis of a coastal seafront renovation project that is required due to the effects of sea storm events (see Section 3.4.2). The quantification considers breakwater zones, sand replacement in affected areas, and sea surge protection. Figure 10 and Table 5 show the cost to maintain the affected areas used for the modelling.

Table 5. Estimated costs in the current and future scenario (RCP 8.5).

Case Study (SW Sea Storm Events)	Current Cost (€)	RCP8.5 Cost (€)
Zone 1. Amadores beach	7,257,857.963	17,613,104.59
Zone 1. Breakwater zone	15,908,075.27	34,863,748.77
Zone 3. Puerto Rico beach	3,661,897.537	16,102,854.49
Zone 3. Breakwater zone 1	4,456,461.924	1,759,701.09
Zone 3. Breakwater zone 2	40,073,569.71	12,382,823.77
Zone 5. Balito beach	3,006,388.264	22,332,505.51
Zone 5. Anfi del Mar	2,789,652.754	7,181,615.099
Zone 5. Breakwater zone 1	13,018,031.7	1,866,736.62
Zone 5. Breakwater zone 2	13,177,390.01	3,359,775.978
Zone 6. Patalavaca beach	4,390,074.539	4,632,867.86
Zone 6. Breakwater zone	14,737,418.15	15,065,843.46
Zone 7. La Carrera beach	3,137,659.948	1,883,187.98
Zone 7. Resort and coast zone	2,744,628.301	491,131.0381

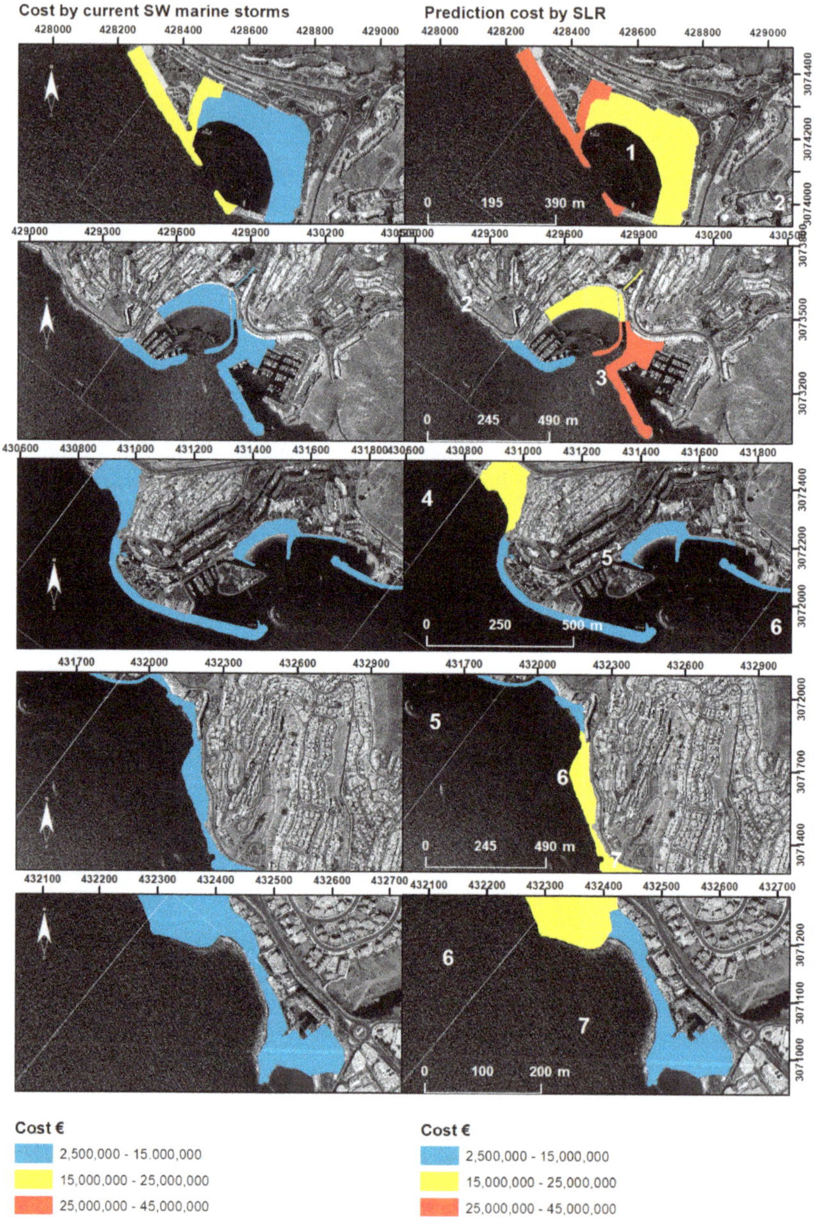

Figure 10. Economic cost of infrastructures and beaches affected by SW sea storm events, with current climate (**left**) and RCP 8.5 scenario (**right**) in the study area for the studied period.

Beaches are one of the principal assets of coastal tourist destinations [49], and their disappearance will entail significant economic losses. The authors of [86] observed that while coastal areas adapt to climate change, the long-term net value of coastal residential properties can fall by as much as 52% when the beach erosion rate triples and the cost of replacement sand quadruples. The present research has focused on an initial approach for the calculation of the repair costs of infrastructures and the recovery of mostly artificial beaches. In future works, calculation of the hedonic price will be key to understanding the real cost that beach loss will entail, including the resulting loss of tourist visitors and hence a vital source of revenue. In all the beach cases, the cost in the RCP 8.5 scenario would be viable if in the future it was decided to eliminate the buildings and construction inland and allow the beach to retreat; otherwise, the beach would completely disappear.

5. General Discussion

In view of the results of the present research, the importance of analysing the effects of SLR through the approach proposed by Pontee (2013) [5] is highlighted, given the conceptual differences between coastal narrowing (effects under current climate conditions) and coastal squeeze (prediction of effects in view of climate change and especially SLR). These conceptual differences also entail physical and environmental differences that allow the corresponding authorities to take effective measures now, in the current scenario, that take into consideration modelling for an RCP 8.5 scenario. Consideration of both scenarios can facilitate territorial planning and recommendations that resolve current environmental problems affecting coastal tourist zones (as considered in the present research) while anticipating the long-term effects of SLR.

This research also highlights the importance of the effects of the SW sea storm events that affect not only the case study area [24], but indeed practically all the archipelagos of the Atlantic Ocean [32–34]. The fact that the frequency and severity of these SW sea storm events have been increasing since the 1990s, affecting tourist areas situated in the S-SW of Atlantic islands, is also highlighted [24]. In Gran Canaria, for example, these S-SW areas are essentially unaffected by the trade winds whose access to the area is blocked by mountainous regions inland, but they are exposed to the impact of SW sea storm events [30].

With respect to the results obtained through the spatialisation of the flood sea levels, it was found that the zones occupied by both infrastructures and protected by breakwaters will be the least affected, although there are exceptions such as Zone 3 (Puerto Rico) and its second breakwater, where the inland entrance of seawater will be easier in some parts of the structure. In general, the current engineering models being implemented in the study area are taking into account the effects of climate change through repair and breakwater reinforcement projects (2018). In this regard, the present research questions the economic viability, especially for the relevant competent bodies, of this type of project as they are likely to become more and more common in view of a phenomenon like climate change that is so difficult to manage. The zones that will practically disappear, however, are the beaches. They are already showing signs of vulnerability to erosive processes (Figure 6). In addition, as explained previously, the beaches in this area have practically zero longshore drift and instead behave like capsules situated in ravine mouths whose sediment/sand supply through erosion processes is terrigenous through a non-river hydrographic basin to which they are associated. Thus, the sediment supply of these beaches is limited and only associated with occasional ravine run-off episodes that do not take place all year round as is the case with river-associated longshore drift. Finally, the combination of beach sediment loss, the SLR, and the complete urbanisation taking place around these systems means that beach retreat inland is impossible and their disappearance is likely [74].

Finally, climate change, as well as causing the anticipated SLR that will affect coastal areas, is causing temperate rises. A mean increase of 1 °C has been predicted by 2025 and of 3 °C by the end of the century [87]. Climate change will result in changes in rain patterns, with an increase in annual precipitation levels (especially in countries north of 50°N) [87] or a drastic reduction in arid and semiarid countries [88], but the most important impact is probably the anticipated increase in rainfall intensity [89]. In this regard, future works will tackle the overall aim of this work with the inclusion of a combination of the effects of SLR and the risk of flooding due to higher flow rates in the water network, in this case of coastal tourist zones situated in the mouths of ravines [54]. This double effect will increase the vulnerability of these coastal spaces.

6. Conclusions

The study presented in this paper analyses the environmental impact of SW sea storm events in coastal tourist areas, including both the effects taking place at the present time and those that are anticipated in view of climate change. For this purpose, a conceptual approach was applied that incorporated the separation of coastal narrowing (current climate) and coastal squeeze (projection of sea level rise) and modelling on the basis of two tidal conditions (half-tide and high tide). The first results obtained offer a justification for this type of research, especially as it was found that sea storm events, especially from the SW, have been increasing in frequency. Such storms can have important impacts on the tourist areas of oceanic islands which are commonly strategically situated in the SW of Atlantic islands because of the optimal levels of comfort and environmental stability that they offer. The calculation and mapping of the flood sea levels of the case study area allowed us to determine the most affected zones at the present time and to predict what would happen in an RCP 8.5 scenario. On the basis of this mapping, it was also possible to determine seafront damage, breakwater damage, seafront and breakwater damage together, and beach erosion. Although the infrastructures generally show a good level of preparation for sea storm events, the beaches have a high level of risk because of the erosion levels detected, including a strong possibility of their future disappearance as the tourist buildings and general urbanisation of the surrounding area is blocking their retreat. As far as infrastructure maintenance is concerned (with the need for constant and expensive repair projects on the part of the local authorities and the need to replace sand to avoid beach erosion), it has been calculated that just maintaining the current situation of the study area would cost 128,359,106.10€ under present conditions and 139,535,896.20€ under an RCP 8.5 scenario. Such amounts would be needed repeatedly and over a long period of time.

Author Contributions: L.G.-R. and P.M.-S., Conceptualization and investigation; L.G.-R., T.C.-G. and J.Á.R.-B., methodology, software, data curation, validation, formal analysis; L.G.-R., writing, writing and editing, and original draft preparation; L.G.-R., Geographical Information Systems; P.M.-S., A.Y.-L. and L.H.-C., review and supervision; P.M.-S. and L.H.-C., project administration, resource and funding acquisition; All authors have read and agreed to the published version of the manuscript.

Funding: This research was supported by a project co-financed with ERDF funds of the cooperation program INTERREG-MAC 2014–2020, project PLANCLIMAC (MAC/3.5b/244).

Data Availability Statement: Not applicable.

Acknowledgments: This work is a contribution of a project co-financed with ERDF funds of the cooperation program INTERREG-MAC 2014–2020, project PLANCLIMAC (MAC/3.5b/244), which allowed a contract to the first author, and the project also co-financed with ERDF funds PID2021-124888OB-I00 (National R&D&i Plan, Spain). Leví García-Romero is a beneficiary of the postdoctoral contract programme 'Catalina Ruiz 2022' of the Canary Islands Government and the European Social Fund (APCR2022010005). We thank Francisco Martínez Castellanos for his support. We are grateful for the work of the three reviewers who improved the original manuscript.

Conflicts of Interest: The authors declare no conflict of interest.

References

1. Kelman, I.; Spence, R.; Palmer, J.; Petal, M.; Saito, K. Tourists and disasters: Lessons from the 26 December 2004 tsunamis. *J. Coast. Conserv.* **2008**, *12*, 105–113. [CrossRef]
2. Nicholls, R.J. Adaptation Options for Coastal Areas and Infrastructure: An Analysis for 2030. 2007. Available online: https://unfccc.int/files/cooperation_and_support/financial_mechanism/application/pdf/nicholls.pdf (accessed on 1 September 2021).
3. Williams, A.T.; Rangel-Buitrago, N.; Pranzini, E.; Anfuso, G. The management of coastal erosion. *Ocean. Coast. Manag.* **2018**, *156*, 4–20. [CrossRef]
4. Doody, J.P. Coastal squeeze and managed realignment in southeast England, does it tell us anything about the future? *Ocean. Coast. Manag.* **2013**, *79*, 34–41. [CrossRef]
5. Pontee, N. Defining coastal squeeze: A discussion. *Ocean. Coast. Manag.* **2013**, *84*, 204–207. [CrossRef]
6. Christy, J. The Future Outlook is Fine. *The Times*, 20 February 2001.

7. Environmental Scientist. *Living in the Greenhouse. Environ. Sci.* **1999**, *8*, 1–3.
8. Giles, J. When doubt is a sure thing. *Nature* **2002**, *418*, 476–478. [CrossRef] [PubMed]
9. World Tourism Organisation. Leading the World's Largest Industry. 2001. Available online: http://www.world-tourism.org/aboutwto.html (accessed on 1 October 2021).
10. Michailidou, A.V.; Vlachokostas, C.; Moussiopoulos, N. Interactions between climate change and the tourism sector: Multiple-criteria decision analysis to assess mitigation and adaptation options in tourism areas. *Tour. Manag.* **2016**, *55*, 1–12. [CrossRef]
11. Phillips, M.R.; Jones, A.L. Erosion and tourism infrastructure in the coastal zone: Problems, consequences and management. *Tour. Manag.* **2006**, *27*, 517–524. [CrossRef]
12. Moreno, A.; Amelung, B. Climate change and tourist comfort on Europe's beaches in summer: A reassessment. *Coast. Manag.* **2009**, *37*, 550–568. [CrossRef]
13. Moreno, A.; Becken, S. A climate change vulnerability assessment methodology for coastal tourism. *J. Sustain. Tour.* **2009**, *17*, 473–488. [CrossRef]
14. Becken, S. Harmonising climate change adaptation and mitigation: The case of tourist resorts in Fiji. *Glob. Environ. Chang.* **2005**, *15*, 381–393. [CrossRef]
15. Santana-Gallego, M.; Ledesma-Rodríguez, F.; Pérez-Rodríguez, J.V. Tourism and trade in small island regions: The case of the Canary Islands. *Tour. Econ.* **2011**, *17*, 107–125. [CrossRef]
16. Tsai, C.H.; Chen, C.W. The establishment of a rapid natural disaster risk assessment model for the tourism industry. *Tour. Manag.* **2011**, *32*, 158–171. [CrossRef]
17. UNISDR Terminología Sobre Reducción del Riesgo de Desastres. *Estrategia Internacional para la Reducción de Desastres de las Naciones Unidas*; United Nations: Geneva, Switzerland, 2009.
18. Carreño, M., Cardona, O.D.; Barbat, A. *Sistema de Indicadores para la Evaluación de Riesgos*; Centre Internacional de Mètodes Numèrics en Enginyeria (CIMNE): Barcelona, Spain, 2002.
19. Oppenheimer, M.; Glavovic, B.; Hinkel, J.; van de Wal, R.; Magnan, A.K.; Abd-Elgawad, A.; Cai, R.; Cifuentes-Jara, M.; Deconto, R.M.; Ghosh, T.; et al. *Sea Level Rise and Implications for Low Lying Islands, Coasts and Communities*; The Intergovernmental Panel on Climate Change: Geneva, Switzerland, 2019.
20. Coller, M.L.F.; Wheeler, P.; Kunapo, J.; Peterson, J. Interactive flood hazard visualization in Adobe Flash. *J. Flood Risk Manag.* **2018**, *11*, 134–146. [CrossRef]
21. IPCC. Climate change 2013: The physical science basis. In *Contribution of Working Group I to the Fifth Assessment Report of the Intergovernmental Panel on Climate Change*; Stocker, T.F., Qin, D., Plattner, G.-K., Tignor, M., Allen, S.K., Boschung, J., Nauels, A., Xia, Y., Bex, V., Midgley, P.M., Eds.; Cambridge University Press: Cambridge, UK, 2013.
22. Medina Santamaría, R. *(Coord.) Impactos en la Costa Española por Efectos del Cambio Climático*; Ministerio de Medio Ambiente: Madrid, Spain, 2004; 423p.
23. Losada, I.; Izaguirre, C.; Diaz, P. *Cambio Climático en la Costa Española*; Oficina Española de Cambio Climático, Ministerio de Agricultura, Alimentación y Medio Ambiente, Gobierno de España: Madrid, Spain, 2014; 133p.
24. Luque, A.Y.; Rodríguez-Báez, J.A.; Suárez, P.M.; Antequera, P.D.; López-Díez, A.; Díaz-Pacheco, J.; Pérez-Chacón, E. Marine storms in coastal tourist areas of the Canary Islands. *Nat. Hazards* **2021**, *109*, 1297–1325. [CrossRef]
25. EN 32007L0060. Directive 2007/60/EC of the European Parliament and of the Council of 23 October 2007 on the Assessment and Management of Flood Risks. European Parliament Council: Strasbourg, France, 2007.
26. Garín-Muñoz, T. Inbound international tourism to Canary Islands: A dynamic panel data model. *Tour. Manag.* **2006**, *27*, 281–291. [CrossRef]
27. INE. Spanish Statistical Institute. 2020. Available online: https://www.ine.es (accessed on 15 September 2021).
28. Hernández, L. *Diagnóstico sobre la Evolución del Sistema de Dunas de Maspalomas (1960–2000)*; Cabildo de Gran Canaria: Las Palmas de Gran Canaria, Spain, 2006.
29. Ferrer-Valero, N.; Hernández-Calvento, L.; Hernández-Cordero, A. Human impacts quantification on the coastal landforms of Gran Canaria Island (Canary Islands). *Geomorphology* **2017**, *286*, 58–67. [CrossRef]
30. Máyer Suárez, P.; Pérez-Chacón Espino, E.; Cruz Avero, N.; Hernández-Calvento, L. Características del viento en el campo de dunas de Maspalomas (Gran Canaria, Islas Canarias, España). *Nimbus* **2012**, *29–30*, 381–397. Available online: https://dialnet.unirioja.es/servlet/articulo?codigo=4376861 (accessed on 23 September 2020).
31. Exceltur. Impactur Canarias 2018. *Estudio del Impacto Económico del Turismo Sobre la Economía y el Empleo de las Islas Canarias. Santa Cruz de Tenerife, Exceltur y Gobierno de Canarias.* 2019. Available online: https://www.exceltur.org/wp-content/uploads/2019/12/IMPACTUR-Canarias-2018.pdf (accessed on 23 September 2020).
32. Lameiras, G.; Fontiela, J.; Borges, P.; Calado, H.; Vieira, O.; Rangel, B.; Gallagher, A. Coastal hazards of Fajã do Calhau (São Miguel—Azores): A first approach. *J. Coast. Res.* **2009**, *1*, 827–831.
33. Arnault, J.; Roux, F. Characteristics of African easterly waves associated with tropical cyclogenesis in the Cape Verde Islands region in July–August–September of 2004–2008. *Atmos. Res.* **2011**, *100*, 61–82. [CrossRef]
34. Moreira, M.E.S. Recreation and conservation of the coastal environment of the Madeira Islands. *Ocean. Shorel. Manag.* **1988**, *11*, 409–425. [CrossRef]
35. Máyer, P. Un siglo de temporales en la prensa de Gran Canaria. *Vegueta* **1999**, *4*, 267–282.

36. Yanes, A.; Marzol, M.V. Los temporales marinos como episodios de riesgo en Tenerife a través de la prensa (1985–2003). *Rev. Soc. Geológica España* **2009**, *22*, 95–104.
37. Yanes, A. Desastres naturales en Canarias. La costa como espacio de riesgo en Tenerife. *Sémata* **2017**, *29*, 67–89. [CrossRef]
38. Máyer, P. *Lluvias e Inundaciones en Gran Canaria, Islas Canarias, España*; Editorial Académica Española: Madrid, Spain, 2011.
39. Nielsen, P.; Hanslow, D.J. Wave runup distributions on natural beaches. *J. Coast. Res.* **1991**, *7*, 1139–1152.
40. Tomas, A.; Medina, R.; Méndez, F.J.; Jaime, F.F.; Castellanos, O.F.; Higuera, P.; Fernandez, F.; Minguez, R.; Diaz-Hernandez, G. *Metodología para la Elaboración de los Mapas de Peligrosidad y RIESGO requeridos por RD 903/2010 en la Costa Española*; Ministerio de Agricultura, Alimentación y Medio Ambiente: Madrid, Spain, 2013; 97p.
41. González, M.; Medina, R.; Gonzalez-Ondina, J.; Osorio, A.F.; Mendez, F.; García, E. An integrated coastal modeling system for analyzing beach processes and beach restoration projects, SMC. *Comput. Geosci.* **2007**, *33*, 916–931. [CrossRef]
42. Kirby, J.T.; Dalrymple, R.A. *Combined Refraction/Diffraction Model REF/DIF 1, Version 2.5: Documentation and User's Manual*; Center for Applied Coastal Research, Department of Civil Engineering, University of Delaware: Newark, DE, USA, 1994.
43. Kirby, J.T.; Dalrymple, R.A.; Shi, F. *Combined Refraction/Diffraction Model. Ref/Dif 1 Version 3.0 Documentation and User's Manual*; University of Delaware: Delaware, DE, USA, 1994.
44. Puertos del Estado, Dirección Técnica. *Red de Mareógrafos del Estado (REDMAR), Resumen de Parámetros Relacionados con el Nivel del mar y la Marea que Afecta a las Condiciones de Diseño y Explotación Portuaria*; Área de Medio Físico: Puertos de Las Palmas, Spain, 2019.
45. McCartney, B.L.; Ahrens, J.P. *Stability of Gobi Block Revetment to Wave Attack*; US Army Corps of Engineers Coastal Engineering Research Center: Washington, DC, USA, 1975.
46. Bolaños-Briganty, J.J.; Martínez-Castellanos, F. *Diseño del Paseo Marítimo Desde Anfi del Mar Hasta Arguineguín*; Escuela de Ingenierías Industriales y Civiles; Universidad de Las Palmas de Gran Canaria: Las Palmas de Gran Canaria, Spain, 2015.
47. Wolf, J.; Woolf, D.; Bricheno, L. Impacts of climate change on storms and waves relevant to the coastal and marine environment around the UK. *MCCIP Sci. Rev.* **2020**, *2020*, 132–157.
48. Bresnan, E.; Baker-Austin, C.; Campos, C.J.A.; Davidson, K.; Edwards, M.; Hall, A.; McKinney, A.; Turner, A.D. Impacts of climate change on human health, HABs and bathing waters, relevant to the coastal and marine environment around the UK. *MCCIP Sci. Rev.* **2020**, *2020*, 521–545.
49. Toimil, A.; Diaz-Simal, P.; Losada, I.J.; Camus, P. Estimating the risk of loss of beach recreation value under climate change. *Tour. Manag.* **2018**, *68*, 387–400. [CrossRef]
50. Ottersen, G.; Planque, B.; Belgrano, A.; Post, E.; Reid, P.C.; Stenseth, N.C. Ecological effects of the North Atlantic oscillation. *Oecologia* **2001**, *128*, 1–14. [CrossRef]
51. Keim, B.D.; Muller, R.A.; Stone, G.W. Spatial and temporal variability of coastal storms in the North Atlantic Basin. *Mar. Geol.* **2004**, *210*, 7–15. [CrossRef]
52. Bertin, X.; Prouteau, E.; Letetrel, C. A significant increase in wave height in the North Atlantic Ocean over the 20th century. *Glob. Planet. Chang.* **2013**, *106*, 77–83. [CrossRef]
53. Paterson, S.K.; Loomis, D.K.; Young, S.E. The human dimension of changing shorelines along the U.S. North Atlantic Coast. *Coast. Manag.* **2014**, *42*, 17–35. [CrossRef]
54. Macías-González, F. La Problemática Ambiental de Urbanizaciones Turísticas Costeras: El Ejemplo de Puerto Rico-Amadores, Gran Canaria (España). Ph.D. Thesis, Universidad de Las Palmas de Gran Canaria, Las Palmas, Spain, 2017.
55. Richmond, B.M.; Fletcher, C.H., III; Grossman, E.E.; Gibbs, A.E. Islands at risk: Coastal hazard assessment and mapping in the Hawaiian Islands. *Environ. Geosci.* **2001**, *8*, 21–37. [CrossRef]
56. Sawaragi, T.; Deguchi, I.; Park, S.K. Experimental study on the function of submerged breakwater to control cross-shore sediment transport on artificially nourished beaches. *Coast. Eng. Jpn.* **1988**, *31*, 121–130. [CrossRef]
57. Lorenzoni, C.; Postacchini, M.; Mancinelli, A.; Brocchini, M. The morphological response of beaches protected by different breakwater configurations. In Proceedings of the 33rd International Conference on Coastal Engineering, Santander, Spain, 1–6 July 2012; p. 2.
58. Dolphin, T.J.; Vincent, C.E.; Bacon, J.C.; Dumont, E.; Terentjeva, A. Decadal-scale impacts of a segmented, shore-parallel breakwater system. *Coast. Eng.* **2012**, *66*, 24–34. [CrossRef]
59. Fletcher, C.H.; Mullane, R.A.; Richmond, B.M. Beach loss along armored shorelines on Oahu, Hawaiian Islands. *J. Coast. Res.* **1997**, *13*, 209–215.
60. Anderson, T.R.; Fletcher, C.H.; Barbee, M.M.; Frazer, L.N.; Romine, B.M. Doubling of coastal erosion under rising sea level by mid-century in Hawaii. *Nat. Hazards* **2015**, *78*, 75–103. [CrossRef]
61. Hunt, I.A. Design of sea-walls and breakwaters. *Trans. Am. Soc. Civ. Eng.* **1961**, *126*, 542–570. [CrossRef]
62. Tainter, S.P. *Bluff Slumping and Stability: A Consumer's Guide*; Michigan Sea Grant: Ann Arbor, MI, USA, 1982; Publication MICHU-SG-82-902; p. 65.
63. Bush, D.M.; Neal, W.J.; Young, R.S.; Pilkey, O.H. Utilization of geoindicators for rapid assessment of coastal-hazard risk and mitigation. *Ocean. Coast. Manag.* **1999**, *42*, 647–670. [CrossRef]
64. Jorgenson, T.; Ely, C. Topography and flooding of coastal ecosystems on the Yukon-Kuskokwim Delta, Alaska: Implications for sea-level rise. *J. Coast. Res.* **2001**, *17*, 124–136.
65. Di Risio, M.; Bruschi, A.; Lisi, I.; Pesarino, V.; Pasquali, D. Comparative analysis of coastal flooding vulnerability and hazard assessment at national scale. *J. Mar. Sci. Eng.* **2017**, *5*, 51. [CrossRef]

66. Lewsey, C.; Cid, G.; Kruse, E. Assessing climate change impacts on coastal infrastructure in the Eastern Caribbean. *Mar. Policy* **2004**, *28*, 393–409. [CrossRef]
67. Bulleri, F.; Chapman, M.G. The introduction of coastal infrastructure as a driver of change in marine environments. *J. Appl. Ecol.* **2010**, *47*, 26–35. [CrossRef]
68. Smythe, T.C. *Assessing the Impacts of Hurricane Sandy on the Port of New York and New Jersey's Maritime Responders and Response Infrastructure*; Natural Hazards Center: St. Trenton, NJ, USA, 2013. [CrossRef]
69. Anderies, J.M.; Janssen, M.A.; Ostrom, E. A Framework to Analyze the Robustness of Social-Ecological Systems from an Institutional Perspective. *Ecol. Soc.* **2004**, *9*, 18. Available online: http://www.ecologyandsociety.org/vol9/iss1/art18 (accessed on 1 October 2021). [CrossRef]
70. Bruun, P. Sea-level rise as a cause of shore erosion. *J. Waterw. Harb. Div.* **1962**, *88*, 117–132. [CrossRef]
71. Cooper, J.A.G.; Masselink, G.; Coco, G.; Short, A.D.; Castelle, B.; Rogers, K.; Anthony, E.; Green, A.N.; Kelley, J.T.; Pilkey, O.H.; et al. Sandy beaches can survive sea-level rise. *Nat. Clim. Chang.* **2020**, *10*, 993–995. [CrossRef]
72. Toimil, A.; Losada, I.J.; Camus, P.; Díaz-Simal, P. Managing coastal erosion under climate change at the regional scale. *Coast. Eng.* **2017**, *128*, 106–122. [CrossRef]
73. Rosati, J.D.; Dean, R.G.; Walton, T.L. The modified Bruun Rule extended for landward transport. *Mar. Geol.* **2013**, *340*, 71–81. [CrossRef]
74. de Santiago, I.; Camus, P.; Gonzalez, M.; Liria, P.; Epelde, I.; Chust, G.; del Campo, A.; Uriarte, A. Impact of climate change on beach erosion in the Basque Coast (NE Spain). *Coast. Eng.* **2021**, *167*, 103916. [CrossRef]
75. Yohe, G. *The Cost of Not Holding Back the Sea: Phase 1 Economic Vulnerability in The Potential Effects of Global Climate Change on the United States*; EPA 230-05-89-052; Report to Congress; Appendix B: Sea Level Rise; U.S. Environmental Protection Agency: Washington, DC, USA, 1989.
76. Yohe, G.; Neumann, J.; Marshall, P.; Ameden, H. The Economic Cost of Greenhouse-Induced Sea-Level Rise for Developed Property in the United States. *Clim. Chang.* **1996**, *32*, 387–410. [CrossRef]
77. Yohe, G.; Schlesinger, M.E. Sea-Level Change: The Expected Economic Cost of Protection or Abandonment in the United States. *Clim. Chang.* **1998**, *38*, 447–472. [CrossRef]
78. Fu, X.; Song, J. Assessing the economic costs of sea level rise and benefits of coastal protection: A spatiotemporal approach. *Sustainability* **2017**, *9*, 1495. [CrossRef]
79. Hanemann, W.M. *What is the Economic Cost of Climate Change?* (No. 1557-2016-132754); University of California, Berkeley: Berkeley, CA, USA, 2008.
80. Heberger, M.; Cooley, H.; Herrera, P.; Gleick, P.H.; Moore, E. *The Impacts of Sea-Level Rise on the California Coast*; California Climate Change Center: San Francisco, CA, USA, 2009.
81. Michael, J. Episodic Flooding and the Cost of Sea-level Rise. *Ecol. Econ.* **2007**, *63*, 149–159. [CrossRef]
82. Kirshen, P.; Knee, K.; Ruth, M. Adaptation to Sea Level Rise in Metro Boston. *Clim. Chang.* **2008**, *90*, 453–473. [CrossRef]
83. U.S. Army Corps of Engineers (USACE). *Economic Guidance Memorandum (EGM) 04-01: Generic Depth-Damage Relationships for Residential Structures with Basements*; USACE: Washington, DC, USA, 2003.
84. King, P.G. *Overcrowding and the Demand for Beaches in Southern California*; Department of Boating and Waterways: San Francisco, CA, USA, 2001.
85. Pendleton, L.; King, P.; Mohn, C.; Webster, D.G.; Vaughn, R.; Adams, P.N. Estimating the potential economic impacts of climate change on Southern California beaches. *Clim. Chang.* **2011**, *109*, 277–298. [CrossRef]
86. Gopalakrishnan, S.; Smith, M.D.; Slott, J.M.; Murray, A.B. The value of disappearing beaches: A hedonic pricing model with endogenous beach width. *J. Environ. Econ. Manag.* **2011**, *61*, 297–310. [CrossRef]
87. Tickell, C. *Global Warming and Its Effects, Engineering for Climate Change*; Institute of Civil Engineers: London, UK, 1993; pp. 9–16.
88. Arnell, N.W. Impacts of Climatic Change on River Flow Regimes in the UK. *Water Environ. J.* **1992**, *6*, 432–442. [CrossRef]
89. Al-Ansari, N.; Abdellatif, M.; Ali, S.; Knutsson, S. Long term effect of climate change on rainfall in northwest Iraq. *Open Eng.* **2014**, *4*, 250–263. [CrossRef]

remote sensing

MDPI

Article

Video-Monitoring Tools for Assessing Beach Morphodynamics in Tidal Beaches

Juan Montes [1,*], Laura del Río [1], Theocharis A. Plomaritis [2], Javier Benavente [1], María Puig [3] and Gonzalo Simarro [4]

[1] Department of Earth Sciences, Faculty of Marine and Environmental Sciences, Instituto Universitario de Investigación Marina (INMAR), University of Cádiz, 11510 Puerto Real, Cádiz, Spain; laura.delrio@uca.es (L.d.R.); javier.benavente@uca.es (J.B.)

[2] Department of Applied Physics, Faculty of Marine and Environmental Sciences, Instituto Universitario de Investigación Marina (INMAR), University of Cádiz, 11510 Puerto Real, Cádiz, Spain; haris.plomaritis@uca.es

[3] TECNALIA, Basque Research and Technology Alliance (BRTA), Astondo Bidea, Edificio 700, 48160 Derio, Bizkaia, Spain; maria.puig@tecnalia.com

[4] Institute of Marine Sciences (CSIC), Passeig Marítim de la Barceloneta 37-49, 08003 Barcelona, Catalonia, Spain; simarro@icm.csic.es

* Correspondence: juan.montes@uca.es

Abstract: Beach behaviour and evolution are controlled by a large number of factors, being susceptible to human-derived pressures and the impacts of climate change. In order to understand beach behaviour at different scales, systematic monitoring programs that assess shoreline and volumetric changes are required. Video-monitoring systems are widely used in this regard, as they are cost-effective and acquire data automatically and continuously, even in bad weather conditions. This work presents a methodology to use the basic products of low-cost IP video cameras to identify both the cross-shore and long-shore variability of tidal beaches. Shorelines were automatically obtained, digital elevation models (DEMs) were generated and validated with real data, and the outputs were combined to analyse beach behaviour from a morphodynamic perspective. The proposed methodology was applied to La Victoria Beach (SW Spain) for the analysis of beach variations over a 5-year period. The combination of shoreline position analysis and data from DEMs facilitates understanding and provides a complete overview of beach behaviour, revealing alongshore differences in an apparently homogeneous beach. Furthermore, the methods used allowed us to inter-relate the different processes occurring on the beach, which is difficult to achieve with other types of techniques.

Keywords: video-monitoring system; shoreline; digital elevation model; EOFs; beach variability

Citation: Montes, J.; del Río, L.; Plomaritis, T.A.; Benavente, J.; Puig, M.; Simarro, G. Video-Monitoring Tools for Assessing Beach Morphodynamics in Tidal Beaches. *Remote Sens.* **2023**, *15*, 2650. https://doi.org/10.3390/rs15102650

Academic Editors: Ramón Blanco Chao, Germán Flor-Blanco and José Juan de Sanjosé Blasco

Received: 15 April 2023
Revised: 17 May 2023
Accepted: 17 May 2023
Published: 19 May 2023

1. Introduction

Beaches are complex natural systems whose evolution and behaviour are conditioned by a large number of factors. Beaches also concentrate population and economic activities, which has led to large urban development in these areas [1]. In addition, beaches are extremely vulnerable to the impacts associated with climate change, such as relative sea level rises or changes in wave direction and/or storm frequency and intensity, as well as to decrease in sediment supply and other human-induced modifications that often lead to localised beach erosion [2]. About half of all sandy beaches in the world could be seriously threatened by erosion by the end of the century because of sea level rise, especially where landward transgression is restricted by infrastructure, although the response is dependent on many local environmental factors [3,4]. Effective management strategies are, therefore, needed for these areas, supported by morphodynamic studies that provide information on local processes.

The analysis of coastal processes and morphological responses is essential to understand beach behaviour at different scales, from event responses to seasonal and interannual changes. Beach evolution analyses of tidal areas add complexity to the study, as tidal cycles shift the oceanographic influence along the beach profile [5]. Therefore, the continuous monitoring of these zones is necessary to understand shoreline and volumetric changes and quantify the morphodynamic processes controlling these systems. On the other hand, the real-time monitoring of the rapid changes generated by high-energy events has traditionally been challenging, making it difficult to understand the processes that actually occur under storm conditions.

In beach-monitoring programs, the spatial and temporal resolution and the area and duration of the survey that can be covered are key parameters, as they determine the type of processes that can be observed, from storm-induced rapid changes to seasonal behaviour or changes driven by global atmospheric circulation patterns [6–8]. In this regard, the RTK-DGPS was an important breakthrough in systematic beach monitoring, making it possible to cover larger areas faster and more accurately than previous techniques, such as the theodolite and the total station [9] methods or low-cost techniques designed to reduce the number of surveyors needed [10]. Terrestrial laser scanners, airborne LiDAR sensors, and unmanned aerial vehicles are widely used for high-resolution surveys, as they are relatively fast and simple [11], although their characteristics and the need for surveyors compromise a temporal resolution. The use of satellite-based observations to study coastal regions is becoming increasingly widespread. Satellites measure a variety of physical parameters and enable the analysis of shorelines, sea levels, waves, and coastal currents, among others [12]. However, a spatiotemporal resolution is still not enough for monitoring some coastal processes in tidal beaches, where errors in satellite-derived shorelines can exceed 30 m [13], with spatial changes close to the satellite pixel resolution and temporal changes close to the satellite revisiting time. The spatial resolution has improved to 30 m for Landsat 4–8 and 10 m for Sentinel-2, while the revisit time is 5 days for Sentinel-2 at the equator, with shorter revisit intervals at higher latitudes [14].

Video-monitoring systems are commonly used to cover the spatiotemporal gap between the more traditional survey techniques and rapidly evolving satellite observations. The use of video-monitoring systems for the study of coastal processes has developed since Holman and Guza [15] applied video techniques to measure wave run-ups and Lippmann and Holman [16] started using these systems to quantify the temporal and spatial variability of sand bar morphologies. Video-monitoring systems are relatively low-cost and acquire data automatically, continuously, and periodically, even in bad weather conditions. These techniques significantly reduce field efforts compared with traditional coastal monitoring systems and allow larger regions to be covered. They can also be used for a wide range of purposes, including the analysis of rip currents [17,18], nearshore hydrodynamics [19], the behaviour of anthropized beaches in storm conditions [20], mesoforms such as bars [21–23] and beach cusps [24–26], run-up measurements [27,28], the estimation of flood processes [29], and to obtain coastal bathymetry [30–32]. The spatial resolution of the video monitoring depends on the system and its position with respect to the monitored area. In low-resolution systems that work with compressed images, as occurs in most IP camera systems, shoreline detection becomes complex, as data compression clusters pixels together, which makes it difficult for algorithms to work properly, mainly in the furthest areas of the system.

The collection of data with high frequency and spatial resolution, as provided by video-monitoring systems, allows for the application of statistical techniques to analyse the conditions controlling beach evolution. Among them, Empirical Orthogonal Function (EOF) analysis is one of the most widely and extensively used methods for decomposing a space–time field into spatial patterns and associated time indices. Winant et al. [33] applied it to study the seasonal changes in cross-shore beach profiles. Since then, it has been used in the analysis of beach morphodynamics (review in Larson et al. [34]). The method is purely statistical and tries to represent the complex field of spatiotemporal variability through a number of basic spatial patterns coupled with time-dependent functions [35]. Although the resulting patterns lack any direct physical meaning, they are often linked with coastal processes and behaviour [36] or can be coupled with oceanographic or sediment transport models to analyse beach changes [37].

The aim of this work is to present and validate a methodology based on the primary video-monitoring products (timex images) using a low-cost IP camera system to calculate morphological parameters that can be analysed to extract the interannual behaviour of tidal beaches. For this purpose, automatic shoreline extraction and the generation of intertidal digital elevation models (DEMs) were implemented. Furthermore, a tool was developed to validate the DEMs with measured data (ground truth). The automatically obtained shorelines were explored with Empirical Orthogonal Function (EOF) techniques in order to identify the spatial and temporal information captured by the video-monitoring station and identify both the cross-shore and long-shore variability of the system. The proposed methodology was applied to La Victoria Beach (SW Spain) for the analysis of beach variations over a 5-year period.

2. Study Zone

The method was developed and applied to a 750 m coastal stretch at La Victoria urban beach, located in the city of Cádiz (SW Spain, Figure 1). The western part of the city, where La Victoria is located, is mainly composed of sandy beaches exposed to the Atlantic Ocean.

Figure 1. (**A**) Location of the study area in SW Spain. (**B**) Position of the video-monitoring system (VMS) in the city of Cadiz. (**C**) General view of the VMS. (**D–F**) Areas covered by each camera.

La Victoria Beach follows an NNW-SSE orientation and presents an intermediate–dissipative profile composed of medium-to-fine sand [38]. Foreshore slope values range between 0.025 and 0.02 with seasonal changes [39]. A wide rocky shore platform extends along the coast, varying in width and depth, affecting the regular bathymetry in the zone [40]. It discontinuously appears around the mean sea level in front of La Victoria, which is considered a reef-supported beach [40]. Several nourishment works were carried out on the beach over the last two decades because of the reduction in the river sediment supply [41,42] in order to prevent erosion and damage to tourism infrastructure.

Regarding the hydrodynamic conditions, storms affecting La Victoria Beach occur mainly between November and March, associated with westerly winds. Easterly winds, although more frequent and more intense, have little impact on wave generation in this area because of their limited fetch [43]. The mean wave height is less than 1 m with associated periods of 5–6 s, although wave height during storms can exceed 4 m [44] with associated periods longer than 8 s. The dominant longshore drift is directed towards the southeast. Tides in this area are semidiurnal and mesotidal. The mean tidal range oscillates between 3.07 m and 1.11 m in mean spring and neap tides, respectively, with a tidal range of 3.89 m during the highest astronomical tide [45].

3. Methodology

The proposed methodology jointly analyses the cross-shore and the long-shore behaviour of the beach. For this purpose, two approaches were combined, namely, shoreline-based 2D analysis, common in coastal studies, and DEM-based 3D analysis, often limited because of the impossibility of obtaining such data with a high temporal resolution. Such a combination of techniques facilitates the comprehension of processes that may be ignored when cross- and longshore behaviours are analysed independently. An overview of the workflow of the proposed analysis is presented in Figure 2.

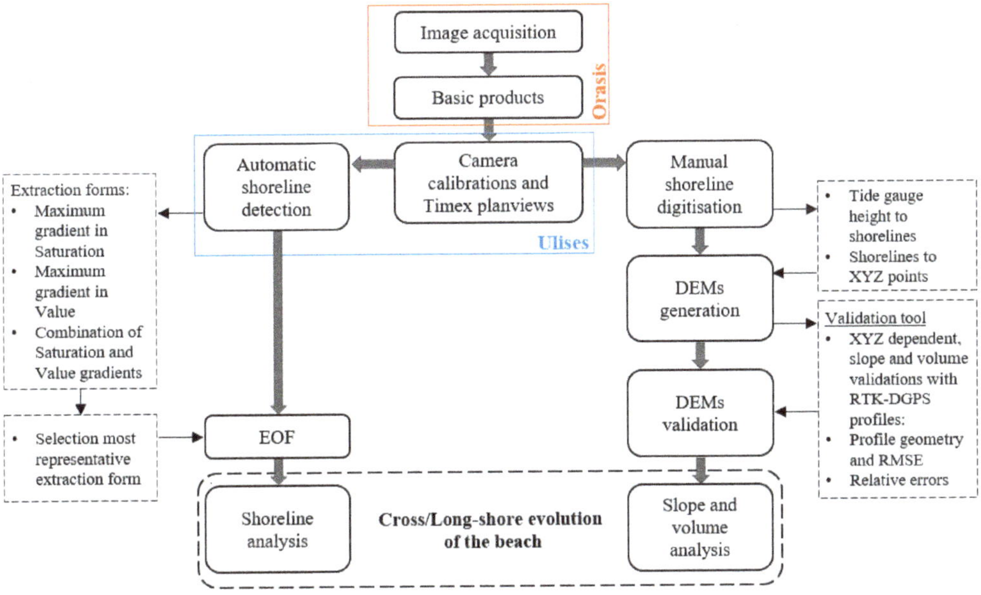

Figure 2. Methodology used to analyse beach cross- and longshore evolution combining the automatic extraction of shorelines and the generation of DEMs.

A coastal video-monitoring system composed of 3 IP cameras of 2 MegaPixel (1600 × 1200 pixels) resolution was used in order to record beach images (Figure 1). The system was installed on top of a residential building (~49 m above the mean sea level) and covers approximately 750 m of shoreline. The system records the first 10 min of each daylight hour with a frequency of 4 Hz [26], i.e., 2400 frames per burst. After recording, three basic products are automatically obtained from each camera by using the ORASIS software [31]: snap images (snapshot image), timex images (time average image of the 2400 frames), and Sigma images (greyscale variance image of the 2400 frames). The interval of images analysed in the present work extends from September 2013 to March 2018, except for small periods in which the system was not operational for technical reasons. Camera calibration, image rectification, and planview generation were undertaken using the ULISES open-source software developed by Simarro et al. [46]. Planviews, both for timex and Sigma images, were generated with a resolution of 0.5 m/pixel. For further details on this video-monitoring system, the generation of basic products, the system calibrations, and the production of planviews, the reader is also referred to Montes et al. [26].

3.1. Automatic Shoreline Detection

Shoreline analyses during the studied period were performed by extracting the shoreline automatically from the planviews (Figure 3). System limitations, mainly derived from the resolution and compression types of the images, constrained automatic shoreline extraction, so different methods were tested to resolve this, and only timex images were used in a similar approach to the one developed by Ribas et al. [47]. The use of Sigma planviews, employed in similar studies [28,31], was discarded because of the poor results obtained in this case. The timex planviews were transformed from Red–Green–Blue (RGB) to Hue–Saturation–Value (HSV) in order to improve the results obtained when extracting the shoreline [48,49]. Three different extraction forms were used: maximum gradient in Saturation, maximum gradient in Value, and a combination of Saturation and Value gradients. The quality of the results for each of the abovementioned procedures was highly dependent on the existing conditions in each planview (sun height, luminosity, presence of shadows, etc.). Therefore, the shorelines for all planviews were generated following Ribas et al. [47] and then manually checked to ignore corrupt shorelines; those showing the best results in each case were selected. To improve the automatic extraction process, a mask was created on the zone of interest in the planview (Figure 3), and the extraction of the shoreline was only conducted during high tide (including the previous and next hour) during daylight conditions (~3 per day). This specific automatic extraction prevents the presence of buildings, roads, people, and intertidal water-filled features that can produce, in some cases, noise that could decrease the quality of the detection [48]. In order to assess shoreline evolution, the land part of the mask was used as a baseline. Moreover, all the obtained shorelines were corrected to their real Z since the planviews were generated with a constant Z.

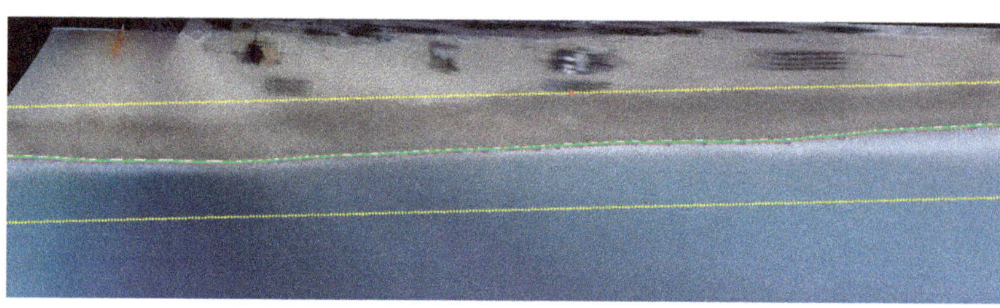

Figure 3. Example of shoreline extraction (green line) using a mask (yellow lines) to avoid features of the beach that might decrease the quality of the extraction. The background planview is in pixel coordinates.

The large number of automatically extracted shorelines obtained with this procedure, approximately 1800, allows us to perform statistical analyses to study the interannual behaviour of the beach. In this work, the spatial and temporal variability in the shoreline position is studied through an analysis of Empirical Orthogonal Functions (EOFs), also known as Principal Component Analysis [50]. Such an analysis can provide useful information on the amount of data obtained with the video-monitoring technique [51]. The EOF analysis can find a set of variables that explain data variability [52]. EOFs are the eigenfunctions of the covariance data, and they represent the dominant spatial patterns, or modes, of the variance, while their associated amplitudes represent the temporal variation, ordered by their absolute values. This method was applied in previous studies at La Victoria Beach to analyse the variations in the beach profile [39,53,54].

3.2. DEM Generation

Intertidal topography was obtained from the coastal video-monitoring system. For this purpose, shorelines of a complete tidal cycle were manually extracted once a month. The digitization of shorelines was not conducted automatically as in Section 3.1 in order to avoid problems in low-tide conditions and to ensure the quality of the extraction. This simple approach has been used by other authors for similar purposes [48,49,55]. The steps followed were (Figure 2) (i) manual digitisation of the shoreline; (ii) assignation to each shoreline of its real height value through the tide level; (iii) decomposition of shorelines into points with X, Y, Z coordinates; and (iv) generation of DEMs from XYZ points. Once generated, the DEMs were validated using the ground truth, and slope and volume values were obtained in different profiles distributed homogeneously over the DEMs (Figure 4).

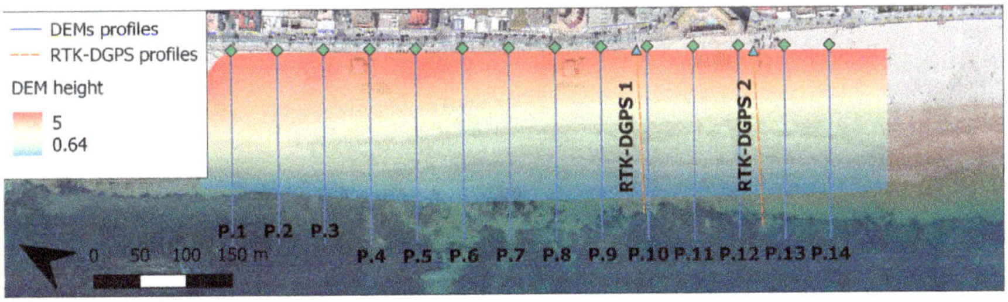

Figure 4. RTK-DGPS profiles used for the validation (orange dashed lines), profiles employed to extract slope and volume (blue lines), and an example of an extracted DEM.

The manually digitised shorelines were projected to their real coordinates via ULISES routines. Subsequently, each shoreline was assigned its corresponding height by using the sea level data from the Cádiz tide gauge [56] located in the city harbour (Figure 1). The ∼6 shorelines of each tidal cycle were decomposed into points every 0.5 m, obtaining around 9000 points with X, Y, Z coordinates for each DEM. Once these points were obtained, they were used to generate the DEMs. Monthly DEMs were built when planviews were available for spring tidal conditions. Low-wave-energy conditions were selected to avoid possible errors in the assignment of a height to each shoreline, e.g., produced by wave run-up. It must be noted that the sea level data obtained from the tide gauge include the influence of atmospheric pressure.

Once DEMs were extracted for the entire study period, an extensive DEM validation was performed using ground-truth data collected by RTK-DGPS. The topographic surveys for the validation were conducted on dates equal to or close to (<4 days) the date chosen to build the DEM in order to ensure a similar beach state (Table 1). The validation was performed using two profiles located in the southern part of the study area (orange profiles in Figure 4), where RTK-DGPS-surveyed profiles and DEM-extracted profiles were compared. In this procedure, three types of variables were computed to carry out the validation: XYZ-dependent variables, slope, and volume.

Table 1. Dates on which the RTK-DGPS profiles were carried out and DEMs were extracted.

Survey Index	RTK-DGPS Survey Date	DEM Date
1	23-09-2013	19-09-2013
2	03-10-2013	06-10-2013
3	03-12-2013	04-12-2013
4	31-01-2014	31-01-2014
5	18-06-2014	14-06-2014
6	09-10-2014	09-10-2014

Table 1. *Cont.*

Survey Index	RTK-DGPS Survey Date	DEM Date
7	25-11-2014	24-11-2014
8	03-12-2014	06-12-2014
9	02-07-2015	04-07-2015

The XYZ-dependent variables analyse the profile geometry, based on Z, with a comparison between the shape of the real (RTK-DGPS) profiles and those extracted from the DEMs. Different parameters commonly employed for the estimation of errors in morphodynamic models were used [57–61]: correlation coefficient (R2), scatter index (SCI), relative bias (RelBias), and the Brier Skill Score (BSS), as well as the Root-Mean-Square Error (RMSE). R2 and SCI provide information about data dispersion. RelBias is a normalised relative measure of the bias. Lastly, BSS relates the variance in the difference between the real and modelled data to the variance in the real data; a BSS of 1 means that both profiles are equal, while 0 means a very bad coupling between the real and computed profiles. A detailed explanation of the parameters can be found in Roelvink et al. [62]. Finally, the RMSE (Equation (1)) estimates the vertical error (in metres), and it has been widely used in the validation of DEMs, such as those generated by UAVs [63,64]. The RMSE was calculated between the points from real surveys and points from DEMs:

$$RMSE = \sqrt{\frac{\sum_{i=1}^{n} (RTKz_i - DEMz_i)^2}{n}} \quad (1)$$

where RTKz is the elevation of a real point measured with the RTK-DGPS, DEMz is the elevation of the DEM at the same coordinates, and n is the total number of RTKz points.

Slope and volume variables were calculated between the minimum and maximum DEM heights in each profile, which is the equivalent of intertidal beach slope and volume. After that, the relative errors were obtained for each one using Equations (2) and (3).

$$\varepsilon_s = \frac{|Slope_{DEM} - Slope_{RTK}|}{Slope_{RTK}} \quad (2)$$

$$\varepsilon_v = \frac{|Volume_{DEM} - Volume_{RTK}|}{Volume_{RTK}} \quad (3)$$

where ε_s is the slope relative error, and ε_v is the volume relative error. Again, RTK refers to ground truth values and DEM refers to values extracted from the DEMs in both equations.

Once DEMs were extracted and validated, slopes and volumes were calculated between September 2013 and March 2018 in 14 cross-shore profiles distributed along the beach and spaced every 50 m (Figure 4). Slopes and volumes were extracted between the minimum and maximum tide levels reached on the day of DEM generation. For each parameter, the evolution in each profile during the study period was obtained, as well as the longshore variation along the beach for each survey.

4. Results

4.1. Automatic Shoreline Extraction and Shoreline Analysis

Approximately 1800 high-tide shorelines were automatically extracted over the study period, with a 46% success rate for all available planviews in high-tide conditions. Although a minimum tidal threshold was established (above 3 m) to reduce the tide influence on the results, certain oscillations directly related to the tidal range above the threshold were observed, which is accentuated under spring tides or by the effect of storm surges. The gaps correspond to periods when the system was not running properly or the shorelines could not be extracted from the planviews.

Regarding shoreline evolution based on the automatic shoreline extraction method applied, it was observed that the dry beach changes (defined from the shoreline to the baseline; see Figure 5A) showed seasonal behaviour, with clear differences between the winter and summer periods (Figure 5B). The mean beach width ranged between 5 and 55 m during this period. Minimum values were observed in winter between October and March, and values below 10 m were observed in January 2015, December 2016, and February/March 2017. Maximum values were mostly recorded in summer. In September 2013 and March 2014, values higher than 50 m were also identified. March 2014 and 2016 were two exceptions to this seasonal behaviour, showing an early increase in the dry beach area, although in both cases, it reduced afterwards, following the same trend observed in other years.

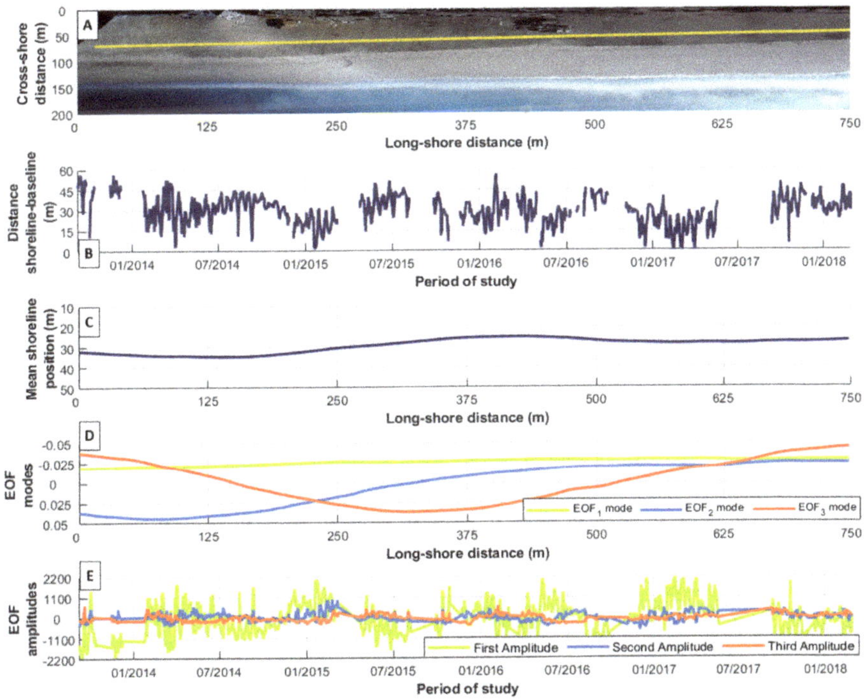

Figure 5. Analysis of shoreline position. (**A**) Planview of La Victoria Beach with the baseline (yellow line) used for the analysis. (**B**) Mean shoreline position with respect to the baseline. (**C**) Result of EOF analysis. (**D**) Evolution of EOF amplitudes along the study period. (**E**) Evolution of mean shoreline position along the study period.

Regarding the alongshore variations, Figure 5C shows the time-averaged beach width with respect to the baseline during the study period. An alongshore differential behaviour was observed and, based on this, the beach can be divided into three sectors: (1) the northern sector, which extends along the first 250 m from the planview origin (N) and presents an average beach width higher than 30 m, with a maximum value of 35 m; (2) the central sector, between 250 and 500 m from the planview origin, which shows a reduction in beach width reaching the minimum value, 25 m, with beach widths always below 30 m; and (3) the southern zone, between 500 and 750 m from the planview origin, which was always below a 30 m beach width like the central zone, although there is a slight increase of 2–3 m, and it shows more homogeneous behaviour.

The results of the EOF analysis are also presented in Figure 5, including the spatial patterns (Figure 5D) and their corresponding amplitudes (Figure 5E), calculated on the average shoreline position. The first three EOFs describe 98% of the total variability of the beach during the study period. EOF1, explaining 89.5% of the variance, represents the accretion/erosion of the beach, depending on the negative/positive amplitude values, and in this case, its spatial mode has a constant negative sign. The values are slightly higher in the northern sector, decreasing in the central zone until they become constant in the southern sector. EOF2, which explains 5.5% of the total variance, represents beach rotation, associated with sediment transport from the zone with positive spatial mode values to the zone with negative spatial mode values, implying sediment transport from the north to the south under specific conditions. Finally, EOF3 explains 3% of the total variance and corresponds to a localised erosion/accretion process under certain conditions (recalling that EOF modes are mathematically forced to be orthogonal). Almost the entire northern and southern zones show negative values (accretion). Conversely, a small part of the northern sector together with the central sector present positive values (erosion).

4.2. DEM Generation and Validation

A total of 48 DEMs were generated for the intertidal zone at La Victoria Beach for the entire study period. All of them were located between the elevations +0.13 and +3.99 m above the tidal datum (hydrographic zero, i.e., the lowest astronomical tide).

The validation results are shown in Tables 2 and 3 and in Figure 6. Regarding the geometry of the beach profiles, all the analysed indexes of the morphodynamic models show satisfactory values (Table 2). The R2 values for both profiles were higher than 97% and in all cases over 95%, and the mean SCI values were lower than 0.13 in both profiles. The relative BIAS parameters were close to zero in most of the validations; only in July 2015 were the values of this parameter higher than 0.10 for both validation profiles (0.13 and 0.16). The BSS parameter indicated a strong similarity between the RTK-DGPS profiles and those generated by DEMs, as in all cases, the value of this parameter was over 0.90, 0.96 being the average for all the validations in both profiles. Finally, the mean RMSE in both profiles was around 25 cm. Only in one case (July 2015) were the values obtained high, greater than 45 cm, while in most cases, the vertical error was lower than 30 cm.

Table 2. Results of the DEM validation using the two RTK-DGPS profiles (P1 and P2, Figure 4) for XYZ-dependent variables where profile geometry was analysed.

Survey Index	Geometry Errors									
	R		SCI		RelBias		BSS		RMSE	
	P1	P2	P1	P2	P1	P2	P1	P2	P1	P2
1	0.98	0.97	0.08	0.09	0.04	0.02	0.97	0.94	0.23	0.26
2	0.98	0.99	0.08	0.1	0.03	−0.02	0.96	0.96	0.2	0.23
3	1	0.97	0.13	0.17	−0.09	−0.12	0.97	0.94	0.3	0.36
4	0.98	0.99	0.14	0.14	0.1	0.13	0.95	0.99	0.31	0.36
5	0.99	0.99	0.05	0.06	−0.02	−0.02	0.99	0.98	0.12	0.14
6	0.96	0.98	0.09	0.08	−0.06	−0.06	0.93	0.97	0.30	0.26
7	0.98	0.97	0.1	0.11	0.01	−0.02	0.96	0.95	0.22	0.23
8	0.99	0.99	0.05	0.04	−0.05	−0.03	0.99	0.99	0.10	0.08
9	0.97	0.97	0.17	0.19	0.13	0.16	0.95	0.96	0.48	0.47
Mean	**0.98**	**0.98**	**0.12**	**0.13**	**0.05**	**0.05**	**0.96**	**0.96**	**0.25**	**0.26**

Regarding the validation of beach slope and volume, low mean relative errors were obtained (Table 3). The mean relative error in slope was 9.9% for profile 1 and 12.7% for profile 2. The mean error for the volume was below 10% in both profiles. Considerably higher error values were only found in the slope calculations for both profiles on the validation of 9 October 2014 (51.5% and 45.3%, respectively), while significantly higher errors (more than 20%) for the volume were only found on the validation of the 4 July 2015 (19.7% and 24.5% for profile 1 and profile 2, respectively).

Table 3. Results of the DEM slope and volume validations using the two RTK-DGPS profiles for the relative slope and volume error.

Survey Index	Relative Errors			
	Slope P1	Slope P2	Volume P1	Volume P2
1	0.01	0.05	0.05	0.02
2	0.02	0.22	0.04	0.02
3	0.02	0.05	0.13	0.19
4	0.13	0	0.12	0.16
5	0.01	0.13	0.05	0.04
6	0.52	0.45	0.12	0.07
7	0.18	0.22	0.04	0
8	0	0	0.08	0.05
9	0	0.02	0.2	0.25
Mean	**0.1**	**0.13**	**0.09**	**0.09**

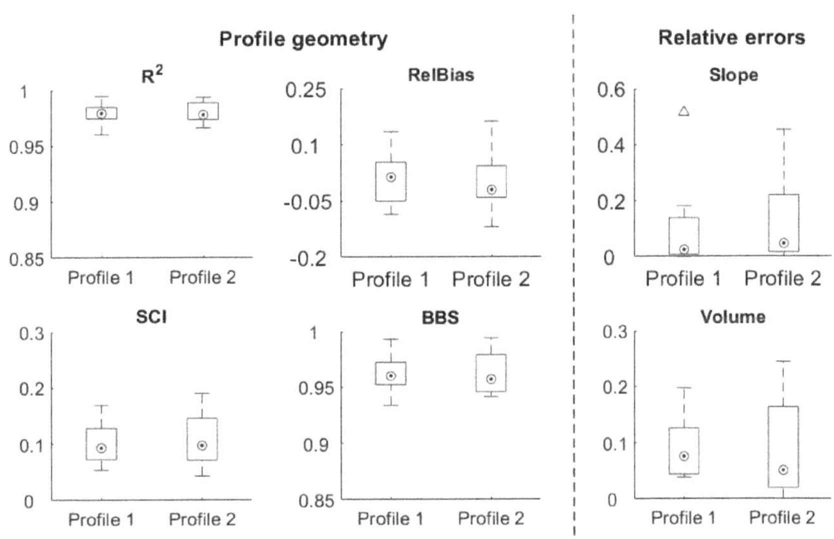

Figure 6. Results of the DEM validation (XYZ-dependent parameters, slope, and volume) using the two RTK-DGPS profiles.

4.3. Slope and Volume Evolution

The DEM-derived slope evolution (Figure 7) shows a seasonal pattern across the entire beach. During winter months, the slope decreases and then increases later during the summer months (upper panel, Figure 7). The variations recorded are small, ranging between 0.02 and 0.07 over the study period. In addition, higher slopes in the northern zone of the beach are observed in all surveys in a detailed profile analysis. In turn, the seasonal oscillations described above are more pronounced in this area, with increased disparities between the northern and southern slopes during the summer period. During winter months, these differences are smaller. In fact, if the values of this parameter are analysed for the beach in each survey (lower panel Figure 7), a higher dispersion of data is observed in the northern zone, while the data scattering is reduced in the southern sector (between 0.025 and 0.055). Furthermore, the study of the average slope behaviour also revealed higher values in the northern part that decrease towards the central zone (profiles 7–8), and the slope remains constant until the southern sector.

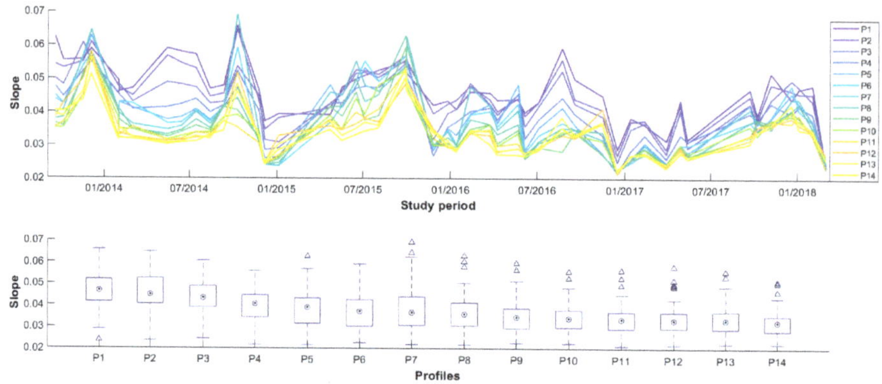

Figure 7. Upper panel: evolution of beach slope in each profile of La Victoria Beach over the studied period, where profile 1 is the northernmost one and profile 14 is the southernmost one. Lower panel: variability in beach slope along the analysed profiles for each survey, where the red lines are the median, the limits of the boxes are the 25th and 75th percentiles, and the markers are the outliers.

The beach volume (Figure 8), normalised according to the analysed profile extension, is relatively homogeneous, as low dispersion is observed between the values of the different profiles for each DEM (upper panel Figure 8). The volume ranges between 1 and 2.5 m^3/m in all profiles. A clear seasonal trend is not detected. Volume variability (lower panel, Figure 8) is slightly lower in the northern area, while the average values per profile show homogeneous behaviour along the beach.

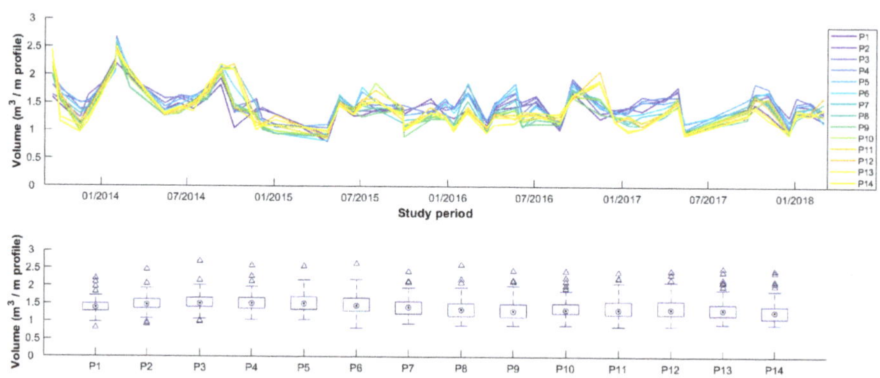

Figure 8. Upper panel: evolution of the normalized volume in each profile of La Victoria Beach over the studied period, where profile 1 is the northernmost one and profile 14 is the southernmost one. Lower panel: variability in the volume along the analysed profiles for each survey, where the red lines are the median, the limits of the boxes are the 25th and 75th percentiles, and the markers are the outliers.

5. Discussion

5.1. Methodological Considerations

The methodological approach developed in this work combines automatic shoreline extraction and DEM generation, validation and extraction of slope and volume data from them. The results demonstrate the validity of this approach, although a number of technical considerations must be noted.

One of the main techniques applied to video-monitoring systems for the automatic extraction of shorelines [65] is the use of Sigma images. This technique was discarded in this work because of the low return rate, despite its widespread use for this purpose [28,31]. Thus, an approximation of the Pixel Intensity Clustering (PIC) model described by Aarninkhof and Roelvink [66] and Aarninkhof [67] was employed in order to improve the success rate of the automatic extraction of the shoreline. The PIC model identifies colour differences between wet and dry sand areas through the transformation of RGB images into HSV images. This model works better on dissipative beaches than, for instance, the Shore Line Intensity Maximum (SLIM) model [68], which is designed for relatively steep beaches and fails in dissipative ones [69]. By applying the PIC model to La Victoria, the shoreline extraction success rate reached 46%. In the future, this rate could be further increased by using Artificial Neural Network systems (ANN) [31,51,69]. ANN systems are trained with manual extractions of shorelines, improving the results for dissipative beaches with complex geometries (i.e., sand bars, inlets, beach cusps, etc.) such as La Victoria Beach [26].

Another major application of video-monitoring systems is topo-bathymetric data extraction, which has been widely studied and applied by different authors [30,31,48,55,70–74]. In this work, intertidal DEMs derived from the video-monitoring system were subjected to extensive validation in order to check their quality by using nine RTK-DGPS topographic surveys. Indexes concerning profile geometry also reveal good results in most of the cases, showing a great similarity between the profiles obtained with RTK-DGPS and the profiles obtained using the DEMs (Table 2 and Figure 6). In morphology, the BSS is considered acceptable when the value is between 0.6–0.8 and excellent when it is higher than 0.8 [59]. In this study, almost every extracted BSS is above 0.95, which indicates an excellent similarity between the measured profiles and those extracted from the DEMs. This is supported by the results of the R2, SCI, and relative bias, which, in both profiles, have values with low scattering between the real and DEM data and a low vertical offset. It is important to note that these parameters are typically used in morphodynamic models to understand the relationship between the measured profiles and the resulting model profiles. Morphodynamic models are complex and require a large amount of input data and a lot of computational time [75], so usually there are processes that are not taken into account and limit the model's ability to estimate reality [76]. In this work, two real profiles were compared, although the one obtained from DEMs was indirectly computed. Under this premise, the threshold to accept the validation results as valid must be more demanding than those considered acceptable for morphodynamic models. Even so, in the validations carried out with these parameters, very satisfactory results were obtained.

For RMSE values, the average in both profiles was around 25 cm. These values are similar to those obtained with other methodologies, often more expensive and with lower temporal resolution. For instance, when using Unmanned Aerial Vehicles (UAVs) to build DEMs, the errors obtained are between 5 and 15 cm [63,64,77]. For LiDAR, the nominal accuracy is 15 cm, although it is only achieved on very flat areas and low flights [78]. In the present study, in some cases, the error is close to the one obtained with DGPS for this purpose, between 5 and 7 cm [79], with the advantage of faster and automated data collection, allowing for higher temporal resolution.

The results obtained also show low relative errors in slope and volume values, as most of the validations for both variables are significantly below a 10% error. These results can be considered a good fit for real data since slope errors are generally greater than volume errors given that integrated properties (volume) are less susceptible to error than derived properties (slope). The high errors obtained in some validations, mainly in the estimation of slope and volume, are associated with errors in the detection of the shoreline position. The presence of steep bars and troughs around the high tide level sometimes prevents the correct tide height estimation of the shoreline (Figure 9) because of uneven flooding in the trough.

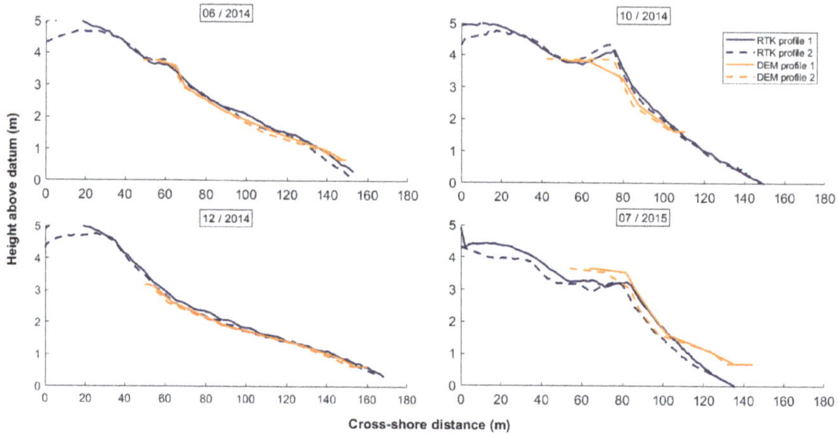

Figure 9. Comparison of beach profiles from RTK-DGPS surveys (blue lines) and DEMs (yellow lines) on different dates.

It must be noted that the effect of wave run-ups on the water level has not been considered for the DEM extraction. The inclusion of this effect could improve the DEM quality, but it would also increase the complexity of the used tool. Furthermore, empirical estimations of wave run-ups could generate some inaccuracies [27], although this approach is widely accepted in coastal engineering. The differential behaviour of the beach slope over the study zone, along with the bathymetric complexity caused by the discontinuity in the rocky platform, hampers the definition of a unique slope value for the whole beach. Finally, low values of relative bias (Figure 6) indicate that the quality of the results is adequate even without considering a wave run-up. This index provides information about vertical variation between the model and measured profiles [62]. In this case, values near zero indicate that shoreline heights are not underestimated for not including the waves.

5.2. Beach Behaviour Analysis Based on Video-Monitoring Tools

The development of a combined methodology for a comprehensive beach analysis and its application to La Victoria Beach has allowed us to obtain results that reveal contrasting behaviour patterns in an apparently homogeneous area. The combined methodology used has made it possible to identify small changes and differential behaviours on both a spatial and temporal scale. As shown in the previous section, the good agreement between the results obtained by the monitoring system and the data obtained by the RTK-DGPS facilitates the long-term analysis of these small variations. This allows us to demonstrate the differential and seasonal behaviour of the beach in the medium term.

In this regard, La Victoria Beach was characterised by marked seasonal behaviour during the study period, previously described in this beach and surrounding ones [38,44,80]. The seasonality was considered to be highly related to the hydrodynamic conditions in the northern area of Cádiz Gulf, in which mean wave height and the peak period are higher during the end of autumn and winter months. This results in two periods with different behaviours: the storm season, between November and March, in which wave energy is higher with more southerly wave directions, and the calm season, between April and October [81]. During the storm period, the beach slope decreases to dissipate the incident energy, in contrast to the calm period, in which the beach recovers and the slope increases [82]. This response was observed in both the shoreline position and the EOF analysis (Figure 5), as was the slope evolution of the entire beach (Figure 7).

From the evolution of the slope in the different transects and the EOF analysis, an alongshore differential behaviour was found in La Victoria Beach. Based on this, the apparently homogeneous beach can be divided into three sectors, defined by using the second and third spatial modes of the EOF analysis (Figure 5D): the northern or protected sector, conditioned by the rocky shore platform supporting the beach profile; the southern or exposed sector, a completely sandy zone; and the central sector or transition zone. The exact position and extension of the transition zone are influenced by the angle of incidence of storms, which can vary between the west and southwest. This results in a longer or shorter extension of the area protected by the rocky platform. The varying angle of storm incidence also has an impact on the presence of cusps and bars in this zone [26].

As demonstrated by several authors, the geological setting has an important role in the shaping of beach profiles [83,84]. In fact, it has special relevance in alongshore differences observed in La Victoria because of the presence of the rocky platform in the northern zone. The recorded differences in slope and shoreline position between the northern and southern sectors are in accordance with those found by Muñoz-Pérez and Medina [40]. The northern zone is a type 1 perched beach according to the classification by Gallop et al. [85], in which the structure or rocky platform is connected to the beach face. Beach response to the presence of structures such as rocky platforms is still poorly understood [86], and it varies based on the specific characteristics of the area (see Gallop et al. [85] for more information and a detailed review). The rocky platform in La Victoria Beach dissipates wave energy, making the slope and the average width of this zone higher than in the exposed one. The protection not only reduces erosion but also the accretion rate [87]. As a consequence of this, the slope shows more variability in the southern zone. This is due to the "locking" effect of the rocky platform on the beach profile, which blocks the cross-shore sediment transport between the lower and upper zone of the rocky platform [88]. Muñoz-Pérez and Medina [40] found that the recovery rate of the exposed zone was three times faster than the protected one. The difference between the erosion/accretion rates in both zones could be responsible for the beach planform rotation observed in the EOF analysis.

Even though the average shoreline position did not suffer remarkable changes over the studied period, there is a slightly decreasing trend in the beach slope. In addition to this, the peak values of the slope during summer, which can be related to beach recovery processes, decreased year after year during the study period. This trend could be amplified in the context of climate change. The methodology used allows for a detailed analysis of the most vulnerable areas in the face of long-term erosion, which will allow us to design interventions to prevent or mitigate erosion in such a way that their effects are maximised and expenses are minimised, as proposed by other authors [89]. Although it is commonly accepted that perched beaches are more protected than exposed beaches, there are some cases in which the effect is the opposite, producing an increase in erosion due to changes in waves and currents generated by the rocky platform [85]. Future studies should analyse the effect of the increasingly deep rocky platform caused by sea level rises and its incidence in the evolution of La Victoria Beach.

6. Conclusions

In this work, specific tools were developed for the exploitation of a 5-year dataset from a beach video-monitoring system. Automatic shoreline extraction and DEM generation, validation and extraction of intertidal slope and volume allowed us to analyse the behaviour of La Victoria Beach. The results show that the combination of shoreline position analysis (2D) and data from DEMs (3D) facilitated the understanding and provided a complete overview of the beach's behaviour. Furthermore, the methods used allowed us to interrelate the different processes occurring on the beach, which is difficult to achieve with other types of techniques.

The validation methodology developed for the generated DEMs is by itself a remarkable output of this work. The combination of the different procedures used allowed us to validate this tool by employing real data. It can provide geometric similarity; data scattering; and vertical errors and associated offset, slope, and volume errors. Moreover, it has wide potential applications, from LiDAR topographic data, UAVs, LaserScan data, satellite-derived bathymetry, etc., to the results of 3D morphodynamic models.

This methodology allowed us to analyse the evolution of a large urban beach, showing its spatial and temporal behaviour. The results reveal a seasonal pattern, with a marked zonation due to the presence of a rocky shore platform. Beach behaviour is defined by three processes that can explain almost all the obtained variability: erosion/accretion, alongshore transport, and beach rotation.

The proposed methodology is cost-effective, obtains data in an automatic mode with high spatiotemporal resolution, and can be easily extrapolated to other areas. This enables the analysis of vulnerable areas and facilitates the management of investments in coastal protection against climate change.

Author Contributions: J.M., conceptualization, formal analysis, investigation, methodology, software, visualization, writing—original draft, and writing—review and editing; L.d.R., conceptualization, funding acquisition, project administration, supervision, and writing—review and editing; T.A.P., formal analysis, funding acquisition, methodology, software, and writing—review and editing; J.B., conceptualization, supervision, and writing—review and editing; M.P., data curation and writing—review and editing; G.S., formal analysis, methodology, software, and writing—review and editing. All authors have read and agreed to the published version of the manuscript.

Funding: This work is a contribution to the ADACOSTA project (CGL2014-53153-R), funded by the Spanish MINECO; the CRUNCH project (FEDER-UCA18-107062), funded by European Union under the 2014-2020 ERDF Operational Programme and by the Department of Economic Transformation, Industry, Knowledge, and Universities of the Regional Government of Andalusia; and the CRISIS project (PID2019-109143RB-I00), funded by the Spanish Ministry of Science and Innovation and the European Union. J.M. was supported by contract BES-2015-073218 and a postdoctoral contract from Margarita Salas at the University of Cadiz from the Ministry of Universities of Spain, funded by the European Union—NextGenerationEU.

Data Availability Statement: Not applicable.

Acknowledgments: This work is a contribution to the research group RNM-328 of the Andalusian Research Plan (PAI).

Conflicts of Interest: The authors declare no conflict of interest.

References

1. Todd, P.A.; Heery, E.C.; Loke, L.H.L.; Thurstan, R.H.; Kotze, D.J.; Swan, C. Towards an urban marine ecology: Characterizing the drivers, patterns and processes of marine ecosystems in coastal cities. *Oikos* **2019**, *128*, 1215–1242. [CrossRef]
2. Cooley, S.; Schoeman, D.; Bopp, L.; Boyd, P.; Donner, S.; Ghebrehiwet, D.Y.; Ito, S.-I.; Kiessling, W.; Martinetto, P.; Ojea, E.; et al. Oceans and Coastal Ecosystems and Their Services. In *Climate Change 2022: Impacts, Adaptation and Vulnerability. Contribution of Working Group II to the Sixth Assessment Report of the Intergovernmental Panel on Climate Change*; Pörtner, H.-O., Roberts, D.C., Tignor, M., Poloczanska, E.S., Mintenbeck, K., Alegría, A., Craig, M., Langsdorf, S., Löschke, S., Möller, V., et al., Eds.; Cambridge University Press: Cambridge, UK; New York, NY, USA, 2022; pp. 379–550. ISBN 9781009325844.
3. Vousdoukas, M.I.; Ranasinghe, R.; Mentaschi, L.; Plomaritis, T.A.; Athanasiou, P.; Luijendijk, A.; Feyen, L. Sandy coastlines under threat of erosion. *Nat. Clim. Chang.* **2020**, *10*, 260–263. [CrossRef]
4. Cooper, J.A.G.; Masselink, G.; Coco, G.; Short, A.D.; Castelle, B.; Rogers, K.; Anthony, E.; Green, A.N.; Kelley, J.T.; Pilkey, O.H.; et al. Sandy beaches can survive sea-level rise. *Nat. Clim. Chang.* **2020**, *10*, 993–995. [CrossRef]
5. Masselink, G.; Hegge, B. Morphodynamics of meso- and macrotidal beaches: Examples from central Queensland, Australia. *Mar. Geol.* **1995**, *129*, 1–23. [CrossRef]
6. Russell, P.E. Mechanisms for beach erosion during storms. *Cont. Shelf Res.* **1993**, *13*, 1243–1265. [CrossRef]
7. Karunarathna, H.; Pender, D.; Ranasinghe, R.; Short, A.D.; Reeve, D.E. The effects of storm clustering on beach profile variability. *Mar. Geol.* **2014**, *348*, 103–112. [CrossRef]
8. Vos, K.; Harley, M.D.; Turner, I.L.; Splinter, K.D. Pacific shoreline erosion and accretion patterns controlled by El Niño/Southern Oscillation. *Nat. Geosci.* **2023**, *16*, 140–146. [CrossRef]

9. Morton, R.A.; Leach, M.P.; Paine, J.G.; Cardoza, M.A. Monitoring beach changes using GPS surveying techniques. *J. Coast. Res.* **1993**, *9*, 702–720.
10. Delgado, I.; Lloyd, G. A Simple Low Cost Method for One Person Beach Profiling. *J. Coast. Res.* **2004**, *204*, 1246–1252. [CrossRef]
11. Guisado-Pintado, E.; Jackson, D.W.T.; Rogers, D. 3D mapping efficacy of a drone and terrestrial laser scanner over a temperate beach-dune zone. *Geomorphology* **2019**, *328*, 157–172. [CrossRef]
12. Laignel, B.; Vignudelli, S.; Almar, R.; Becker, M.; Bentamy, A.; Benveniste, J.; Birol, F.; Frappart, F.; Idier, D.; Salameh, E.; et al. Observation of the Coastal Areas, Estuaries and Deltas from Space. *Surv. Geophys.* **2023**. [CrossRef]
13. Castelle, B.; Masselink, G.; Scott, T.; Stokes, C.; Konstantinou, A.; Marieu, V.; Bujan, S. Satellite-derived shoreline detection at a high-energy meso-macrotidal beach. *Geomorphology* **2021**, *383*, 107707. [CrossRef]
14. Bergsma, E.W.J.; Almar, R. Coastal coverage of ESA' Sentinel 2 mission. *Adv. Space Res.* **2020**, *65*, 2636–2644. [CrossRef]
15. Holman, R.A.; Guza, R.T. Measuring run-up on a natural beach. *Coast. Eng.* **1984**, *8*, 129–140. [CrossRef]
16. Lippmann, T.C.; Holman, R.A. The spatial and temporal variability of sand bar morphology. *J. Geophys. Res.* **1990**, *95*, 11575. [CrossRef]
17. Austin, M.; Scott, T.; Brown, J.; Brown, J.; MacMahan, J.; Masselink, G.; Russell, P. Temporal observations of rip current circulation on a macro-tidal beach. *Cont. Shelf Res.* **2010**, *30*, 1149–1165. [CrossRef]
18. Bracs, M.A.; Turner, I.L.; Splinter, K.D.; Short, A.D.; Lane, C.; Davidson, M.A.; Goodwin, I.D.; Pritchard, T.; Cameron, D. Evaluation of Opportunistic Shoreline Monitoring Capability Utilizing Existing "Surfcam" Infrastructure. *J. Coast. Res.* **2016**, *319*, 542–554. [CrossRef]
19. Chickadel, C.C.; Holman, R.A.; Freilich, M.H. An optical technique for the measurement of longshore currents. *J. Geophys. Res. Ocean.* **2003**, *108*, 3364. [CrossRef]
20. Sancho-García, A.; Guillén, J.; Ojeda, E. Storm-induced readjustment of an embayed beach after modification by protection works. *Geo-Mar. Lett.* **2013**, *33*, 159–172. [CrossRef]
21. Armaroli, C.; Ciavola, P. Geomorphology Dynamics of a nearshore bar system in the northern Adriatic: A video-based morphological classi fi cation. *Geomorphology* **2011**, *126*, 201–216. [CrossRef]
22. Pellón, E.; Garnier, R.; Medina, R. Intertidal finger bars at El Puntal, Bay of Santander, Spain: Observation and forcing analysis. *Earth Surf. Dyn.* **2014**, *2*, 349–361. [CrossRef]
23. Splinter, K.D.; Gonzalez, M.V.G.; Oltman-shay, J.; Rutten, J.; Holman, R. Observations and modelling of shoreline and multiple sandbar behaviour on a high-energy meso-tidal beach. *Cont. Shelf Res.* **2018**, *159*, 33–45. [CrossRef]
24. Almar, R.; Coco, G.; Bryan, K.R.; Huntley, D.A.; Short, A.D.; Senechal, N. Video observations of beach cusp morphodynamics. *Mar. Geol.* **2008**, *254*, 216–223. [CrossRef]
25. Vousdoukas, M.I. Erosion/accretion patterns and multiple beach cusp systems on a meso-tidal, steeply-sloping beach. *Geomorphology* **2012**, *141–142*, 34–46. [CrossRef]
26. Montes, J.; Simarro, G.; Benavente, J.; Plomaritis, T.A.; Del Río, L. Morphodynamics Assessment by Means of Mesoforms and Video-Monitoring in a Dissipative Beach. *Geosciences* **2018**, *8*, 448. [CrossRef]
27. Vousdoukas, M.I.; Wziatek, D.; Almeida, L.P. Coastal vulnerability assessment based on video wave run-up observations at a mesotidal, steep-sloped beach. *Ocean Dyn.* **2012**, *62*, 123–137. [CrossRef]
28. Simarro, G.; Bryan, K.R.; Guedes, R.M.C.; Sancho, A.; Guillen, J.; Coco, G. On the use of variance images for runup and shoreline detection. *Coast. Eng.* **2015**, *99*, 136–147. [CrossRef]
29. Sancho-García, A.; Ruessink, B.G.; Guillén, J. Storm-surge inundation along a multibarred beach. *J. Coast. Res.* **2011**, *SI 64*, 1911–1915.
30. Stockdon, H.F.; Holman, R.A. Estimation of wave phase speed and nearshore bathymetry from video imagery. *J. Geophys. Res. Ocean.* **2000**, *105*, 22015–22033. [CrossRef]
31. Vousdoukas, M.I.; Ferreira, P.M.; Almeida, L.P.; Dodet, G.; Psaros, F.; Andriolo, U.; Taborda, R.; Silva, A.; Ruano, A.; Ferreira, Ó. Performance of intertidal topography video monitoring of a meso-tidal reflective beach in South Portugal. *Ocean Dyn.* **2011**, *61*, 1521–1540. [CrossRef]
32. Simarro, G.; Calvete, D.; Plomaritis, T.A.; Moreno-Noguer, F.; Giannoukakou-Leontsini, I.; Montes, J.; Durán, R. The Influence of Camera Calibration on Nearshore Bathymetry Estimation from UAV Videos. *Remote Sens.* **2021**, *13*, 150. [CrossRef]
33. Winant, C.D.; Inman, D.L.; Nordstrom, C.E. Description of seasonal beach changes using empirical eigenfunctions. *J. Geophys. Res.* **1975**, *80*, 1979–1986. [CrossRef]
34. Larson, M.; Capobianco, M.; Jansen, H.; Różyński, G.; Southgate, H.N.; Stive, M.; Wijnberg, K.M.; Hulscher, S. Analysis and Modeling of Field Data on Coastal Morphological Evolution over Yearly and Decadal Time Scales. Part 1: Background and Linear Techniques. *J. Coast. Res.* **2003**, *19*, 760–775.
35. Kroon, A.; Larson, M.; Möller, I.; Yokoki, H.; Rozynski, G.; Cox, J.; Larroude, P. Statistical analysis of coastal morphological data sets over seasonal to decadal time scales. *Coast. Eng.* **2008**, *55*, 581–600. [CrossRef]
36. Harley, M.D.; Turner, I.L.; Short, A.D.; Ranasinghe, R. A reevaluation of coastal embayment rotation: The dominance of cross-shore versus alongshore sediment transport processes, Collaroy-Narrabeen Beach, southeast Australia. *J. Geophys. Res.* **2011**, *116*, F04033. [CrossRef]
37. Zhang, R.; Chen, L.; Liu, S.; Zhang, H.; Gong, W.; Lin, G. Shoreline evolution in an embayed beach adjacent to tidal inlet: The impact of anthropogenic activities. *Geomorphology* **2019**, *346*, 106856. [CrossRef]

38. Del Río, L.; Plomaritis, T.A.; Benavente, J.; Valladares, M.; Ribera, P. Establishing storm thresholds for the Spanish Gulf of Cádiz coast. *Geomorphology* **2012**, *143–144*, 13–23. [CrossRef]
39. Benavente, J.; Plomaritis, T.A.; Del Río, L.; Puig, M.; Valenzuela, C.; Minuzzi, B.; Del Río, L.; Puig, M.; Valenzuela, C.; Minuzzi, B. Differential short- and medium-term behavior of two sections of an urban beach. *J. Coast. Res.* **2014**, *70*, 621–626. [CrossRef]
40. Muñoz-Pérez, J.J.; Medina, R. Comparison of long-, medium- and short-term variations of beach profiles with and without submerged geological control. *Coast. Eng.* **2010**, *57*, 241–251. [CrossRef]
41. Benavente, J.; Borja, F.; Gracia, F.J.; Rodríguez, A. Introduction to the Gulf of Cadiz coast. In *Geomorphology of the South-Atlantic Spanish Coast. Guide for Fieldtrip A4, Proceedings of the 6th International Conference on Geomorphology, Zaragoza, Spain, 7–11 September 2005*; Gracia, F.J., Ed.; Universidad de Zaragoza: Zaragoza, Spain, 2005; pp. 1–11.
42. Del Río, L.; Gracia, F.J.; Benavente, J. Morphological and evolutionary classification of sandy beaches in Cadiz coast (SW Spain). *J. Coast. Res.* **2013**, *65*, 2113–2118. [CrossRef]
43. Plomaritis, T.A.; Benavente, J.; Del Río, L.; Reyes, E.; Dastis, C.; Gómez, M.; Bruno, M. Storm early warning system as a last plug-in of a regional operational oceanography system: The case of the gulf of Cádiz. *Int. Conf. Coastal. Eng.* **2012**, *1*, management.54. [CrossRef]
44. Puig, M.; Del Río, L.; Plomaritis, T.A.; Benavente, J. Contribution of storms to shoreline changes in mesotidal dissipative beaches: Case study in the Gulf of Cádiz (SW Spain). *Nat. Hazards Earth Syst. Sci.* **2016**, *16*, 2543–2557. [CrossRef]
45. Instituto Hidrográfico de la Marina. Anuario de Mareas. Available online: https://armada.defensa.gob.es/ArmadaPortal/page/Portal/ArmadaEspannola/cienciaihm1/prefLang-es/02ProductosServicios--08InfoInteres--04infoAnuario (accessed on 15 May 2021).
46. Simarro, G.; Ribas, F.; Álvarez, A.; Guillén, J.; Chic, Ò.; Orfila, A. ULISES: An Open Source Code for Extrinsic Calibrations and Planview Generations in Coastal Video Monitoring Systems. *J. Coast. Res.* **2017**, *335*, 1217–1227. [CrossRef]
47. Ribas, F.; Simarro, G.; Arriaga, J.; Luque, P. Automatic shoreline detection from video images by combining information from different methods. *Remote Sens.* **2020**, *12*, 3717. [CrossRef]
48. Aarninkhof, S.G.J.; Turner, I.L.; Dronkers, T.D.T.; Caljouw, M.; Nipius, L. A video-based technique for mapping intertidal beach bathymetry. *Coast. Eng.* **2003**, *49*, 275–289. [CrossRef]
49. Huisman, C.E.; Bryan, K.R.; Coco, G.; Ruessink, B.G. The use of video imagery to analyse groundwater and shoreline dynamics on a dissipative beach. *Cont. Shelf Res.* **2011**, *31*, 1728–1738. [CrossRef]
50. Lorenz, E.N. *Empirical Orthogonal Functions and Statistical Weather Prediction*; Massachusetts Institute of Technology, Department of Meteorology: Cambridge, MA, USA, 1956.
51. Álvarez-Ellacuría, A.; Orfila, A.; Gómez-Pujol, L.; Simarro, G.; Obregon, N. Decoupling spatial and temporal patterns in short-term beach shoreline response to wave climate. *Geomorphology* **2011**, *128*, 199–208. [CrossRef]
52. Hannachi, A.; Jolliffe, I.T.; Stephenson, D.B. Empirical orthogonal functions and related techniques in atmospheric science: A review. *Int. J. Climatol.* **2007**, *27*, 1119–1152. [CrossRef]
53. Muñoz-Pérez, J.J.; Medina, R. Profile changes due to a fortnightly tidal cycle. In Proceedings of the International Conference on Coastal Engineering (ASCE), Sydney, Australia, 16–21 July 2000; pp. 3063–3075.
54. Muñoz-Pérez, J.J.; Tejedor, B. Las funciones empíricas ortogonales y los cambios en el perfil de playa a corto, medio, y largo plazo. *Física Tierra* **2001**, *13*, 139–166.
55. Uunk, L.; Wijnberg, K.M.; Morelissen, R. Automated mapping of the intertidal beach bathymetry from video images. *Coast. Eng.* **2010**, *57*, 461–469. [CrossRef]
56. Red Mareógrafica del IEO. Available online: http://indamar.ieo.es/mareas/formulario_datos.htm (accessed on 15 May 2021).
57. Ferreira, Ó.; Plomaritis, T.A.; Costas, S. Effectiveness assessment of risk reduction measures at coastal areas using a decision support system: Findings from Emma storm. *Sci. Total Environ.* **2019**, *657*, 124–135. [CrossRef]
58. McCall, R.T.; van Thiel de Vries, J.S.M.; Plant, N.G.; Van Dongeren, A.R.; Roelvink, J.A.; Thompson, D.M.; Reniers, A.J.H.M. Two-dimensional time dependent hurricane overwash and erosion modeling at Santa Rosa Island. *Coast. Eng.* **2010**, *57*, 668–683. [CrossRef]
59. van Rijn, L.C.; Walstra, D.J.R.; Grasmeijer, B.; Sutherland, J.; Pan, S.; Sierra, J.P. The predictability of cross-shore bed evolution of sandy beaches at the time scale of storms and seasons using process-based profile models. *Coast. Eng.* **2003**, *47*, 295–327. [CrossRef]
60. Smallegan, S.M.; Irish, J.L.; van Dongeren, A.R.; Den Bieman, J.P. Morphological response of a sandy barrier island with a buried seawall during Hurricane Sandy. *Coast. Eng.* **2016**, *110*, 102–110. [CrossRef]
61. Sutherland, J.; Peet, A.H.; Soulsby, R.L. Evaluating the performance of morphological models. *Coast. Eng.* **2004**, *51*, 917–939. [CrossRef]
62. Roelvink, D.; Reniers, A.; van Dongeren, A.R.; van Thiel de Vries, J.S.M.; McCall, R.; Lescinski, J. Modelling storm impacts on beaches, dunes and barrier islands. *Coast. Eng.* **2009**, *56*, 1133–1152. [CrossRef]
63. Casella, E.; Rovere, A.; Pedroncini, A.; Stark, C.P.; Casella, M.; Ferrari, M.; Firpo, M. Drones as tools for monitoring beach topography changes in the Ligurian Sea (NW Mediterranean). *Geo-Mar. Lett.* **2016**, *36*, 151–163. [CrossRef]
64. Talavera, L.; Del Río, L.; Benavente, J.; Barbero, L.; López-Ramírez, J.A. UAS & SfM-based approach to Monitor Overwash Dynamics and Beach Evolution in a Sandy Spit. *J. Coast. Res.* **2018**, *85*, 221–225. [CrossRef]

65. Armaroli, C.; Ciavola, P.; Balouin, Y.; Gatti, M. An Integrated Study of Shoreline Variability Using GIS and ARGUS Techniques. *J. Coast. Res.* **2004**, *39*, 473–477.
66. Aarninkhof, S.G.J.; Roelvink, J.A. Argus-based monitoring of intertidal beach morphodynamics. In Proceedings of the Coastal Sediments Conference, Long Island, NY, USA, 21–23 June 1999; Edge, B.E., Ed.; ASCE: New York, NY, USA, 1999.
67. Aarninkhof, S.G.J. *Nearshore Bathymetry Derived from Video Imagery*; Delft University of Technology: Delft, The Netherlands, 2003.
68. Plant, N.G.; Holman, R.A. Intertidal beach profile estimation using video images. *Mar. Geol.* **1997**, *140*, 1–24. [CrossRef]
69. Plant, N.G.; Aarninkhof, S.G.J.; Turner, I.L.; Kingston, K.S. The Performance of Shoreline Detection Models Applied to Video Imagery. *J. Coast. Res.* **2007**, *233*, 658–670. [CrossRef]
70. Aarninkhof, S.G.J.; Ruessink, B.G.; Roelvink, J.A. Nearshore subtidal bathymetry from time-exposure video images. *J. Geophys. Res.* **2005**, *110*. [CrossRef]
71. Holman, R.; Plant, N.G.; Holland, T. cBathy: A robust algorithm for estimating nearshore bathymetry. *J. Geophys. Res. Ocean.* **2013**, *118*, 2595–2609. [CrossRef]
72. Simarro, G.; Calvete, D.; Luque, P.; Orfila, A.; Ribas, F. UBathy: A new approach for bathymetric inversion from video imagery. *Remote Sens.* **2019**, *11*, 2722. [CrossRef]
73. Bergsma, E.; Conley, D.; Davidson, M.; O'Hare, T.; Almar, R. Storm Event to Seasonal Evolution of Nearshore Bathymetry Derived from Shore-Based Video Imagery. *Remote Sens.* **2019**, *11*, 519. [CrossRef]
74. Santos, D.; Abreu, T.; Silva, P.A.; Santos, F.; Baptista, P. Nearshore Bathymetry Retrieval from Wave-Based Inversion for Video Imagery. *Remote Sens.* **2022**, *14*, 2155. [CrossRef]
75. Poelhekke, L.; Jäger, W.S.; van Dongeren, A.R.; Plomaritis, T.A.; McCall, R.; Ferreira, Ó. Predicting coastal hazards for sandy coasts with a Bayesian Network. *Coast. Eng.* **2016**, *118*, 21–34. [CrossRef]
76. Donnelly, C.; Kraus, N.; Larson, M. State of Knowledge on Measurement and Modeling of Coastal Overwash. *J. Coast. Res.* **2006**, *224*, 965–991. [CrossRef]
77. Duo, E.; Chris Trembanis, A.; Dohner, S.; Grottoli, E.; Ciavola, P. Local-scale post-event assessments with GPS and UAV-based quick-response surveys: A pilot case from the Emilia-Romagna (Italy) coast. *Nat. Hazards Earth Syst. Sci.* **2018**, *18*, 2969–2989. [CrossRef]
78. Aguilar, F.J.; Mills, J.P.; Delgado, J.; Aguilar, M.A.; Negreiros, J.G.; Pérez, J.L. Modelling vertical error in LiDAR-derived digital elevation models. *ISPRS J. Photogramm. Remote Sens.* **2010**, *65*, 103–110. [CrossRef]
79. Baptista, P.; Bastos, L.; Bernardes, C.; Cunha, T.; Dias, J. Monitoring Sandy Shores Morphologies by DGPS—A Practical Tool to Generate Digital Elevation Models. *J. Coast. Res.* **2008**, *246*, 1516–1528. [CrossRef]
80. Benavente, J.; Gracia, F.J.; López-Aguayo, F. Empirical model of morphodynamic beachface behaviour for low-energy mesotidal environments. *Mar. Geol.* **2000**, *167*, 375–390. [CrossRef]
81. Plomaritis, T.A.; Benavente, J.; Laiz, I.; Del Río, L. Variability in storm climate along the Gulf of Cadiz: The role of large scale atmospheric forcing and implications to coastal hazards. *Clim. Dyn.* **2015**, *45*, 2499–2514. [CrossRef]
82. Benavente, J.; Del Río, L.; Anfuso, G.; Gracia, F.J.; Reyes, J.L. Utility of Morphodynamic Characterisation in the Prediction of Beach Damage by Storms. *J. Coast. Res.* **2002**, *36*, 56–64. [CrossRef]
83. Muñoz-Pérez, J.J.; Tejedor, L.; Medina, R. Equilibrium beach profile model for perched beaches. *J. Coast. Res.* **1999**, *15*, 950–957. [CrossRef]
84. Jackson, D.W.T.; Cooper, J.A.G.; Del Río, L. Geological control of beach morphodynamic state. *Mar. Geol.* **2005**, *216*, 297–314. [CrossRef]
85. Gallop, S.L.; Bosserelle, C.; Pattiaratchi, C.; Eliot, I. Rock topography causes spatial variation in the wave, current and beach response to sea breeze activity. *Mar. Geol.* **2011**, *290*, 29–40. [CrossRef]
86. Ranasinghe, R.; Turner, I.L.; Symonds, G. Shoreline response to multi-functional artificial surfing reefs: A numerical and physical modelling study. *Coast. Eng.* **2006**, *53*, 589–611. [CrossRef]
87. Gallop, S.L.; Bosserelle, C.; Eliot, I.; Pattiaratchi, C.B. The influence of limestone reefs on storm erosion and recovery of a perched beach. *Cont. Shelf Res.* **2012**, *47*, 16–27. [CrossRef]
88. Vousdoukas, M.I.; Velegrakis, A.F.; Plomaritis, T.A. Beachrock occurrence, characteristics, formation mechanisms and impacts. *Earth-Sci. Rev.* **2007**, *85*, 23–46. [CrossRef]
89. López-Dóriga, U.; Jiménez, J.A.; Bisaro, A.; Hinkel, J. Financing and implementation of adaptation measures to climate change along the Spanish coast. *Sci. Total Environ.* **2020**, *712*, 135685. [CrossRef] [PubMed]

remote sensing

MDPI

Article

Global Distribution and Morphodynamic Patterns of Paired Spits Developed at the Mouths of Interdistributary Bays of Deltas and within Coastal Channels

Javier Alcántara-Carrió [1,*], Ángela Fontán-Bouzas [2,3], Ana Caicedo Rodríguez [4], Rogério Portantiolo Manzolli [1] and Luana Portz [1]

1 Department of Geology and Geochemistry, Faculty of Sciences, Autonomous University of Madrid, C/Tomás y Valiente 7, 29049 Madrid, Spain; rogerio.manzolli@uam.es (R.P.M.); luana.portz@uam.es (L.P.)
2 Department of Geodynamics, Stratigraphy and Paleontology, Faculty of Geological Sciences, Complutense University of Madrid, C/José Antonio Novais 12, 28040 Madrid, Spain; anfontan@ucm.es
3 CIM–UVIGO GEOMA, University of Vigo, 36310 Vigo, Spain
4 Seccional Oceanografía, Universidad of Antioquia, C/104 #110-2a, Cra. 25 #110120, Turbo 057867, Colombia; ana.caicedo@udea.edu.co
* Correspondence: javier.alcantara@uam.es

Abstract: Previously, paired spits have been described at the mouths of bays, estuaries, and deltas. This study analyzed the worldwide distribution and morphodynamic patterns of paired spits located at the mouths of interdistributary bays of deltas (three systems) and within coastal channels (24 systems). The methodology was based on the detailed analysis of satellite images, nautical charts, and tidal-range databases. The paired spits found were mainly located on microtidal coasts at high or mid latitudes. Waves were the main factor controlling convergent progradation and breaching of the spits, while the hydraulic blockage for the development of these paired spits was mainly due to tide-induced currents, as well as minor fluvial outlets in the interdistributary bays. Three morphodynamic patterns were identified: (i) stable, with low progradation rates, generally without breaching or degradation of any of the spits; (ii) stationary, with high progradation rates, alternating degradation or breaching of any of the spits with the formation of new spits or closure of the breaches; and (iii) instable or ephemeral, which included three subtypes, the severe erosion of one or both spits, the joining of the head of the two spits forming a single barrier, and the merging of each with its channel margin.

Keywords: geomorphology; baymouth spits; double spits; progradation; erosion; degradation; breaching

Citation: Alcántara-Carrió, J.; Fontán-Bouzas, Á.; Caicedo Rodríguez, A.; Manzolli, R.P.; Portz, L. Global Distribution and Morphodynamic Patterns of Paired Spits Developed at the Mouths of Interdistributary Bays of Deltas and within Coastal Channels. *Remote Sens.* **2023**, *15*, 2713. https://doi.org/10.3390/rs15112713

Received: 14 April 2023
Revised: 18 May 2023
Accepted: 19 May 2023
Published: 23 May 2023

1. Introduction

Sandy and gravel spits are elongated barriers that extend laterally through the successive construction of progradational beach ridges [1–3]. Wave-induced processes are the main factor controlling spit development [4]. Bays and estuaries are often partially enclosed by spits [5]. Intense littoral drift leads to the formation of spits in the outer parts of the bays, which can evolve into systems of sand bars and coastal lagoons, with a mouth maintained by tidal currents or rivers flowing into the bay [6,7]. Similarly, littoral drift also forms barrier spits at estuarine river mouths, which can break up due to intense river discharge events [8–11], storms [12–14], tsunami events [15], the temporal decrease of sediment supply [16] or artificial breaching [17]. Spits also develop in asymmetric and very asymmetric (deflected) wave-dominated deltas [18–21], where they can be extremely dynamic [22].

On exceptional occasions, the bays, estuaries, and deltas are confined by systems of paired spits (PS), i.e., two spits with a converging longshore drift that partially or almost completely enclose the bay or river mouth. The formation of paired spits has been described by five morphogenetic models: (i) coastal barrier break [23,24]; (ii) convergent longshore drift in a narrow bay [25]; (iii) bidirectional longshore drift with hydraulic blockage [26]; (iv) cutting a detached spit due to ebb-tidal currents [27] or by high energy events such as storms [28], hurricanes [29] or high fluvial discharge [30]; and (v) the convergence of two estuary mouths and the associated spits [30]. In previous studies about paired spits, they are also referred to as baymouth spits, double spits or baymouth barriers. The term paired spits is also used for gravel or sandy points growing within coastal lagoons [31], with countless examples worldwide [32–36].

The formation and evolution of convergent baymouth spits is influenced by the hydrodynamics of waves, tides, and river discharge. In narrow bays, the convergent progradation of both spits may be due to unidirectional frontal waves, which generate a convergent littoral drift from both ends of the bay [25]. On the other hand, on open coasts, the paired convergent spits are related to a seasonal alternation of bidirectional waves. At each time of the year, the waves favor the development of one of the spits, while the hydraulic blockage, generated by tidal currents [29,37] or intense fluvial discharge [27,30,38], favors the net progradation of the countercurrent spit, by preventing or at least minimizing its erosion by the dominant waves. Usually, paired spits have been described at the mouth of bays and rivers [39]. However, the development of paired spits constraining the mouths of interdistributary bays of deltas has only been described for the Po River delta (N Adriatic Sea) [40–44], and the development of paired spits within channels have only been very briefly reported [45–47]. Therefore, determining the global distribution, geomorphological characteristics and morphodynamic evolution of paired spits located both at the interdistributary bays of deltas and within coastal channels was the main objective of this study.

2. Materials and Methods

The analysis of open mean resolution of satellite LANDSAT imagery with worldwide and five-decades historical coverage, supported by high performance platforms such as Google Earth Pro, is a very useful tool to investigate coastal processes, such as shoreline changes [48–50] and spit morphodynamics [12,38,51] on a planetary scale and in an economical way. The identification of the global distribution and a description of the four-decades geomorphological evolution of all the paired spits in the world found at the mouths of interdistributary bays of deltas and within coastal channels were obtained by analyzing annual series of LANDSAT satellite imagery (1984–2020), supported by Google Earth™ [52]. Exceptionally, for a small system within the Safaga Strait (NW Red Sea), data were analyzed using Satellite Quickbird imagery, with a higher spatial resolution; data were only available since 2003 in Google Earth Pro. This platform does not provide access to the original multispectral data and therefore a supervised shoreline extraction could not be realized.

Initially, the shoreline of each spit was digitized using the Google Earth ruler tool. The traced lines can be considered as representations of the mean littoral zone because tidal variations are not taken into account in these images [53]. However, monitoring the length of the shoreline from the root to the head is not useful when determining spit progradation, because this length also increases due to small cuspate forelands, beach cusps or local erosion, for instance at the neck of the spit. Therefore, a reference line was defined joining the root of the two spits, along the axis of each spit up to the head (considering the head of the spit for each year to follow its progradation), passing through the inlet between the heads and following the longitudinal axis of the possible islands (Figure 1). This reference line was defined following [29], but it is not straight now, because the morphology and head progradation of the spits are not straight either. The 1984–2020 evolution of the distance, on this reference line, from the root to the head of the spits, as well as to the ends of the developed islands, were measured using the Google Earth ruler tool.

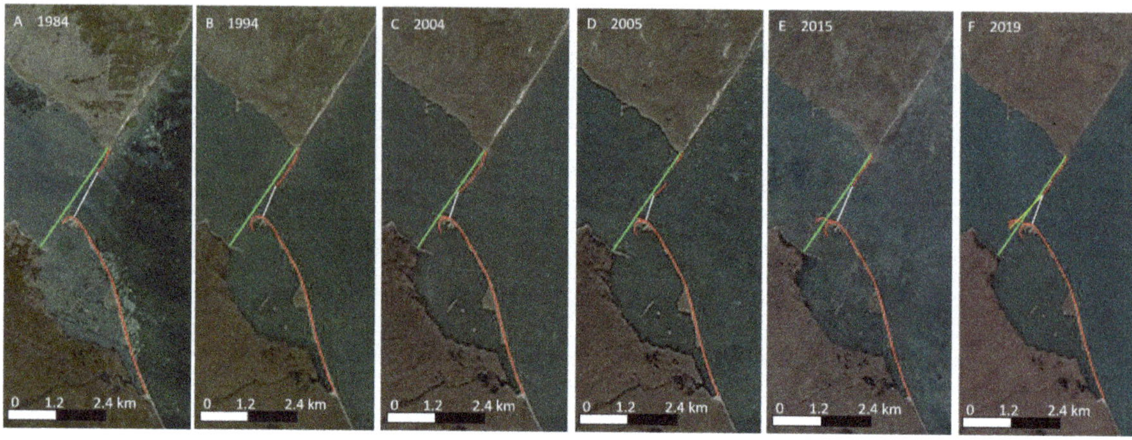

Figure 1. Plot and measurement of annual axial lines (red lines) from the root to the head of each spit for the period from 1984 (**A**) to 2020. The length of successive hooked ridges (**B**,**E**) was measured in order to determine head progradation, but not the changes in neck curvature (**C**). Formation of islands by spit breaching (**D**) and the new formation of spits (**D**–**F**) were also monitored. The reference line considers the longest axial line for each spit and the distance (yellow line) between the heads of the two spits (**F**). The minimum width of the inlet (white line) varied from year to year, but the width at the entrance of the channel or at the mouth of the interdistributary bay (green line) was considered constant throughout the study period.

The distances obtained were plotted using Grapher^{TM} (Golden Software) and the head progradation rates (m/yr) for the spits were determined on these graphs. In the case of spit degradation, such as on the western margin of the Scardovari lagoon and on the eastern margin of the Goro lagoon (both in the SW sector of the Po River delta), they were monitored using the old spit before its partial degradation as the new spit that developed afterwards.

The long-term scale decrease in width, due to constriction by the development of paired spits, was defined as the ratio of the length of the main inlet to the length of the original mouth of the interdistributary bay or of the original entrance to the channel without considering the paired spits. The minimum width of the main inlet was measured on the digitized shorelines of the spits, varying over the study period 1984–2020. The width of the original baymouth or the channel entrance, which was considered constant for the study period, was measured directly from satellite imagery.

The maximum depth of the inlet between the two spits, for each paired spits system, was determined by analysis of online bathymetric charts supported by Navionics® Chart Viewer [54], except for the Goro lagoon, which was obtained from [55]. Bathymetric data supported by Google Earth, based on the bathymetric model of [56] and its updates, were initially considered but later discarded, due to inconsistencies between these data and both the nautical charts and the spits morphology observed in the satellite images.

The offshore data series near each paired spits system of the mean wave direction, peak period (Tp) and significant height (Hs) of combined wind, waves, and swell were obtained from the ERA5 reanalysis [57] for the period 2018–2021 (except for PS-17 which were for the period 2010–2013). For each data series, the percentage of the year with data (i.e., a proxy for the period free of ice sheet formation at high latitudes) was calculated, and directional histograms (wave roses) were plotted by Grapher^{TM} (Golden Software) to identify the uni-, bi- or multi-directional wave regimes. Then, for the first and second significant directions identified considering the wave approach and coastal orientation, the mean annual and 95% percentile of Hs (m), and the predominance of wind sea (Tp < 6 s), swell (Tp > 9 s) or intermediate peak periods (6 to 9 s) were determined. Offshore wave propagation by numerical models was not carried out because high-resolution bathymetry is not available for many of the study sites, due to the fact that they are located in remote areas and have very dynamic systems, and therefore bathymetric charts are not accurate. Satellite-derived bathymetry can help solve this problem for future detailed research of each study area, even though it also has limitations, such as turbidity or the short window of opportunity for many of them due to ice-sheet formation [58].

Tidal range (m) was obtained using the free online software WXTIDE [59], except for the paired spits located in Canada, where it was replaced by data available online from the tidal data stations supported by the Canadian Hydrographic Service [60]. Each paired spits system was then classified as micro- (0–2), meso- (2–4), macro- (4–6) or mega-tidal (>6 m), according to criteria of [61,62]. The development of ice plates at high-latitude coastal areas, which impedes the morphodynamic evolution of the paired spits, was confirmed by observation of monthly LANDSAT-8 satellite imagery, supported by the USGS GloVis online repository [63].

The presence of local human activities was obtained from the analysis of satellite images and then cross-checked with bibliographic information. The spits were classified as natural, rural, semi-urban, urban, or artificial coasts following [64]. The rural coast included agricultural uses and small harbors for fishing activities, as well as spits on natural coast with the presence of coastal defense structures.

In summary, the methodology applied in this study included: the extraction of the shoreline of the paired spits by analyzing satellite images; the determination of the geomorphological characteristics of the spits; the plotting of the evolution of the head of the spits and the ends of the islands; the calculation of both the rates of head progradation and the decrease in width at the mouth of the interdistributary delta bays or at the entrance to the channels; the determination of the maximum inlet depth and the tidal range for each system; the description of morphodynamic events (spit breaching, development of spit islands, closure of breaches, degradation of spits, and formation of new spits); the analysis of hydrodynamic control, sediment availability and sedimentary interactions; the definition of morphogenetic models and natural morphodynamic patterns; and the analysis of human disturbances. It is summarized in the workflow (Figure 2).

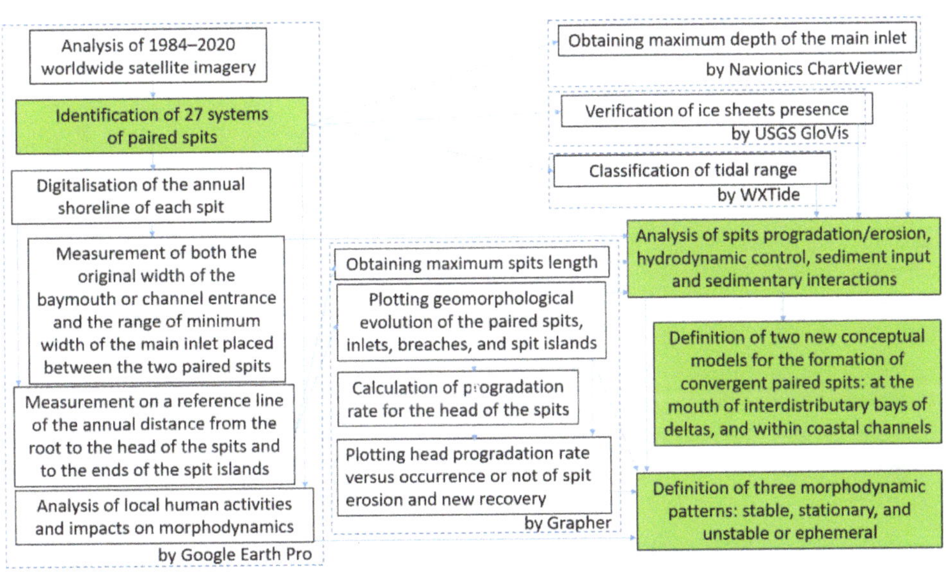

Figure 2. Workflow. The main results are indicated in green boxes.

3. Results

3.1. Global Distribution and Geomorphological Description

Twenty-seven systems of paired spits at the mouth of interdistributary bays of deltas and within coastal channels were identified (Figures 3 and 4). They are mainly located on microtidal coasts of high or mid latitudes, although they have also been found on mesotidal coasts and at low latitudes (Table 1). All of them were found in the northern hemisphere, except for one ephemeral system located on the southwest coast of Angola (PS-27). Three paired spits systems were identified at the mouths of interdistributary bays of deltas. PS-1 is located at the mouth of Demarcation Bay (South of the Beaufort Sea), an interdistributary bay between the Kongakut and Clarence River deltas. The western spit of PS-1 developed from the barrier island off the Kongakut River delta, and the eastern spit is a barrier spit developed from the Clarence River delta. PS-2 and PS3, are located, respectively, at the mouth of the Scardovari and Goro lagoons, in two interdistributary bays of the Po River delta,. The other twenty-four paired spits systems were identified at the entrance to coastal channels, except for: PS-24, which is in the middle of the Safaga Strait (Red Sea); PS-10 in the middle of the Litke Strait, between Karaginsky Island and the Kamchatka Peninsula (Bering Sea); and PS-17 in the northern sector within the Strait of Canso (Northeastern of the Nova Scotia Peninsula) (Table 1). PS-6 is located at the entrance to the channel between Booth and Fiji Islands (South of the Beaufort Sea), and there is another spit in the northern margin, within the channel. Similarly, eastward of PS-7 (West of the Parry Peninsula, South of the Beaufort Sea), the nautical chart shows that there was also a second spit within the channel, on the northern margin, but satellite imagery showed that it was almost completely degraded before 1985. Another particular coastal configuration are channels with paired spits at both entrances, which is the case for PS-12 and PS-13 in the Tugidak Passage, between Tugidak and Sitkinak islands (West of Gulf of Alaska), and PS-25 and PS-26 in Almejas Bay, between Santa Margarita Island and the western coast of Baja California Peninsula. In addition, three entrances to Nantucket Sound (Northwest Atlantic Ocean) are limited by the development of paired spits, one of them between Monomoy and Nantucket islands (PS-20), and two others in the Muskeget Channel (PS-21 between Tuckernuck and Muskeget islands, and PS-22 between Nantucket and Tuckernuck islands).

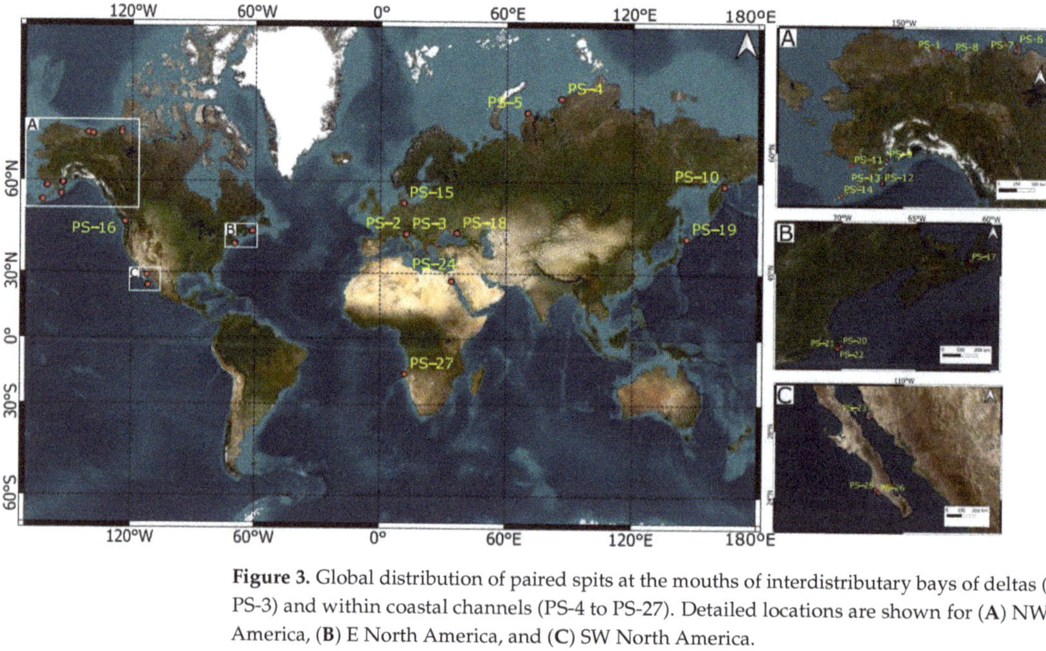

Figure 3. Global distribution of paired spits at the mouths of interdistributary bays of deltas (PS-1 to PS-3) and within coastal channels (PS-4 to PS-27). Detailed locations are shown for (**A**) NW North America, (**B**) E North America, and (**C**) SW North America.

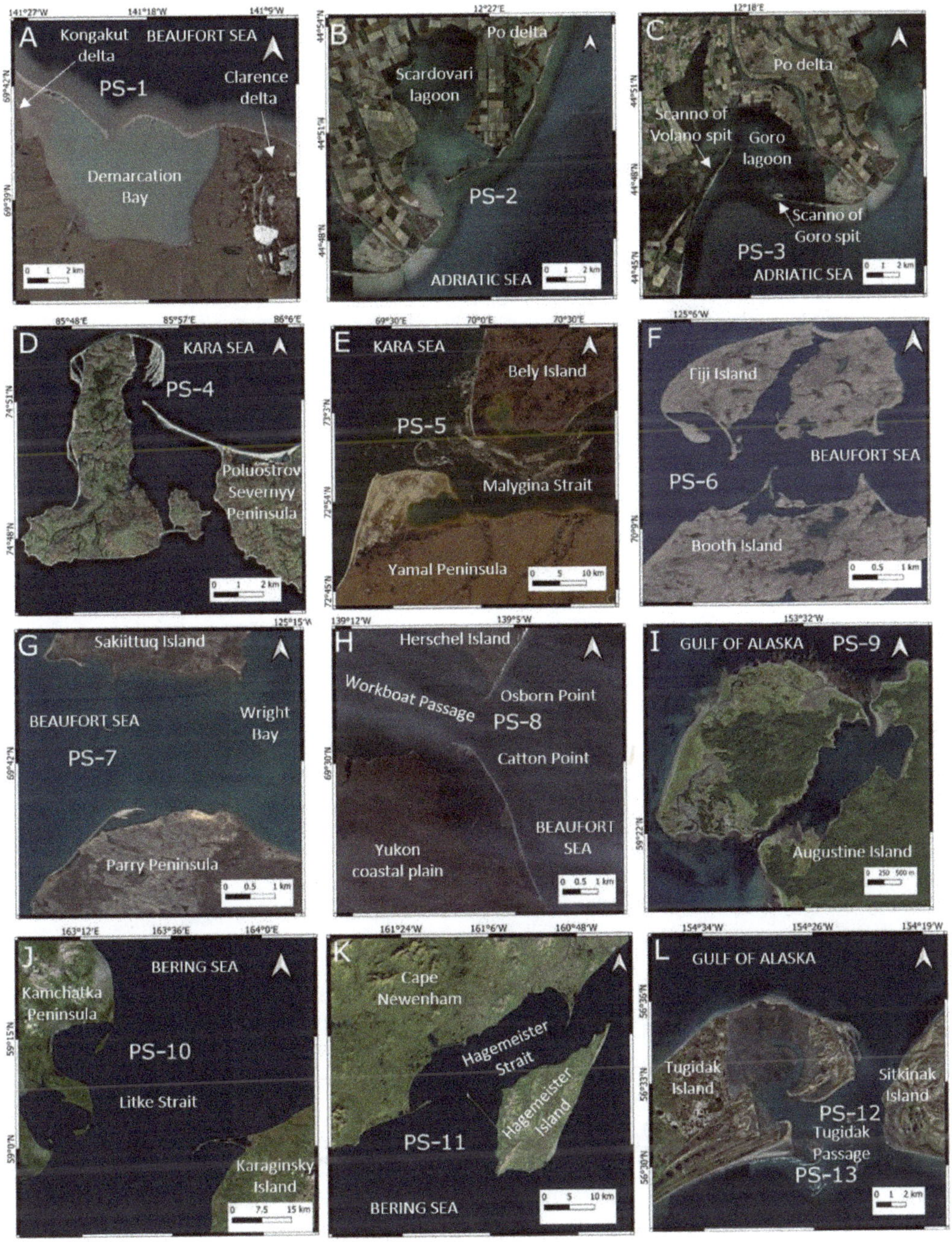

Figure 4. *Cont.*

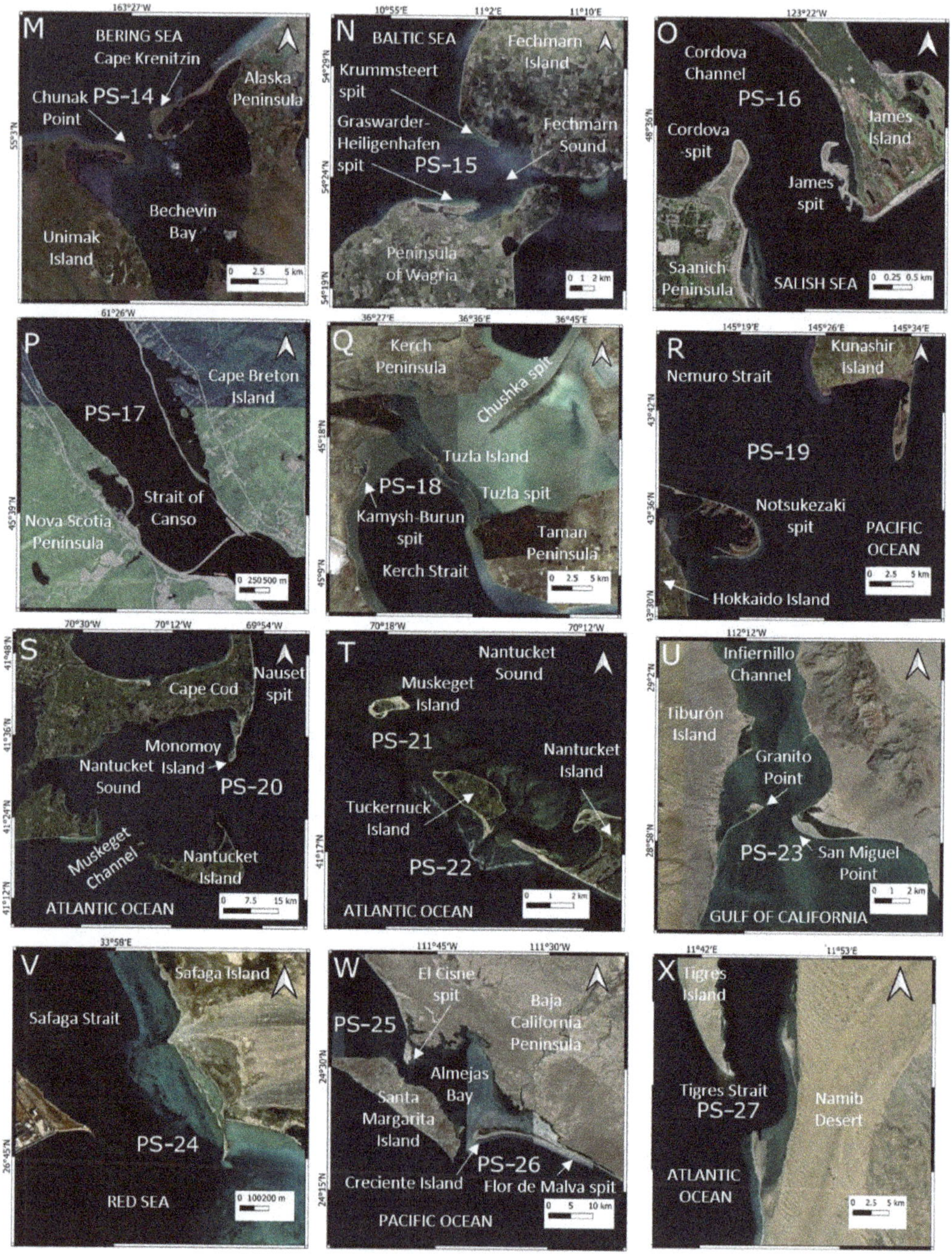

Figure 4. Paired spits (PS) located at the mouths of interdistributary bays of deltas (PS-1 to PS-3) and within coastal channels (PS-4 to PS-27). (Source of the images: ESRI Satellite, Google Satellite, and Bing Satellite). (**A**) PS-1, (**B**) PS-2, (**C**) PS-3, (**D**) PS-4, (**E**) PS-5, (**F**) PS-6, (**G**) PS-7, (**H**) PS-8, (**I**) PS-9, (**J**) PS-10, (**K**) PS-11, (**L**) PS-12 and PS-13, (**M**) PS-14, (**N**) PS-15, (**O**) PS-16, (**P**) PS-17, (**Q**) PS-18, (**R**) PS-19, (**S**) PS-20, (**T**) PS-21 and PS-22, (**U**) PS-23, (**V**) PS-24, (**W**) PS-25 and PS-26, (**X**) PS-27.

Table 1. Location, tidal range, and morphological characteristics (1984–2020) of the paired spits identified at the mouth of the interdistributary bays of deltas (PS-1 to PS-3) and within coastal channels (PS-4 to PS-27). Their morphology of the paired spits is parameterized by the maximum length (m) of the longest and the shortest spit, the ratio between them, the maximum depth (m) of the inlet between the head of both spits, and the decrease in width, i.e., the ratio of the width of the main inlet to the original width of the mouth of the interdistributary bay or channel.

Paired Spits	Latitude	Longitude	Place	Region	Tidal Range	Longest Spit	Shortest Spit	Length Ratio	Inlet Depth	Width Decrease
PS-1	69.68°N	141.34°W	Demarcation Bay	Beaufort Sea	micro–	5965	5647	1.06	5.2	0.02–0.14
PS-2	44.82°N	12.43°E	Scardovari lagoon	Adriatic Sea	micro–	2393	2188	1.09	2.2	0.21–0.41
PS-3	44.80°N	12.30°E	Goro lagoon		micro–	8313	6741	1.23	3.5	0.17–0.42
PS-4	74.84°N	85.95°E	NW Poluostrov Severnyy Peninsula	Kara Sea	micro–	3978	3460	1.15	<5	0.05–0.18
PS-5	72.98°N	69.92°E	Malygina Strait		micro–	31,329	6466	4.85	<5	0.35
PS-6	70.16°N	125.07°W	Channel between Booth and Fiji islands	Beaufort Sea	micro–	1181	379	3.12	13.5	0.40
PS-7	69.70°N	125.36°W	W of Parry Peninsula		micro–	2435	361	6.75	0.3	0.71–0.72
PS-8	69.51°N	139.11°W	Workboat Passage		micro–	5626	1472	3.82	6.4	0.19–0.44
PS-9	59.38°N	153.52°W	Augustine Island	Gulf of Alaska	meso–	487	182	2.68	0.3	0.33
PS-10	59.10°N	163.50°E	Litke Strait	Bering Sea	micro–	16,831	9272	1.82	5	0.77
PS-11	58.68°N	161.23°W	Hagemeister Strait		meso–	8990	5131	1.75	20.1	0.42
PS-12	56.58°N	154.49°W	Tugidak Passage	Gulf of Alaska	meso–	17,941	2326	7.71	31	0.15–0.16
PS-13	56.52°N	154.40°W			meso–	15,934	5780	2.77	31	0.289
PS-14	55.05°N	163.44°W	NW Bechevin Bay	Bering Sea	micro–	11,860	6119	1.94	23.8	0.17
PS-15	54.41°N	10.99°E	Fehmarn Sound	Baltic Sea	micro–	4931	2624	1.88	12.3	0.74–0.75
PS-16	48.59°N	123.36°W	Cordova Channel	Salish Sea	meso–	1177	1083	1.09	32	0.62
PS-17	45.65°N	61.43°W	Strait of Canso	Gulf of St. Lawrence	micro–	2640	582	4.53	40	0.58
PS-18	45.24°N	36.54°E	Kerch Strait	Black Sea	micro–	4795	2186	2.19	6.7	0.46
PS-19	43.61°N	145.44°E	Nemuro Strait	Sea of Okhotsk	micro–	24,613	1810	13.60	20	0.68
PS-20	41.47°N	70.03°W	Main Channel	Nantucket Sound, NW Atlantic Ocean	micro–	15,443	8282	1.86	17.0	0.46–0.47
PS-21	41.32°N	70.29°W	Muskeget Channel		micro–	2047	887	2.31	1.2	0.72–1
PS-22	41.28°N	70.24°W			micro–	2765	1444	1.91	7.3	0.53–0.67
PS-23	28.97°N	112.18°W	Infiernillo Channel	Gulf of California	micro–	4797	3585	1.34	13.4	0.19
PS-24	26.75°N	33.96°E	Safaga Strait	Red Sea	micro–	887	340	2.61	0.5	0.59–0.89
PS-25	24.51°N	111.83°W	Almejas Bay	NE Pacific Ocean	micro–	15,231	1149	13.25	26.6	0.09–0.10
PS-26	24.37°N	111.67°W			micro–	11,408	3339	10.66	13.2	0.08
PS-27	16.73°S	11.75°E	Tigres Strait	SE Atlantic Ocean	micro–	15,346	2961	5.18	9.7	——

The maximum length along the shore of the spits, from root to head, ranges from 31,359 m for the southern spit of PS-5 in the Malygina Strait, which separates Bely Island from the Yamal Peninsula (Northwest of Siberia), to 182 m for the western spit of PS-9, located in a small channel cutting through the volcanic Augustine Island (West of Gulf of Alaska). The maximum length of both spits is similar in the interdistributary bays, but the length of the longest spit within channels varies from the same order to more than thirteen times the length of the shortest spit (Table 1).

The formation of paired spits reduces the width at the mouth of the interdistributary bays or channels. Maximum closure is identified for PS-1, in Demarcation Bay, and for PS-4, in a channel between Tsirkul Island and the Poluostrov Severnyy Peninsula (E Kara Sea), remaining around 2–5% of the original width. It was not possible to determine the decrease in width for PS-27 because the width of the channel progressively increased due to high erosion at the southern end of the Tigres island.

The maximum depth between the head of the paired spits ranges from less than 1 m to 40 m (Table 1). In some systems, the local depth increases significantly just between the two spits, along the entire inlet, as is the case on the Hagemeister Strait (PS-11), the Tugidak Passage (PS-12, PS-13), and the Infiernillo Channel (PS-23), or only right next to one of the spits, as is the case of the eastern spit of PS-14 in the NW entrance to Bechevin Bay, and the Notsukezaki spit (PS-19), in the western margin of the southern entrance to the Nemuro Strait.

3.2. *Natural Morphodynamics*

Most of the paired spits are very dynamic, with high progradation rates reaching up to 156.93 m/yr (PS-20). In some paired spits, the longest spits show higher progradation rates that the shortest ones, but in others, the opposite is true. Progradation in some recurved spits is not identified from the head, but from the middle of the spit, allowing new hooked ridges to develop. The eastern recurved spit of PS-19 shows progradation from the middle and the head of the spit. In contrast, other paired spits show very low progradation rates, which cannot be measured due to the resolution of the satellite images used, without other significant geomorphological changes; these progradation rates must be lower than 1 m/yr, i.e., below the detection threshold of the analysis carried out (Table 2). The the migration of some of the spits from the surrounding coast to the mouth of the channel (northern spit of PS-20), or alongside of the channel towards the central sector (southern spit of PS-7, western spit of PS-22, eastern spit of PS-24, and western spit of PS-25), has been observed, as well as a rotation of the spits from the root (both spits of PS-21, western spits of PS-24 and PS-26).

Progradation is usually combined with erosion, which results in the degradation or breaching of some of the spits, developing minor inlets and spit islands. Subsequently, some of these spit islands eroded completely (PS-3, PS-8), but in other cases the breaches closed, and they joined the previous spit (PS-22, PS-25), or continued as islands by their natural evolution (PS-26) or because the limits of the spit islands and the head of the spit was fixed by coastal defense structures (PS-2; PS-18). It is also possible that the spit islands migrated to the mainland, developing new spits (PS-20). Extreme erosion due to the development of new inlets reaches a shoreline retreat of up to −167.65 m/yr, for the head of the northern spit of PS-27, and has even led to total degradation of one of the spits (northern spit of PS-20, western spit of PS-21, and northern spit of PS-27) or severe root retreat of the northern spit of PS-27 (Figure 5). Cannibalization of the root of the western spit of PS-12 was also identified, while coastal defenses were built close to the root of other spits (e.g., the southern spit of PS-15, the two spits of PS-17, western spit of PS-19) to avoid their erosion.

The development of the paired spits constrains the interdistributary bay or channel (Figure 4; Table 1). Positive progradation rates of the spits were observed in the study period (Table 2) as well as a decrease in the width of the main inlet (Figure 5), although in several cases, this decrease was very small compared to the overall dimensions of the system (PS-20, PS-26). However, in one exceptional case (PS-24), there was an increase in the width of the main inlet, due to the 40° clockwise rotation of the western spit, i.e., towards its channel margin. At PS-27, similar rates of progradation of the southern spit and erosion of the northern spit (Table 2) resulted in a northward migration of the inlet with a slight increase in its width (Figure 5L). The main inlet of PS-22 also showed a westward migration, combined with an alternately decreasing, increasing, and again decreasing width (Figure 5I).

Table 2. Natural morphodynamic processes of the paired spits located at the mouths of interdistributary bays of deltas (PS-1 to PS-3) and within coastal channels (PS-4 to PS-27): average progradation rate (m/yr; negative values for erosion), presence of hooked ridges, breaching of the spit developing a spit island or degradation of the spit by erosion, merging of the island with the rest of the spit or formation of a new spit, and classification of the natural morphodynamic pattern.

Paired Spits	Progradation Rate			Hooked Ridges	Spit Breaching or Degradation	Spits Merging or Formation of a New Spit	Morphodynamic Pattern
	Study Period	Longer Spit	Shorter Spit				
PS-1	1984–2020	10.88	16.83	Yes	Yes	Yes	Ephemeral
PS-2	1984–1989	104.54	106.88	Yes	Yes	Yes	Stationary
PS-3	1984–2020	75.26	Very low	Yes	Yes	Yes	Stationary
PS-4	1984–2020	3.74	8.25	Yes	No	No	Ephemeral
PS-5	1984–2020	Very low		No	No	No	Stable
PS-6	1984–2020	Very low		No	No	No	Stable
PS-7	1984–2020	12.48	Very low	Yes	Yes	Yes	Stationary
PS-8	1984–2004	6.16	11.24	Yes	Yes	Yes	Stationary
	2005–2020		17.42				
PS-9	1984–2020	Very low		No	No	No	Stable
PS-10	1984–2020	Very low		No	No	No	Stable
PS-11	1984–2020	Very low		No	No	No	Stable
PS-12	1984–2020	63.65	Very low	Yes	No	No	Stationary
PS-13	1984–2020	31.98	6.44	Yes	No	No	Stationary
PS-14	1984–2020	Very low		Yes	No	No	Stable
PS-15	1984–2020	9.46	17.69	Yes	No	No	Stationary
PS-16	1984–2020	Very low		No	No	No	Stable
PS-17	1985–2019	2.68	Very low	No	Yes	No	Ephemeral
PS-18	1984–2003		17.90	Yes	Yes	Yes	Stationary
	1984–2020	3.29					
PS-19	1984–2020	10.46	15.13	Yes	No	No	Stationary
PS-20	1992–2006	156.93		Yes	Yes	Yes	Stationary
	2007–2012	−13.29					
	2013–2016	−135.30					
	1984–2020		1.92				
PS-21	1984–1991	51.99	21.69	Yes	Yes	Yes	Ephemeral
PS-22	1986–2006	55.72	−17.70	Yes	Yes	Yes	Stationary
	2007–2014	−79.98					
	2015–2020	42.06					
PS-23	1984–2020	Very low		No	No	No	Stable
PS-24	2003–2020	3.08	−9.27	No	Yes	Yes	Ephemeral
PS-25	1984–2012	35.49	3.58	Yes	Yes	Yes	Stationary
	2012–2020		6.91				
PS-26	1984–2020	1.13	1.67	Yes	Yes	Yes	Stationary
PS-27	1984–2020	134.92	−167.65	Yes	Yes	No	Ephemeral

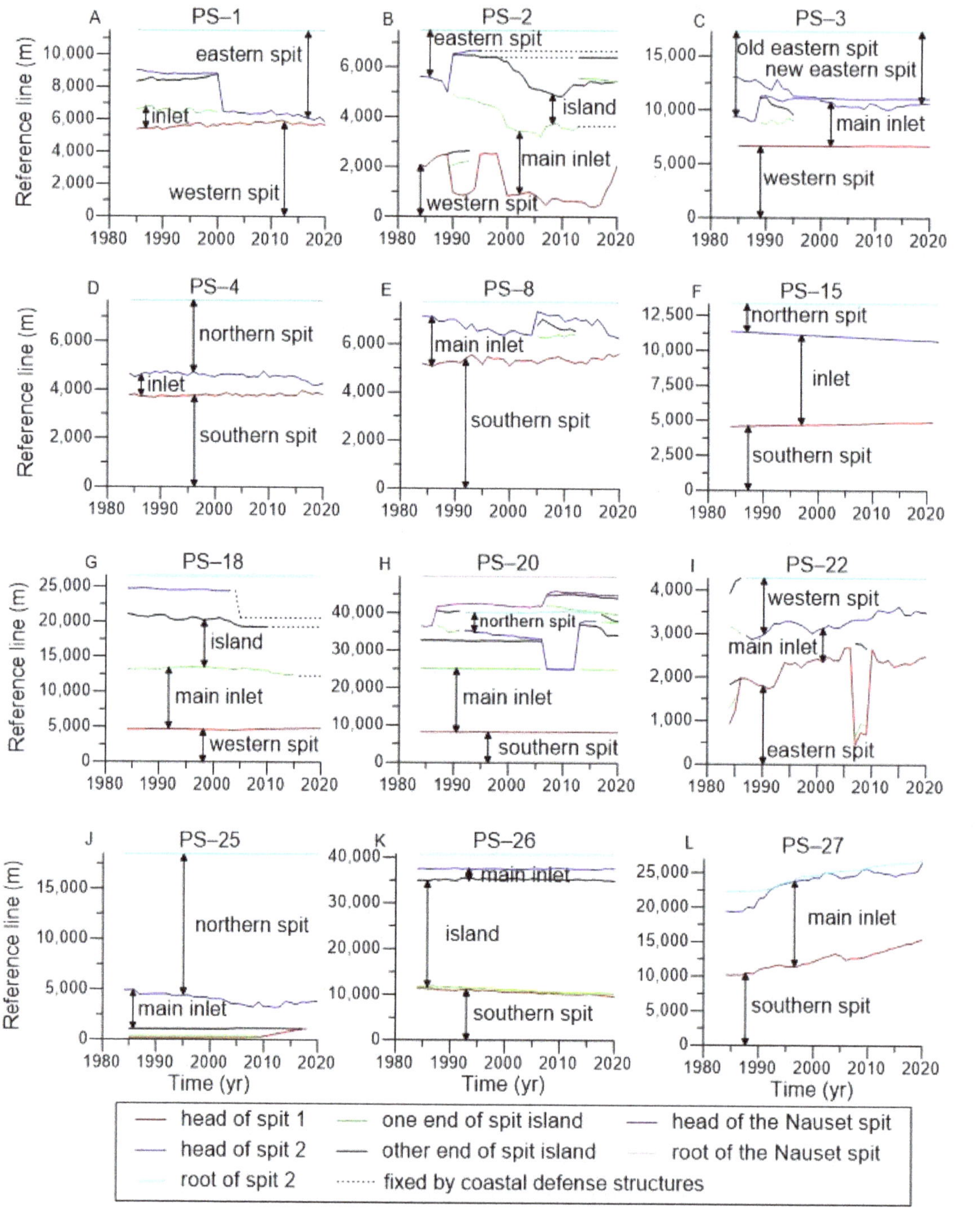

Figure 5. Geomorphological evolution (1984–2020) of some of the paired spits identified at the mouths of interdistributary bays of deltas (PS-1 to PS-3) and within channels (PS-4 to PS-27). Annual distances (m) for the head of the spits and ends of island are plotted on a reference line defined from the root of the two spits. (**A**) PS-1, (**B**) PS-2, (**C**) PS-3, (**D**) PS-4, (**E**) PS-8, (**F**) PS-15, (**G**) PS-18, (**H**) PS-20, (**I**) PS-22, (**J**) PS-25, (**K**) PS-26, (**L**) PS-27.

Three morphodynamic patterns for paired spits can be defined following the geomorphological evolution in the last decades of the twenty-seven systems analyzed (Table 2 and Figure 6). The first pattern corresponds to the stable systems, which are the eight systems of paired spits with very low progradation rates, and imperceptible geomorphological changes in the study period. Among them, PS-6 shows evidence of previous breaching and degradation of the central sector of the southern spit, but its progradation during the study period is very low. Another set of thirteen paired spits can be classified as stationary systems, with higher progradation rates and successive processes within a few decades of breaching or degradation of some spit, followed by the formation of new spits. PS-7 shows evidence of a previous degradation of the northern spit and the formation of the new spit in this margin of the channel is very slow. PS-6 can be considered an intermediate case between stable and stationary, as it shows evidence of previous breaching of the southern spit in its central sector, but the present progradation rate are very low.

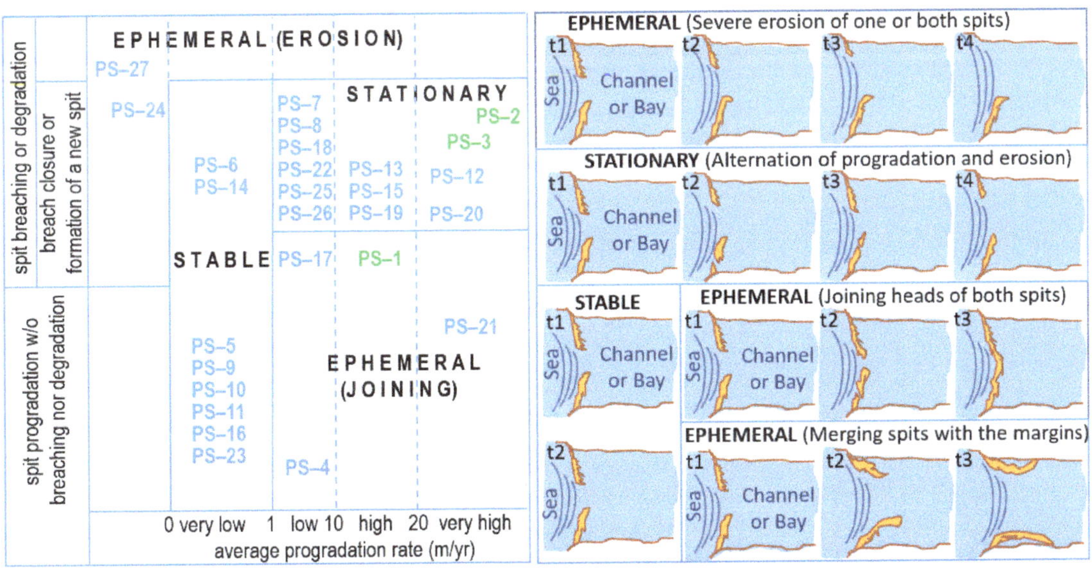

Figure 6. Morphodynamic patterns. Left: classification of stable, stationary, or ephemeral morphodynamic patterns of paired spits at the mouths of interdistributary bays of deltas (PS-1 to PS-3, green font) and within coastal channels (PS-4 to PS-27, blue font). Right: Synoptic description of the evolution for each morphodynamic pattern; the ephemeral (i.e., unstable) morphodynamic pattern includes three subtypes.

The third morphodynamic pattern corresponds to six unstable systems. They include PS-1 and PS-4, because the progradation of their heads will determine the near merging of the two spits and, consequently, these paired spits are ephemeral (Table 2, Figure 5A,D). Another type of unstable configuration occurs when the heads of the paired spits join to the respective sides of the channel. Thus, the two small spits observed in 1984 between Tuckernuck and Muskeget islands (PS-21) progressively rotated, erosion broke the western spit in 1991, and the heads of both spits merged their respective islands in 1994. Thus, the paired spits configuration of PS-21 disappeared. Similarly, the progradation of the two spits of PS-17 trended towards their respective margins, in the interior of the Strait of Canso. The progradation of the western spit towards the mainland of the Nova Scotia Peninsula was closing the lagoon entrance, decreasing its width to only about 4 m in 2019, while the head progradation of the eastern spit was towards the mainland of Cape Breton Island, with a lagoon entrance in 2020 of only 22 m in width (Figure 4P). The third type of unstable morphodynamic is due to high and continuous erosion, ultimately leading to the permanent disappearance of one of the spits. The neck of the Tigres spit was breached in March 1962. Later, converging paired spits PS-27 developed on both sides of the inlet, and the southern spit tied to the mainland. However, the northern spit progressively eroded (negative progradation rate, Table 2) until it disappeared completely in 1993. Subsequently, since 2003, an incipient and very narrow spit started to develop, but disappeared again in 2019 due to severe erosion (Figure 5L). Therefore, the paired spits system has become ephemeral, while two other spits are developing northward in the mainland. Similarly, the length of the western spit of PS-24 within the Safaga Strait is decreasing; it virtually disappeared in 2011 and thereafter does not reach its previous size.

3.3. *Hydrodynamics*

The paired spits analyzed are located in a wide range of tide and wave regimes. Most of them (81.48%) are on microtidal coasts, except PS-9, PS-11, PS-12, PS-13 and PS-16, which are on mesotidal ones (Table 1). The wave regimes affecting the paired spits, considering their offshore wave pattern and shoreline orientation, includes unidirectional (51.85%), bidirectional (37.04%) and even multi-directional (11.11%) cases. The mean annual offshore Hs for the dominant component ranges between 0.45 and 1.84 m, and between 0.44 and 2.08 m for the second component. Peak periods include wind waves (53.84%), swell (28.21%) and intermediate (17.95%) waves (Table 3). Most of them are located on wave-dominated coasts, but there are also some others on mixed-energy or tide-dominated coasts.

PS-16, located in the sheltered Salish Sea, presents a mesotidal range (2.26 m) and offshore unidirectional wind waves with mean offshore Hs of 0.51 m from the SE; therefore, it can be classified as a tide-dominated environment. Progradation rates of the two spits are very low and the morphodynamic pattern is stable, without spit degradation or breaching. The remaining paired spits on mesotidal coasts (Table 1) are located in SW Alaska and are either tide-dominated (PS-9) or mixed-energy environments (PS-11, PS-12, PS-13). Offshore Hs are lower than 1.2 m for PS-9 and PS-11, which show a stable morphodynamic pattern. In contrast, the swell in PS-12 and PS-13, with a mean offshore Hs above 1.5 from the SSW (and above 2 m for the second component from the NW for PS-12), determine that the progradation rates are very high for the western spits of both systems, developed eastward from Tugidak Island, but low and very low for the eastern spits, developed westward from Sitkinak Island (Table 2 and Figure 4L).

Another important factor, along with waves and tides, is the formation of ice sheets in winter near the coast, as well as coastal snowfall and freezing, which were observed in satellite imagery and confirmed by gaps in ERAS5 wave data series for 51.85% of the paired spits analyzed in this study. This factor is most relevant for paired spits located in the Beaufort Sea (PS-1, PS-7 and PS-8) and the Kara Sea (PS-4 to PS-6), which are under wave attack for only 30–40% of the year, but was also identified for paired spits located in the Bering Sea (PS-10, PS-11 and PS-14), Gulf of Alaska (PS-9, PS-12 and PS-13), Gulf of St. Lawrence (PS-17) and the Sea of Okhotsk (PS-19), where exposure to wave influence is greater. The progradation rates of this set of paired spits range from very low to very high (Table 2, Figure 6), implying that they show stable (PS-5, PS-6, PS-9, PS-10, PS-11, PS-14), stationary (PS-7, PS-8, PS-12, PS-13, PS-19) and even ephemeral (PS-1, PS-4, PS-17) morphodynamic patterns. Wave and tide action in the period of the year free of ice sheets and coastal freeze-up determines the morphodynamic pattern.

Six other paired spits are located in sheltered seas with microtidal range and unidirectional or bidirectional wind waves of mean annual offshore Hs of 0.77 m or less (Table 3), as is the case of the paired spits found in the Adriatic Sea (PS-2, PS-3), Baltic Sea (PS-15), Black Sea (PS-18), Red Sea (PS-24) and Gulf of California (PS-21). Their progradation rates vary from low to very high and they are characterized by a stationary or ephemeral (erosional) morphodynamic pattern, with alternating erosion (degradation, breaching) and spit accretion (closure of inlets, formation of new spits), except in the Gulf of California where low energy waves explain that PS-21 shows low progradation rates, without significant morphological changes, and consequently a stable pattern (Table 2).

Table 3. Offshore wave (2018–2021): Percentage of the year free of ice-sheet formation; unidirectional, bidirectional, or multi-directional wave regime; frequency (%), mean annual and 95% percentile of Hs (m), approach direction and Tp (s) of the first and second component. The Tp includes wind waves (w), swell (sw) and intermediate (i) types.

Paired Spit	Wave Data Point		% Year	Wave Regime	1st Component					2nd Component				
	Latitude	Longitude			Freq	Hs	P95	Dir	Tp	Freq	Hs	P95	Dir	Tp
PS-1	69.71°N	141.3°W	33.21	Bi	21.71	0.85	2.12	ENE	w	11.34	0.80	1.77	NW	w
PS-2	44.75°N	12.28°E	100.00	Uni-	37.45	0.45	1.23	SE	w					
PS-3	44.79°N	12.47°E	100.00	Bi-	45.77	0.57	1.66	ENE	w	37.09	0.44	1.15	SE	w
PS-4	74.94°N	86.05°E	31.76	Bi-	9.27	0.74	1.67	NNE	w	9.04	0.79	1.98	NW	i
PS-5	73.09°N	69.50°E	40.93	Uni'-	18.71	1.03	2.47	WSW	w					
PS-6	70.16°N	125.31°W	31.07	Bi-	14.00	0.83	1.85	NW	w	10.01	0.72	1.56	ENE	
PS-7	69.69°N	125.37°W	31.07	Uni-	14.51	0.66	1.54	NNW	w					
PS-8	69.50°N	138.98°W	37.98	Uni-	17.72	0.53	1.07	ENE	w					
PS-9	59.45°N	153.50°W	100.00	Bi-	25.06	1.11	2.49	WNW	w	21.59	1.42	2.88	NE	w
PS-10	59.38°N	164.57°E	79.36	Uni-	29.20	1.64	3.75	E	sw					
PS-11	58.59°N	161.24°W	91.51	Bi-	40.41	1.06	2.35	SW	sw	15.76	1.31	2.69	ESE	w
PS-12	56.64°N	154.33°W	100	Bí-	57.22	1.55	3.13	SSW	sw	15.41	2.05	3.88	NW	i
PS-13	56.41°N	154.43°W	100	Uni-	57.68	1.77	3.57	SSW	sw					
PS-14	55.14°N	163.48°W	100	Uni-	65.79	1.42	3.10	W	sw					
PS-15	54.43°N	10.94°E	100	Bi-	34.47	0.68	1.54	WNW	w	18.07	0.76	1.61	SW	w
PS-16	48.52°N	123.28°W	100	Uni-	15.63	0.51	1.12	SE	w					
PS-17	45.74°N	61.53°W	96.44	Uni-	38.54	1.07	2.60	NNW	w					
PS-18	45.05°N	36.56°E	100	Bi-	40.11	0.77	1.79	NNE	w	36.73	0.68	2.60	SW	w
PS-19	43.69°N	145.32°E	92.13	Bi-	37.48	0.78	1.54	SE	sw	21.93	1.13	2.20	NNW	sw
PS-20	41.45°N	70.01°W	100	Multi-	26.91	0.89	1.86	SSE	i	18.95	1.35	2.99	ENE	sw
PS-21	41.32°N	70.27°W	100	Multi-	47.45	1.35	2.26	SSW	i	12.69	0.97	1.90	ESE	i
PS-22	41.25°N	70.26°W	100	Multi-	47.60	1.18	2.43	SSW	i	12.86	1.04	2.01	ESE	i
PS-23	28.80°N	112.16°W	100	Uni-	29.31	0.48	0.99	SSE	w					
PS-24	27.00°N	34.20°E	100	Uni-	100	0.68	1.31	NN	w					
PS-25	24.55°N	111.87°W	100	Uni-	44.06	1.19	1.60	SW	sw					
PS-26	24.30°N	111.66°W	100	Uni-	45.81	1.16	1.57	SW	sw					
PS-27	16.77°S	11.68°E	100.00	Uni-	100	1.84	2.80	SSW	sw					

The rest of the paired spits are located on microtidal open coasts of the Pacific and Atlantic oceans, under influence of swell or intermediate (6 to 9 s) waves with mean annual Hs varying from higher than 1.15 m to 2.8 m (Table 3), i.e., they are the paired spits under the most energetic waves, and consequently they are highly dynamic systems, with stationary or ephemeral morphodynamic patterns. The paired spits in Nantucket Sound (PS-20, PS-21, PS-22) are the only ones below an offshore multi-directional wave regime, but their different coastal orientation determines that PS-20 is mainly affected by ENE swell and SSE intermediate waves, while PS-21 and PS-22 are more influenced by SSW and ESE intermediate waves. Paired spits in Almejas Bay (PS-25, PS-26; Figure 4W) are below the influence of SW swell waves, with similar offshore Hs for both systems, but very high progradation rates for El Cisne spit, and low rates for the other three spits (Table 2). Finally, PS-27 lies under the strongest waves, with a unidirectional approach direction from the

SSW swell and annual offshore Hs of 2.8 m (Table 3), which determines the intense erosion of this ephemeral system (Figure 5L).

3.4. Anthropization of the Paired Spits

Most of the paired spits identified in this study are found in natural coasts, with no direct anthropogenic disturbance. In fact, some of them are nature reserves. Thus, for example, Demarcation Bay (PS-1), as well as Tugidak and Sitkinak islands, where are located PS-12 and PS-13, are Alaska National Wildlife Refuges, and for the PS-15, the Krummsteert spit and the outer half of the Graswarder–Heiligenhafen spit are two nature reserves. However, other paired spits analyzed in this study show the impact of local human activities (Table 4).

Table 4. Local human impacts and consequent classification of the paired spits at the mouth of interdistributary bays of deltas (PS-1 to PS-3) and within coastal channels (PS-4 to PS-27) from natural, to rural, semi-urban, urban, or artificial coasts.

Paired Spits	Longer Spit				Shorter Spit			
	Houses & Roads	Bridges	Seawalls	Type	Houses & Roads	Bridges	Seawalls	Type
PS-1	No	No	No	Natural	No	No	No	Natural
PS-2	No	No	Yes	Rural	No	No	Yes	Rural
PS-3	No	No	No	Rural	Yes	No	No	Semi-urban
PS-4	No	No	No	Natural	No	No	No	Natural
PS-5	No	No	No	Natural	No	No	No	Natural
PS-6	No	No	No	Natural	No	No	No	Natural
PS-7	No	No	No	Natural	No	No	No	Natural
PS-8	No	No	No	Natural	No	No	No	Natural
PS-9	No	No	No	Natural	No	No	No	Natural
PS-10	No	No	No	Natural	No	No	No	Natural
PS-11	No	No	No	Natural	No	No	No	Natural
PS-12	No	No	No	Natural	No	No	No	Natural
PS-13	No	No	No	Natural	No	No	No	Natural
PS-14	No	No	No	Natural	No	No	No	Natural
PS-15	Yes	No	Yes	Semi-urban	No	No	No	Natural
PS-16	Yes	No	No	Rural	Yes	No	Yes	Rural
PS-17	No	Yes	Yes	Rural	Yes	No	Yes	Semi-urban
PS-18	Yes	No	No	Urban	Yes	Yes	Yes	Urban
PS-19	Yes	No	Yes	Semi-urban	No	No	No	Natural
PS-20	No	No	No	Natural	No	No	No	Natural
PS-21	No	No	No	Natural	No	No	No	Natural
PS-22	Yes	No	No	Semi-urban	No	No	No	Natural
PS-23	No	No	No	Natural	No	No	No	Natural
PS-24	Yes	No	No	Natural	No	No	No	Natural
PS-25	No	No	No	Natural	Yes	No	No	Semi-urban
PS-26	No	No	No	Natural	No	No	No	Natural
PS-27	No	No	No	Natural	No	No	No	Natural

The Scardovari lagoon, in the Po delta, presented in 1985 natural paired spits (PS-2) at the mouth, with a stationary morphodynamic pattern including breaching of the eastern spit and formation of a baymouth barrier island. However, since 1995 several stone reefs have progressively fixed the shoreline of the two spits and the central island; the channel trough the eastern spit was also fixed to keep this lagoon open (Figure 5B).

The Goro lagoon (PS-3) is located to the west of the Scardovari lagoon. On the eastern side, a small bridge and some other minor structures were built in the head of the oldest spit, which remains stable after previous high erosion from 1989 to 1995, while a new spit developed seaward (Figure 5C). The western spit is occupied in the middle sector by buildings for tourist activity.

The Graswarder–Heiligenhafen spit (PS-15) shows high human occupation on the neck, including buildings, a marina on the inner side and coastal defense structures on the seaward side to prevent erosion. On the other side of the channel, human occupation of the Krummsteert spit (S of Fehmarn Island) is very low, with the only presence of a lighthouse, a few houses, and some agricultural land close to the beginning of the spit.

The James spit (PS-16), in the eastern margin of Cordova Channel, contains a road, a small marina in the inland, close to the head, while a stone seawall was constructed to avoid erosion of the neck. On the opposite side, Cordova spit shows no significant human occupation, but only a dirt road and isolated house.

In the Strait of Canso, the neck and some other areas near the head of the eastern spit of PS-17 are secured by boulders to prevent shore erosion. However, in 2007, a small breach opened in the middle of the spit. There is also a small bridge from the head of this eastern spit to the mainland of Cape Breton Island. The western spit, developed from the Ontario Peninsula, contains only a few buildings and a large electricity tower at the neck of the spit.

The western spit of PS-18, in the Kerch Strait, shows high human occupation with an axial road, buildings and tourism infrastructures by the seaside, as well as an industrial harbor in the inner part. In the northern sector of the Tuzla spit island, there was, from before 1984, a small harbor and a village that grew progressively, while the southern end of the island was attached by coastal defense structures since 2007. The Tuzla spit, on the eastern margin of the Kerch Strait, showed no local human alterations until 2003, when it was lengthened by artificial nourishment, the northern end was fixed by the construction of two stone-filled blocks in 2004, and construction of a motorway began on it. In 2015, the motorway and a railway linked the Tuzla spit to Tuzla Island by a bridge and another bridge linked the island to the eastern margin of the Kerch Peninsula. Therefore, both spits and the central spit island currently show high human disturbance.

At the southern end of the Nemuro Strait (also named Notsuke Strait or Kunashirsky Strait) is located PS-19. On the western margin, the Notsukezaki spit, from Hokkaido Island, is the largest sand spit in Japan, and shows strong human interventions, including an axial road, scattered buildings, several groins, and a seawall that is virtually continuous along its entire length, strongly fixing the shoreline except at the far end of the head. On the opposite side, two spits at the southern end of Kunashir Island (Kuril Archipelago) show no human disturbance, the smaller of which is also at the southern entrance to the Nemuro Strait.

The western spit of PS-24, in the Safaga Strait (Red Sea) developed from the head of a cuspate foreland with high human occupation related to the expansion of the urbanization of the city of Safaga in recent decades, prevents sediment input to the western margin of the strait. This cuspate foreland includes a small marina to the north, which is locally trapping sediments, and an industrial harbor to the south. On the root of the western spit, there is a house and a small marina. In contrast, the eastern spit of PS-24, on Safaga Island, is free of local human disturbance, except a dirt road.

At the northern entrance to Almejas Bay, near the head of the western spit of PS-25 there are remains of some abandoned buildings and a deserted small harbor. In contrast, the eastern spit of PS-25 and the two spits of PS-26 at the southern entrance to Almejas Bay are pristine coasts.

4. Discussion

4.1. Hydrodynamic Control

Wave-induced currents and associated littoral drift are the main factors controlling spit morphodynamics [7,65,66]; however, tidal currents can be decisive for megatidal coasts [16]. The paired spits identified in this study were more frequent on microtidal coasts but can also be found on mesotidal coasts (Table 1). This distribution is similar to that of paired spits at the mouths of bays, estuaries and deltas, which mainly correspond to microtidal coasts, while only a few studies refer to paired spits on the mesotidal coasts of England [62–72], Ireland [37,73] and Spain [74], on the macrotidal coasts of England [23,24,75] or the megatidal coasts of France [16,76]. Considering the relative energy of waves versus tides, it was deduced that the formation of paired spits mainly occurs on wave-dominated coasts, but also on tide-dominated (PS-9, PS-16) and mixed-energy (PS-11, PS-12, PS-13) coasts. This is probably because spits are like other coastal barriers, which develop best in microtidal environments with intermediate conditions between wave-dominated and river-dominated coasts [77].

The formation of paired spits has been previously described for unidirectional [25,37] and bidirectional [27,30,38] wave regimes. In this study, convergent longshore drift leading to progradation of the two spits was identified under uni-, bi- and multidirectional offshore wave regimes (Table 3). In the case of deltas, the intense refraction due to their prominent morphology may explain that offshore bidirectional wave regimes give rise to unidirectional wave regimes for paired spits in interdistributary bays, as in the case of the Goro lagoon (PS-3) in the Po delta [78]. In other interdistributary bays of deltas, such as Demarcation Bay, the offshore bidirectional wave regime [79] explains the convergent progradation of the paired spits (PS-1).

The development of spits in the mouths of deltas is typical of wave-dominated deltas, and these spits tend to evolve rapidly [80], as is also the case of the paired spits analyzed in the interdistributary bays of deltas (PS-1, PS-2, and PS-3). Fluvial discharge is primarily responsible for sediment supply and the formation of successive delta lobes and interdistributary bays [81,82]. The natural tendency for sedimentary filling of the interdistributary bay by fine-grained organic-rich and clastic sediments [83–85], together with the high convergent progradation of the paired spits in the mouths of interdistributary bays (Table 2), suggest that these systems are rare because they are ephemeral under natural conditions (PS-1) and the inlet only remains open in the deltas because of high human intervention (PS-2 and PS-3). In fact, the breaching of the eastern spit of PS-1 was observed ot the beginning of the study period, as well as plotted in the nautical chart, but this minor inlet closed in 2000 and the convergent progradation continued (Figure 5A). The final closure or not of this interdistributary bay will be determined by the hydraulic blockage between the two spits due to tidal currents or minor Clarence River outfalls that flow directly into the interdistributary bay.

Accumulative features, such as spits and barriers, are common along many Arctic coasts [86], which explains why some of the paired spits identified in this study are also found on the Arctic coast (Figure 1 and Table 1). The development of ice shelves limits the effects of winds, waves, tides, and river outflows for nine months of an average year on Arctic coasts, i.e., except for the open-water season, typically from July to September [87,88], and similarly for other high-latitude regions. Analysis of monthly LANDSAT images and offshore-wave data series (Table 3) has shown that ice-sheet development and coastal freezing impede or delay the morphodynamic evolution of the paired spits in the Arctic Ocean, the Bering Sea, the Gulf of Alaska, the Kamchatka Peninsula, the Gulf of St. Lawrence, and the Sea of Okhotsk (PS-1, PS-4 to PS-11, PS-14, PS-17 and PS-19). This is a new factor not previously considered in the scientific literature on the hydrodynamic agents controlling the morphodynamics of paired spits.

The hydraulic blockages that allow the entrance between the convergent paired spits within channels to remain open, and even the degradation and development of new spits on behalf of their connection into a single barrier, is due to tidal currents flowing through the channels [45,89]. The increase in local depth observed on the nautical charts of some study sites (Table 1), just between the two spits, is evidence of these strong currents through the channels. Most of the paired spits located at the entrance to the channels are oriented inwards of the channel, indicating the incoming dominance of the longshore drift, with the only exception being the Nemuro Strait (S Sea of Okhotsk), where the paired spits PS-19 are oriented outwards, i.e., related to the outgoing currents of the channel. In the Safaga Strait (Figure 4V), the paired spits SP-20 developed in the middle of the channel; the morphology, orientation and morphodynamics of these paired spits show that they are clearly related to a southward current flowing through the channel [90].

With regard to long-term evolution, sea-level oscillations will determine the evolution of these paired spits. The Holocene transgression favored the occurrence of offshore islands and, consequently, the development within the channels of tombolos [91] and paired spits. Similarly, on the Arctic coast, the Holocene transgression induced the formation of coastal bays, due to breaching of thermokarst lakes, and consequently favored the subsequent formation of barrier islands and spits [92], including paired spits. However, the present sea-level rise due to the impact of the anthropogenic climate change is a clear hazard for the future evolution of Arctic coasts [93] and the Po River delta [94], which include erosion risks for the paired spits developed at their interdistributary bays, as well as for the spits and paired spits developed within coastal channels [95].

4.2. Sediment Availability and Sedimentary Interactions

Good sediment availability is also necessary for the development of spits [29,87,96] and paired spits [97]. The dimensions of the spits (Figure 4 and Table 1), the intensity of shoreline progradation and the ability to develop new spits or regenerate old ones after degradation or breaching (Figure 5 and Table 2) are evidence of this sediment availability. Dimensions are similar in the mouths of the interdistributary bays of deltas, but usually very different at the entrance and within the channels, which indicate the different intensities of the longshore drift and sediment inputs in the latter.

In the interdistributary bays of deltas, the fluvial source of sediments for the paired spits can be from different rivers (PS-1, PS-3) or from the same river (PS-2). In fact, the sediment source for the development of the two spits of the Scardovari lagoon (PS-2), as well as the Scanno of Goro spit, in the eastern flank of the Goro lagoon (PS-3) are related to the modern Po River delta [40,42,98]. Progradation rates are similar for these three spits (Table 2), which is consistent with a similar fluvial discharge supply for the southern mouths of the delta [99,100]. The two paired spits constraining the Goro lagoon show similar lengths (Table 1, Figures 4C and 5C, but the sediment input for the Scanno of Volano spit, on the western flank, is lower and mainly due to littoral drift from the south [42,78], which is consistent with its very low progradation rate (Table 2).

In the medium-term temporal scale, another factor determining the evolution of deltas is subsidence, which allows for the erosion of the abandoned delta lobes, including their distal spits and islands [101]. The Scardovari and Goro lagoons are not related to the abandoned lobes of Po River delta, quite the contrary; however, these lobes have shown the highest subsidence rates of the entire Po River delta from 1967 to 2017. The zone of highest subsidence in the delta has been shifting from north to south since at least 1957 [102], implying an erosion risk for PS-2 and PS-3.

In coastal channels with good sediment availability, the development of cuspate forelands and tombolos in the wave-shadow central sector is common [103–106], whereas sediment scarcity favors the formation of long, narrow, and deep coastal channels [107], even though the channel morphology predicts the presence of cuspate forelands or tombolos [105,106,108]. Tombolos and cuspate forelands have developed as a result of sediment inputs from both channel entrances [4,25,104]. Paired spits within channels are also cuspate bedforms but are related to convergent longshore drift and sediment input toward a single channel entrance. Sediment supply for progradation of paired spits constraining channels can be due to cliff erosion upstream, as in the case of the Graswarder–Heiligenhafen spit [109,110] of PS-15, at the western entrance to the Fehmarn Sound (Baltic Sea), or to littoral drift from spits, beaches, and dunes upstream [45,111]. Nevertheless, many times, sediment supply for spit progradation is also due to cannibalization of its own neck [112–114] as can be observed in the western spit of PS-12, or material scoured from the inlet can be the dominant source for spit accretion, as shown by [115] for the northern spit of PS-20.

Paired spits within coastal channels are, in general, of different lengths (Table 1). The high sediment availability in the Magdalena coastal plain (W coast of the Baja California Peninsula) is related to the development of large Pleistocene coastal dune sheets [116] and large Holocene spits constrain both entrances to Almejas Bay, i.e., the El Cisne spit (PS-25) at the northern entrance, as well as the Flor de Malva spit and the Creciente spit island (PS-26) at the southern entrance [117]. The dimensions and average progradation rates (1984–2020) of the El Cisne spit are higher than those for Creciente Island (Table 2), indicating a greater littoral drift at the northern spit. On the other hand, the small inlet between Creciente Island and the Flor de Malva spit is not closing and it shows a migration rate of 44.50 m/yr (1984–2020) to the southeast. This implies that the dominant longshore drift at the root of the spit is also to the southeast, i.e., similar to the direction of the inlet migration [2,118,119], because updrift migration of inlets only occurs for shorter periods, lasting months or years [115,120]. Therefore, this suggests the current existence of a divergent littoral drift from the middle of the Flor de Malva spit, as the progradation of the head of the Creciente spit island is westward (Figure 5K and Table 2). On the opposite channel margin, the western spits of PS-25 and PS-26, linked to Santa Margarita Island, are very short (Table 1), likely due to low sediment availability from the island, and progradation rates are also low (Table 2).

Spit migration has been described for single spits [19,22] and paired spits [121,122]. In this study, obtained by analysis of satellite images from 1984 to 2022, spit migration was identified in some paired spits systems (PS-7, PS-20, PS-22, PS-24, PS-25), but may also have occurred in others before the start of the study period. Therefore, spit migration is a possible process for sediment input and the development of paired spits at the entrance to channels or within them, due to the breaching of nearby spits and their migration along the shore towards the channel.

In any case, the source area of sediments for paired spit bounding interdistributary bays of deltas or channels is usually different for each spit, i.e., sediment supply is due to longshore drift from different coastal fringes for each margin of the bay or channel. However, sedimentary interactions between them can be expected when they are located on shallow coasts, the channel width is low and the development of the paired spits implies a significant decrease on the width of the interdistributary bay (PS-1) or channel (PS-4) (Table 1, Figure 4A,D,I), favoring sediment bypass through the inlet by several possible mechanisms [27]. On the other hand, the increase in depth in some of the inlets between the two paired spits is evidence of local intensification of the hydrodynamics due to constriction of the channel entrance (PS-11, PS-12, PS-13, PS-23). These currents and increased depth prevent the two spits from merging, but may generate increased head progradation rates, and consequently cannibalization of the neck of the spits.

Sedimentary interactions due to channel constriction can also be identified between nearby paired spits systems, such as PS-20, PS-21 and PS-22, located in different entrances of Nantucket Sound (NW Atlantic Ocean). The largest entrance to Nantucket Sound is located to the east, between Cape Cod and Nantucket Island. Outer Cape Cod is composed of two divergent spits, the Provincetown spit northward and the Nauset spit southward, due to Holocene reworking of glacial outcrop sediments by divergent longshore drift [123]. A historic inlet near the northern end of the Nauset spit is migrating downward, i.e., southward, even though it has temporarily migrated updrift developing paired spits [115,124]. The nodal point for the divergence of the longshore drift between the Provincetown and Nauset spits is also migrating southward [125,126], and, consequently, sediment input to the Nauset spit decreased. An extratropical storm in 1987 allows formation of a new inlet in the Nauset spit off Chatham Light [127–129] and a spit island that subsequently migrated and elongated southward. In 1992, the northern end of the island elongated updrift and connected to the mainland, becoming a new spit (the northern spit of Figure 5H). In 2007, the southern end of this new spit merged with Monomoy Island [130], located at the northern margin of the main entrance to Nantucket Sound. Thus, this new spit together with the northern spit (named Great Point) of Nantucket Island are paired spits (PS-20), which bound the main entrance to Nantucket Sound (Figure 4S).

The northern spit of PS-20 breached in 2013 to become an island and was subsequently completely degraded by wave erosion (Figure 5H). Nevertheless, since 2007, a new inlet and spit island developed again by breaching of the Nauset spit off Minister's Point (Figure 5H); therefore, this island will probably join the mainland and/or Monomoy Island in the coming decades. A graphical reconstruction of the 1984–2020 morphodynamic evolution of the Nauset spit, the new spit southward, and Monomoy Island was reported by the NASA [131]. The reconstruction of the coastal evolution since 1846 has shown that this process of extension of the Nauset Spit to connect to Monomoy Island and the subsequent break-up is cyclical, repeats approximately every 150 years, and therefore, Monomoy Island is related to former spits and spit islands formed from the breaching of the Nauset spit [111,132]. The southward progradation rate of Monomoy Island is much lower than that for the southern spit developed from the breaching of the Nauset spit (Table 2), with values of 7.64 (1984–2006), 13.29 (2007–2012), and 6.36 m/yr (2013–2021). In 1971, a decrease of southward progradation of Monomoy Island was predicted, due to encroachment of the southern end of the island into two deep basins to the northeast and southeast [133]. However, this was not identified in the study period, and the increase of the progradation rate between 2007 and 2012, when Monomoy Island merged with the new spit developed from breaching of the Nauset spit, is remarkable. Perhaps, the decrease in southward progradation will occur in the near future.

On the southern margin of Nantucket Sound, two other paired spits (PS-21 and PS-22) were identified in Muskeget Channel, westward of Nantucket Island (Figure 4T), with recurved spits and the development of flood- and ebb-tidal deltas, denoting that they are controlled by both wave and tidal currents. The maximum narrowing of the main entrance to Nantucket Sound occurred from 2007 to 2012 (Figure 5H), when the new spit joined Monomoy Island (PS-20). This basically corresponds to the period when the head of the two spits of PS-22 eroded (2007–2014) and even developed a small inlet (2007–2009) from the breach of the eastern spit (Figure 5I). Water circulation on Nantucket Sound is tidally dominated [89]. Therefore, narrowing of the main entrance by the development of PS-20 generated the intensification of tidal currents through Muskeget Channel, where PS-21 and PS-22 are located. Subsequently, when the northern spit of PS-20 became an island again and was progressively eroded away (Figure 5H), widening the inlet between Monomoy Island and the Cape Cod mainland, the hydrodynamic outflow through the Muskeget Channel decreased and the paired spits of PS-22 again showed convergent progradation (Figure 5I). Contrary to the erosion of PS-22, the eastern spit of PS-21 developed a new ridge from 2007 to 2012, due to sediment input from the degradation of PS-22, but later eroded drastically when PS-22 progressively regained its convergent progradation. The western spit of PS-21 only developed in the first decades of the study period, but in the 1990s it migrated, rotated, and merged with its island, which has shown severe erosion and remodeling throughout the study period. Therefore, the combined analysis of PS-20, PS-21 and PS-20 shows that hydrodynamic and sedimentary interactions occur not only between the two paired spits of the same system, but even with other closer paired spits in the same channel. This provides evidence that they are not independent spits, but a complex system including several paired spits.

In contrast, in the case of channels with deep and very wide inlets between the paired spits, there is no evidence of sedimentary interactions between the two paired spits, such as PS-19 in the Nemuro Strait, where they are far apart and bathymetry only shows an intensification of currents very close to the western spit. Similarly, sedimentary interactions are not clear between the two spits of PS-15, at the entrance to Fehmarn Sound. The Graswarder–Heiligenhafen spit, located in the north of the Peninsula of Wagria, has grown eastward by longshore transport for about 3000 years, with the addition of successive hooked ridges developed in intervals of about 150 years [134,135]. It shows a current progradation rate of 9.46 m/yr (1984–2022, Table 2), which is significantly higher than previous rates of 3.5 m/yr (1950–1985) after [109] or 2–3 m/yr suggested by [110]. In the same PS-15, the Krummsteert spit (S of Fehmarn Island) shows an even higher progradation rate (Table 2). However, both spits developed almost parallel to the channel margins, so that a merging of the two spits in the near future, or any other sedimentary interaction, is unlikely. All the same, the development of both spits implies a decrease to 75% of the original width at the mouth of the Fehmarn Sound, and consequently an intensification of the currents through it.

4.3. Morphogenetic Models and Morphodynamic Patterns of Paired Spits

Five conceptual models for the formation of paired spits were previously described at the mouth of bays and rivers (Figure 7): (i) the break of a coastal barrier, not necessarily a spit, by new fluvial outlets [68,74], hurricanes [23] or tsunamis [136,137]; (ii) convergent longshore drift in a narrow bay [25]; (iii) bidirectional longshore drift with hydraulic blockage by tidal currents or river flow in a bay or river mouth [26,120]; (iv) cutting of an estuarine detached spit by high-energy events [27–30]; and (v) convergence of two estuarine mouths and the associated spits [30]. In addition, previous studies in the Po River delta have already defined the presence of two spits in the mouths of both the Goro and the Scardovari lagoons [40–44], although none of these studies consider them as paired spits or relate their development to other baymouth spits. The presence of paired spits within channels has been succinctly reported [45–47]. Therefore, in this study, two new environments and conceptual models for the formation of convergent paired spits were described in detail and monitored, at the mouths of the interdistributary bays between two delta lobes of the same or different rivers (Figure 7F), and at the two margins of a channel, in the entrance or within it, usually for incoming, or exceptionally for outgoing, longshore drift (Figure 7G).

The development of paired spits in the interdistributary bays of deltas and within coastal channels was described for uni-, bi- and multidirectional wave regimes (Table 3). Therefore, they are under similar hydrodynamic patterns to conceptual models ii and iii but are located in different coastal environments. In fact, conceptual models i, iv, and v have only been reported in a few places [23,27–30,68,74,136,137]; therefore, they can be considered as unusual.

In addition to the new morphogenetic models for paired spits (Figure 7F,G), three morphodynamic patterns (stable, stationary and ephemeral) were defined for the subsequent geomorphological evolution of paired spits (Figure 6, Table 2). For stationary systems, the degradation and development of new spits is the key that allows them to maintain the paired spit configuration, i.e., to achieve a stationary equilibrium, oscillating between eroded to accretionary states, which resembles seasonal beach morphodynamics [138]. A stationary pattern was described for single spits, due to high erosion after breaching and subsequent closure of the inlet e.g., [14,139] or even total and fast degradation by tsunamis followed by the formation of a new spit [140]. Similarly, the eastern spit of PS-3 has undergone intense morphodynamics changes, including phases of rapid longshore growth, hooked ridges development, cannibalization, overwash, and breaching [98]. There are also intermediate cases between stable and stationary systems. For instance, a historical analysis of nautical charts shows a minor inlet developed in PS-14 from 1949 until before 1974 [47], but can currently be considered stable because average migration rates in recent decades (1984–2020) are very low (Figure 6). Some other paired spits show an unstable or ephemeral morphodynamic pattern. The filling of the interdistributary bay on the backshore of PS-1 and the disappearance of this coastal configuration with paired spits should be expected on a medium-term scale, similar to the closure in previous centuries of the Sinoe lagoon, an interdis tributary bay in the southwest of the Danube delta [141,142].

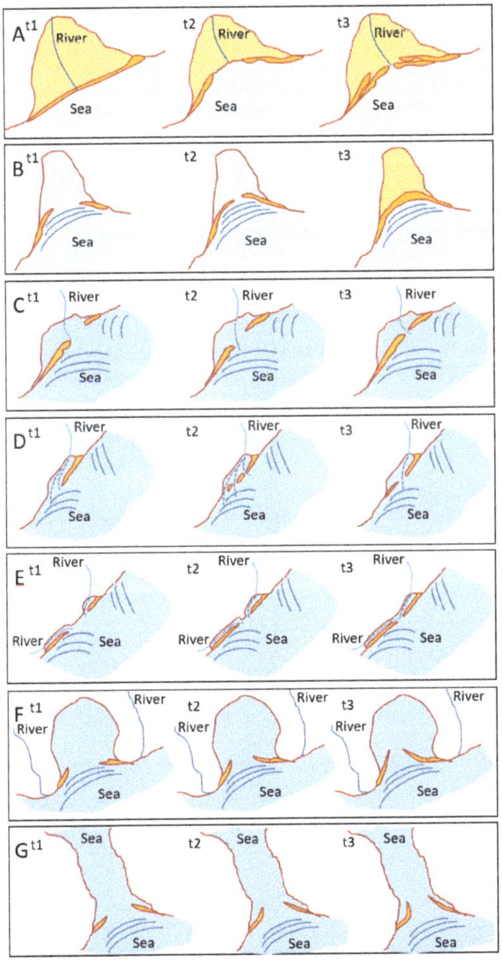

Figure 7. Conceptual models for formation of paired spits: (**A**) coastal barrier break; (**B**) convergent longshore drift in a narrow bay; (**C**) bidirectional longshore drift with hydraulic blockage; (**D**) cutting of a estuarine detached spit by high energy events; (**E**) convergence of two estuarine mouths and the associated spits; (**F**) convergent progradation from two delta lobes to the interdistributary bay (PS-1 to PS-3); and (**G**) convergent longshore drift from both margins to a channel entrance (PS-4 to PS-27) (Modified from [36]).

The three morphodynamic patterns adescribed above can also be applied to characterize the morphodynamics of paired spits at the mouths of bays and rivers. Most of them correspond to systems with a stable geomorphological configuration, while only a few studies have referred to spit breaching or degradation and the development of new spits [29,30,74,122,143]. The paired spits at the mouth of the Guadalquivir estuary (SW Spain), showed an initial stationary pattern, with degradation and the new formation of the southern spit (6900 to 2300 BP); subsequent intense progradation of the northern spit (2300 to 1000 BP) evolved to a stable pattern, where currently the southern spit is fixed by the sedimentary fill of the estuary mouth [74]. Another interesting case is that of the paired spit PS-17, in the Strait of Canso, because in the late 19th century there was an inlet in the middle of the western spit (Church, 1878; in [144]), but nowadays this inlet does not exist, and the head of the spit is very close to the western margin of the channel. Therefore, it evolved from a stationary to an ephemeral pattern. Some other studies have described paired spits at river mouths that later disappeared, i.e., were ephemeral [28–30,145]. In summary, it can be deduced that paired spits can evolve from stationary to ephemeral or stable morphodynamic patterns and can even become relict bedforms from a stable pattern.

4.4. Anthropization of the Paired Spits

Most of the paired spits identified at the mouths of interdistributary bays of deltas and within channels have shown processes of degradation and the development of new spits; some other paired spits tended to develop a continuous sand barrier, or even to disappear (Table 2, Figure 6). Their intense sedimentary dynamics and geomorphological evolution, together with the remote location of many of them, explain why they have not been described before, whereas paired spits at the mouth of bays, estuaries and deltas have been described more extensively for the anthropized coast of England [23,24,67–72,75,146,147], Ireland [73,148,149], France [16,76], Spain [74], Russia [25], India [26,34,38], China [121,122], United States of America [29], Egypt [143], Senegal [150], Colombia [151], Brazil [30,39,152], and New Zealand [28,145]. Nevertheless, paired spits have also been described in natural coasts, at the mouths of bays in the Kamchatka Peninsula [25], and at the mouth of the Volta River delta, although the latter is heavily affected by the Akosombo dam [150,153].

The morphodynamics of many single spits are affected by natural climatic oscillations, such as ENSO [36,154–156] and NAO [76,157], as well as by human disturbances, ranging from global climate change, including sea-level rise and ice melting between other effects [158–161], to local impacts. Coastal defense structures have been built in many spits to avoid erosion, usually on the seashore [10], and more rarely on the bayside [162]. At other times, the artificial opening of channels for fishermen's navigation [163] or to prevent flooding [17] has generated severe erosion of the spits.

Similarly, paired spits at the mouths of bays, estuaries and deltas are often affected by local human activities, such as urbanization, dam construction, coastal defense structures, and dredging to keep channels open for navigation [143,147,149,150,153,164,165]. Elsewhere, degradation of baymouth spits by harbor development or other human disturbances is even more severe [122]. In contrast, most of the paired spits identified in this study are found on natural coasts (Table 4), with no direct anthropogenic disturbance. Therefore, they are ideal locations for environmental conservation plans. They are also very interesting places for the analysis of the impact of both natural climatic oscillations such as ENSO [30,121] and climate change, since in these paired spits ice melting, coastal erosion and flooding due to sea-level rise are not affected by interactions with other regional or local anthropogenic factors that may mask the impact of natural climatic oscillations and global climate change.

Other paired spits analyzed in this study show local human disturbances of very different intensities (Table 4). The paired spits with a higher human influence on their morphodynamics are PS-2 and PS-3, the two interdistributary bays located in the SW of the Po River delta. In fact, the historical and current evolution of the delta is strongly conditioned by human activities [40,166]. The modern delta was induced 400 years ago by the excavation of an artificial channel [167]. In the southern part of the delta, the development of the paired spits at the mouth of the Scardovari lagoon (PS-2) is relatively recent. The lagoon developed in the early 19th century, without paired spits at the mouth, and acquired its present shape in 1955 [168]. Later, the construction of reef stones after 1985 not only protected the coast from erosion, but even generated high sedimentation [169], keeping the lagoon open by artificial inlets. The Goro lagoon has progressively migrated southwards since the 16th century. The nautical charts show the presence of paired spits on the mouth of the Goro lagoon since the late 19th century [41]. The human impact on these paired spits (PS-3) is lower than those for the Scardovari lagoon (PS-2), despite the mouth of the outlets has been also fixed by stone reefs. However, the bathymetry near the main entrance to the lagoon is very shallow [99] and dredging has been suggested [55].

The fixation of the spits by coastal defense structures, roads, railways, and big bridges are strong human disturbances on the geomorphology of the paired spits. Coastal defense structures are usually located on the neck of the spits (southern spit of PS-15, eastern spit of PS-16, both spits of PS-17) but sometimes they are almost along the whole seaside of the spits, as is the case with the Tuzla spit (PS-18) in the eastern margin of the Kerch Strait [170] and the Notsukezaki spit (PS-19) in the western margin of Nemuro Strait [171,172]. Fixing of the shoreline avoids breaching and the formation of spit islands. While head progradation of the spits can continue (PS-19), it can be drastically reduced due to the lack of sediment inputs from the neck (PS-18), or it can even increase, as indicated above for PS-2 [169]. Therefore, some paired spits with natural stationary morphodynamic patterns have evolved into a human-induced stable morphodynamic pattern (PS-18) or a human-induced ephemeral pattern by joining each spit with its margin of the channel (PS-17). Tuzla Island was formed due to breaching of the Tuzla spit (PS-18) in 1925 [173], which is characteristic of a stationary pattern. However, the shoreline of both Tuzla Island and the Tuzla spit has been progressively fixed since 2004 (Figure 5G) [170]. In other cases, there are only small villages or a few houses, which implies a change from natural to rural or semi-urban coast, but these human activities do not affect the natural morphodynamic pattern (eastern spit of PS-22, western spit of PS-25).

In an exceptional case, the coastal region has changed from high human occupation to a natural or nearly natural environment. Tigres Island was initially a 40-km long spit, with a well-established fishing village to the northeast. However, when the neck of the spit broke in 1962 and became an island, the village was abandoned [45], and currently is only visited by tourist excursions. The ephemeral (erosive) morphodynamic pattern of PS-27 has not been influenced by the previous urban development or any other human activity.

Coastal management, including implementation of coastal conservation areas and coastal defense structures, are usually only focused on one spit of the paired spits systems [147,165,174]. However, analysis of the hydrodynamics and sedimentary interactions, and models of the formation and morphodynamic patterns show that, rather than independent spits, the paired spits constitute a system, which should be taken into account for future coastal management plans. Mitigation measures, such as the construction of sand barriers and sediment replenishment, can help protect these vulnerable areas [14]. Nevertheless, at present, certain management approaches involve the removal of hard coastal defense structures that were previously used [162,175]

Spits are important coastal formations that protect coastal areas. However, the dynamics of these formations are extremely complex and can be influenced by multiple factors. To understand this complexity, it is essential for researchers and coastal managers to be aware of the different behaviors of spits. This may include analyzing their geometry, studying sedimentation and erosion on their surfaces, and identifying sediment transport processes, among other aspects. By emphasizing the complexities of these behaviors, these professionals can be equipped with the necessary tools to properly manage these coastal barriers. This will enable them to better decide how best to protect coastal areas and the communities living in them.

In the future, high-resolution monitoring will be integrated with operational wave models to create accurate systems for predicting coastal evolution at various temporal and spatial scales. In addition, combining advances in satellite monitoring with machine learning and data-assimilated modeling will help to resolve questions about the causes and effects of coastal erosion at large scales [176].

5. Conclusions

This study presented the first analysis of the global distribution and geomorphological description of paired spits developed both at the mouths of interdistributary bays of deltas (three systems) and within channels (24 systems). Two new environments and conceptual models for the formation of paired spits were presented, in addition to the five morphogenetic models described in previous studies. The identified paired spits were mainly located at high or mid latitudes, and on microtidal coasts, although they have also been found at low latitudes and on mesotidal coasts. Longshore drift and wave erosion are the main control factors in their formation and development. Paired spits can be generated under multidirectional to unidirectional approach waves. The hydraulic blockage necessary for the development of these paired spits is mainly due to tide-induced currents, as well as minor fluvial outlets in the interdistributary bays. However, the formation of ice sheets at high latitudes prevents wave action and, therefore, delays the morphodynamic evolution of paired spits. The sediment source for the paired spits at the mouths of the interdistributary bays can be from the same river or from different rivers. The sediment source for the spit at each channel margin is from different coastal strips. Nevertheless, sedimentary interactions between the two paired spits can be identified, particularly when they are located on shallow coasts and when the heads of the spits are close together. These hydrodynamic and sedimentary interactions make clear that paired spits are a system, rather than independent spits. Hydrodynamic and sedimentary interactions were also identified between systems of paired spits located at different entrances of the same channel, which therefore constitute a complex system of paired spits.

The main morphodynamic processes that characterized the evolution of the paired spit analyzed were: convergent progradation from the head and/or from the middle of the spits, developing new hooked ridges; possible migration of the spits to the entrance or within the channel, as well possible rotation of the spits and even joining of the spits to their margin of the channel; spit erosion by degradation of the spits or the formation of breaches on the spits, where subsequently the spit islands can be degraded or the breach closed. As a result of convergent progradation, the width of the inlet between the heads of the two spits usually decreases, but sometimes it only migrates and sometimes it even increases. Three morphodynamic patterns were defined as a function of these processes: (i) a stable pattern, which corresponds to systems with average progradation rates below 1 m/yr, generally without breaching or the degradation of any of the spits; (ii) a stationary pattern, for systems with higher average progradation rates and alternating degradation or breaching of one of the spits with formation of new spits or closure of the breaches; and (iii) an unstable or ephemeral pattern, which includes three subtypes, the first due to severe erosion of one or both spits, the second due to the joining of the head of the two spits forming a single barrier, and the third when each spit merges with its channel margin. These three morphodynamic patterns can also be applied to paired spits in the mouths of bays or rivers. Furthermore, the morphodynamic pattern identified for each paired spits system in the decadal scale analysis may evolve in the medium and long term, as evidence historical charts and stratigraphic studies indicates.

Most of the paired spits are located on natural coasts. The geological uniqueness, wide geographical distribution, good ecologic conservation, and low human occupation of many of the paired spits analyzed allow them to be defined as sites of high scientific interest which are suitable as field laboratories for the analysis of the effect of natural climatic oscillations and climate change, as well as potential nature reserve areas. In other cases, anthropization of the spits ranges from agricultural and fishing activities to semi-urban and even urban development. Human actions also include the construction of coastal defense structures to avoid erosion in several of the spits.

Author Contributions: Conceptualization, J.A.-C.; Methodology, J.A.-C., Á.F.-B. and R.P.M.; validation, J.A.-C., A.C.R. and R.P.M.; formal analysis, J.A.-C., Á.F.-B., A.C.R. and L.P.; investigation, J.A.-C. and A.C.R.; resources, J.A.-C., Á.F.-B., R.P.M. and L.P.; data curation, J.A.-C. and A.C.R.; writing—original draft preparation, J.A.-C.; writing—review & editing, J.A.-C., Á.F.-B., R.P.M. and L.P.; visualization, J.A.-C.; supervision, J.A.-C. All authors have read and agreed to the published version of the manuscript.

Funding: This research received no external funding.

Data Availability Statement: The data presented in this study are available on request from the corresponding author.

Acknowledgments: We are grateful to reviewers who significantly contributed to the improvement of this paper. Wave data series have been supported by the European Centre for Medium-Range Weather Forecasts.

Conflicts of Interest: The authors declare no conflict of interest.

References

1. Bowman, G.; Harvey, N. Geomorphic evolution of a Holocene beach-ridge complex, LeFevre Peninsula, South Australia. *J. Coast. Res.* **1986**, *2*, 345–362.
2. Davis, R.A.; FitzGerald, D.M. *Beaches and Coasts*; Blackwell Publishing: Oxford, UK, 2004; 419p.
3. Shawler, J.L.; Hein, C.J.; Obara, C.A.; Robbins, M.G.; Huot, S.; Fenster, M.S. The effect of coastal landform development on decadal-to millennial-scale longshore sediment fluxes: Evidence from the Holocene evolution of the central mid-Atlantic coast, USA. *Quat. Sci. Rev.* **2021**, *7*, 107096. [CrossRef]
4. Evans, O.F. The origin of spits, bars and related structures. *J. Geol.* **1942**, *50*, 846–865. [CrossRef]
5. Otvos, E.G. Coastal barriers—Nomenclature, processes, and classification issues. *Geomorphology* **2012**, *139–140*, 39–52. [CrossRef]
6. Davis, R.A. Barrier Island Systems—A Geologic Overview. In *Geology of Holocene Barrier Island Systems*; Davis, R.A., Ed.; Springer: Berlin/Heidelberg, Germany, 1994; pp. 1–46. [CrossRef]
7. Ashton, A.D.; Nienhuis, J.; Ells, K. On a neck, on a spit: Controls on the shape of free spits. *Earth Surf. Dyn.* **2016**, *4*, 193–210. [CrossRef]
8. Cartera, R.W.G.; Orford, J.D.; Jennings, S.C. The recent transgressive evolution of a paraglacial estuary as a consequence of coastal barrier breakdown: Lower Chezzetcook Inlet, Nova Scotia, Canada. *J. Coast. Res.* **1990**, *9*, 564–590.
9. Kraus, N.C.; Patsch, K.; Munger, S. Barrier beach breaching from the lagoon side, with reference to Northern California. *Shore Beach* **2008**, *76*, 33–43.
10. Bastos, L.; Bio, A.; Pinho, J.L.S.; Granja, H.; da Silva, A.J. Dynamics of the Douro estuary sand spit before and after breakwater construction. *Estuar. Coast. Shelf Sci.* **2012**, *109*, 53–69. [CrossRef]
11. Behrens, D.K.; Bombardelli, F.A.; Largier, J.L.; Twohy, E. Episodic closure of the tidal inlet at the mouth of the Russian River—A small bar-built estuary in California. *Geomorphology* **2013**, *189*, 66–80. [CrossRef]
12. Kumar, A.; Narayana, A.C.; Jayappa, K.S. Shoreline changes and morphology of spits along southern Karnataka, west coast of India: A remote sensing and statistics-based approach. *Geomorphology* **2010**, *120*, 133–152. [CrossRef]
13. Green, A.; Cooper, J.A.G.; LeVieux, A. Unusual barrier/inlet behaviour associated with active coastal progradation and river-dominated estuaries. *J. Coast. Res.* **2013**, *69*, 35–45. [CrossRef]
14. Bateman, M.D.; McHale, K.; Bayntun, H.J.; Williams, N. Understanding historical coastal spit evolution: A case study from Spurn, East Yorkshire, UK. *Earth Surf. Process. Landf.* **2020**, *45*, 3670–3686. [CrossRef]
15. Hoang, V.C.; Tanaka, H.; Mitobe, Y. Estuarine morphology recovery after the 2011 Great East Japan earthquake tsunami. *Mar. Geol.* **2018**, *398*, 112–125. [CrossRef]
16. Robin, N.; Levoy, F.; Anthony, E.J.; Monfort, O. Sand spit dynamics in a large tidal-range environment: Insight from multiple LiDAR, UAV and hydrodynamic measurements on multiple spit hook development, breaching, reconstruction, and shoreline changes. *Earth Surf. Process. Landf.* **2020**, *45*, 2706–2726. [CrossRef]
17. Gunasinghe, G.P.; Ruhunage, L.; Ratnayake, N.P.; Ratnayake, A.S.; Samaradivakara, G.V.I.; Jayaratne, R. Influence of manmade effects on geomorphology, bathymetry and coastal dynamics in a monsoon-affected river outlet in Southwest coast of Sri Lanka. *Environ. Earth Sci.* **2021**, *80*, 238. [CrossRef]
18. Bhattacharya, J.P.; Giosan, L. Wave-influenced deltas: Geomorphological implications for facies reconstruction. *Sedimentology* **2003**, *50*, 187–210. [CrossRef]
19. Van Maren, D.S. Barrier formation on an actively prograding delta system: The Red River delta, Vietnam. *Mar. Geol.* **2005**, *224*, 123–143. [CrossRef]
20. Ashton, A.D.; Giosan, L. Wave-angle control of delta evolution. *Geophys. Res. Lett.* **2011**, *38*, L13405. [CrossRef]
21. Anthony, E.J. Wave influence in the construction, shaping and destruction of river deltas: A review. *Mar. Geol.* **2015**, *361*, 53–78. [CrossRef]
22. Dan, S.; Stive, M.J.; Walstra, D.J.R.; Panin, N. Wave climate, coastal sediment budget and shoreline changes for the Danube Delta. *Mar. Geol.* **2009**, *262*, 39–49. [CrossRef]
23. Ward, E.M. *English Coastal Evolution*; Methuen & Co. Ltd.: London, UK, 1922; 105p.
24. Lovegrove, C.H. Old shorelines near Camber Castle. *Geog. J.* **1953**, *119*, 200. [CrossRef]
25. Zenkovich, V.P. *Processes of Coastal Development*; Wiley-Interscience: New York, NY, USA, 1967; 751p.
26. Kunte, P.D.; Wagle, B. Spit evolution and shore drift direction along South Karnataka Coast, India. *G. Giol.* **1991**, *153*, 71–80.
27. FitzGerald, D.M.; Kraus, N.C.; Hands, E.B. *Natural Mechanisms of Sediment Bypassing at Tidal Inlets, Coastal and Hydraulics Engineering Technical Note IV–30*; United States Army Engineer Research and Development Center: Vicksburg, MS, USA, 2000; 10p.
28. Hume, T.M.; Herdendorf, C.E. Factors congtrolling tidal inlet characteristics on low drift coasts. *J. Coast. Res.* **1992**, *8*, 355–375.

29. Aubrey, D.G.; Gaines, A.G. Rapid formation and degradation of barrier spits in areas with low rates of littoral drift. *Mar. Geol.* **1982**, *49*, 257–277. [CrossRef]
30. Alcántara-Carrió, J.; Dinkel, T.M.; Portz, L.; Mahiques, M.M. Two new conceptual models for formation and degradation of baymouth spits by longshore drift and fluvial discharge (Iguape, SE Brazil). *Earth Surf. Process. Landf.* **2018**, *43*, 695–709. [CrossRef]
31. Zenkovitch, V.P. On the genesis of the cuspate spits along lagoon shores. *J. Geol.* **1959**, *67*, 269–277. [CrossRef]
32. Jensen, J.B.; Stecher, O. Paraglacial barrier-lagoon development in the late Pleistocene Baltic Ice Lake, southwestern Baltic. *Mar. Geol.* **1992**, *107*, 81–101. [CrossRef]
33. Boggs, D.A.; Boggs, G.S.; Eliot, I.; Knott, B. Regional patterns of salt lake morphology in the lower Yarra Yarra drainage system of Western Australia. *J. Arid Environ.* **2006**, *64*, 97–115. [CrossRef]
34. Mahanty, M.M.; Mohanty, P.K.; Pradhan, S.; Samal, R.N.; Ranga Rao, V. Spit and inlet morphodynamics of a tropical coastal lagoon. *Mar. Geod.* **2019**, *42*, 130–165. [CrossRef]
35. Rosen, P.S. Origin and processes of cuspate spit shorelines. In *Estuarine and Research. Vol II Geology and Engineering*; Academic Press: New York, NY, USA, 1975; pp. 77–92.
36. Alcántara-Carrió, J.; Caicedo, A.C.; Hernández, J.; Jaramillo-Velez, A.; Manzolli, R.P. Sediment bypassing from the new human-induced lobe to the ancient lobe of the Turbo delta (Gulf of Urabá; southern Caribbean Sea). *J. Coast. Res.* **2019**, *35*, 196–209.
37. O'Shea, M.; Murphy, J. Predicting and monitoring the evolution of a coastal barrier dune system postbreaching. *J. Coast. Res.* **2013**, *29*, 38–50. [CrossRef]
38. Avinash, K.; Deepika, B.; Jayappa, K.S. Evolution of spit morphology: A case study using a remote sensing and statistical based approach. *J. Coast. Conserv.* **2013**, *17*, 327–337. [CrossRef]
39. Alcántara-Carrió, J.; Mahiques, M.M.; Portz, L. Paired baymouth spits. In *Encyclopedia of Coastal Sciences*; Finkl, C.W., Makowski, C., Eds.; Springer International Publishing: Cham, Switzerland, 2018; pp. 1–7. [CrossRef]
40. Cencini, C. Physical process and human activities in the evolution of Po River delta, Italy. *J. Coast. Res.* **1998**, *14*, 774–793.
41. Gabbianelli, G.; Del Grande, C.; Simeoni, U.; Zamariolo, A.; Calderoni, G. Evoluzione dell'area di Goro negli ultimi cinque secoli (Delta del Po). *Studi Costieri* **2000**, *2*, 45–63.
42. Simeoni, U.; Fontolan, G.; Tessari, U.; Corbau, C. Domains of spit evolution in the Goro area, Po River delta, Italy. *Geomorphology* **2007**, *86*, 332–348. [CrossRef]
43. Simeoni, U.; Fontolan, G.; Dal Cin, R.; Calderoni, G.; Zamariolo, A. Dinamica sedimentaria dell'area di Goro (Delta del Po). *Studi Costieri* **2000**, *2*, 139–151.
44. Simeoni, U.; Dal Cin, R.; Fontolan, G.; Tessari, U. Morfogenesi ed evoluzione dello Scanno di Goro (Delta del Po). *Studi Costieri* **2000**, *2*, 5–20.
45. Guilcher, A. Angola. In *Encyclopedia of the World's Coastal Landforms*; Bird, E.C.F., Ed.; Springer: Dordrecht, The Netherlands, 2010; pp. 963–967. [CrossRef]
46. Bird, E.C.F. (Ed.) New Brunswick and Nova Scotia. In *Encyclopedia of the World's Coastal Landforms*; Springer: Dordrecht, The Netherlands, 2010; pp. 141–154.
47. Zimmermann, M.; Prescott, M.M. False Pass, Alaska: Significant changes in depth and shoreline in the historic time period. *Fish. Oceanogr.* **2021**, *30*, 264–279. [CrossRef]
48. Gad, F.K.; Hatiris, G.A.; Loukaidi, V.; Dimitriadou, S.; Drakopoulou, P.; Sioulas, A.; Kapsimalis, V. Long-term shoreline displacements and coastal morphodynamic pattern of north Rhodes Island, Greece. *Water* **2018**, *10*, 849. [CrossRef]
49. Moghaddam, E.I.; Allahdadi, M.N.; Ashrafi, A.; Chaichitehrani, N. Coastal system evolution along the southern Caspian Sea coast using satellite image analysis: Response to the sea level fall during 1994-2-15. *Arab. J. Geosci.* **2021**, *14*, 771. [CrossRef]
50. Jose, F.; Carlin, F. Storm-Driven Morphodynamics of a Sandy Beach in Florida. *J. Coast. Res.* **2022**, *38*, 896–907. [CrossRef]
51. Taveneau, A.; Almar, R.; Bergsma, E.W.; Sy, B.A.; Ndour, A.; Sadio, M.; Garlan, T. Observing and predicting coastal erosion at the Langue de Barbarie sand spit around Saint Louis (Senegal, West Africa) through satellite-derived digital elevation model and shoreline. *Remote Sen.* **2021**, *13*, 2454. [CrossRef]
52. Google Earth Pro. 2023. Available online: http://www.earth.google.com (accessed on 11 April 2023).
53. Harvey, N.; Gross, A.M.; Jose, F.; Savarese, M.; Missimer, T. Geomorphologic impact of Hurricane Irma on Marco Island, southwest Florida. *Nat. Hazards* **2021**, *3*, 1–17. [CrossRef]
54. Navionic ChartViewer. 2023. Available online: https://webapp.navionics.com (accessed on 11 April 2023).
55. Maicu, F.; Alessandri, J.; Pinardi, N.; Verri, G.; Umgiesser, G.; Lovo, S.; Turolla, S.; Paccagnella, T.; Valentini, A. Downscaling with an unstructured coastal-ocean model to the Goro Lagoon and the Po River delta branches. *Front. Mar. Sci.* **2021**, *8*, 647781. [CrossRef]
56. Smith, W.H.F.; Sandwell, D.T. Global sea floor topography from satellite altimetry and ship depth soundings. *Science* **1997**, *277*, 1956–1962. [CrossRef]
57. Hersbach, H.; Bell, B.; Berrisford, P.; Biavati, G.; Horányi, A.; Muñoz Sabater, J.; Nicolas, J.; Peubey, C.; Radu, R.; Rozum, I.; et al. *ERA5 Hourly Data on Single Levels from 1940 to Present*; Copernicus Climate Change Service (C3S) Climate Data Store (CDS): Reading, UK, 2023. [CrossRef]

58. Pe'eri, S.; Keown, P.; Snyder, L.P.; Gonsalves, M.; Nyberg, J. Reconnaissance surveying of Bechevin Bay, AK using satellite-derived bathymetry. In Proceedings of the US Hydrographic Conference 2015, National Harbor, MD, USA, 16–19 March 2015.
59. Flater, D. WXTide32. 2007. Available online: http://www.wxtide32.com (accessed on 11 April 2023).
60. Canadian Hydrographic Service. 2023. Available online: https://tides.gc.ca/en/stations (accessed on 11 April 2023).
61. Davies, J.L. A morphogenetic approach to world shorelines. *Z. Geomorphol.* **1964**, *8*, 1–42.
62. Woodroffe, C.D. *Coasts: Form, Process and Evolution*; Cambridge University Press: Cambridge, UK, 2003; 640p.
63. *GloVis*; 2023. Available online: http://glovis.usgs.gov (accessed on 11 April 2023).
64. Alcántara-Carrió, J.; Fontán Bouzas, A.; Albarracín, S.; Correa, I.D.; Montoya Montes, I.; Mahiques, M. Geomorphological coastal classification after natural processes and human disturbance. *Oceanography* **2014**, *2*, e108. [CrossRef]
65. Petersen, D.; Deigaard, R.; Fredsøe, J. Modelling the morphology of sandy spits. *Coast. Eng.* **2008**, *55*, 671–684. [CrossRef]
66. Kim, D.; Jo, J.; Nam, S.I.; Choi, K. Morphodynamic evolution of paraglacial spit complexes on a tide-influenced Arctic fjord delta (Dicksonfjorden, Svalbard). *Mar. Geol.* **2022**, *447*, 106800. [CrossRef]
67. Carr, A.P. Aspects of spit development and decay: The estuary of the River Ore, Suffolk. *Field Stud.* **1972**, *3*, 633–653.
68. Carr, A.P. The estuary of the river Ore, Suffolk: Three decades of change in a longer-term context. *Field Stud.* **1986**, *6*, 439–458.
69. Fox, D.; Pontee, N.I.; Fisher, E.; Box, S.; Rogers, J.R.; Reeve, D.E.; Chadwick, A.J.; Sims, P. Spits and flood risk: The Exe estuary. In Proceedings of the Flood and Coastal Management Conference, Manchester, UK, 1–3 July 2008; pp. 1–10.
70. Harlow, D.A. The littoral sediment budget between Selsey Bill and Gilkicker Point, and its relevance to coast protection works on Hayling Island. *Q. J. Eng. Geol. Hydrogeol.* **1979**, *12*, 257–265. [CrossRef]
71. Kidson, C. Dawlish Warren: A study of the evolution of the sand spits across the mouth of the River Exe in Devon. *Trans. Papers Inst. Br. Geogr.* **1950**, *16*, 69–80. [CrossRef]
72. Kidson, C. The growth of sand and shingle spits across estuaries. *Z. Geomorphol.* **1963**, *7*, 1–22.
73. Devoy, R.J.N. The development and management of the Dingle Bay spit-barriers of Southwest Ireland. In *Sand and Gravel Spits*; Randazzo, G., Jackson, D., Cooper, A., Eds.; Coastal Research Library 12; Springer: Cham, Switzerland, 2015; pp. 139–180. [CrossRef]
74. Rodríguez-Ramírez, A.; Morales, J.A.; Delgado, I.; Cantano, M. The impact of man on the morphodynamics of the Huelva coast. *J. Iber. Geol.* **2008**, *34*, 313–327.
75. Robinson, A.H.W. The harbor entrances of Poole, Christchurch and Pagham. *Geog. J.* **1955**, *121*, 33–50. [CrossRef]
76. Levoy, F.; Monfort, O.; Anthony, E.J. Multi-decadal mobility of a managed sandy tidal coast (Normandy, France): Behavioural variability in a context of sea-level rise and increasing storm intensity. *Reg. Stud. Mar. Sci.* **2023**, *62*, 102973. [CrossRef]
77. Boyd, R.; Dalrymple, R.W.; Zaitlin, B.A. Classification of coastal sedimentary environments. *Sedim. Geol.* **1992**, *80*, 139–150. [CrossRef]
78. Ciavola, P.; Corbau, C.; Cibin, U.; Perini, L. Mapping of the coastal zone of the Emilia-Romagna region using geographical information systems. In Proceedings of the Sixth International Conference on the Mediterranean Coastal Environment, Ravenna, Italy, 7–11 October 2003; Volume 3, pp. 2363–2374.
79. Short, A.D. Beach dynamics and nearshore morphology of the Alaskan Arctic coast, 2498. In *LSU Historical Dissertations and Theses*; Louisiana State University and Agricultural & Mechanical College: Baton Rouge, LA, USA, 1973; 140p. Available online: https://digitalcommons.lsu.edu/gradschool_disstheses/2498 (accessed on 20 March 2023).
80. Anthony, E.J.; Marriner, N.; Morhange, C. Human influence and the changing geomorphology of Mediterranean deltas and coasts over the last 6000 years: From progradation to destruction phase? *Earth Sci. Rev.* **2014**, *139*, 336–361. [CrossRef]
81. Coleman, J.M.; Gagliano, S.M. Cyclic sedimentation in the Mississippi river deltaic plain. *Gulf Coast Ass. Geol. Societ. Trans.* **1964**, *14*, 67–80.
82. Wu, X.; Bi, N.; Kanai, Y.; Saito, Y.; Zhang, Y.; Yang, Z.; Fan, D.; Wang, H. Sedimentary records off the modern Huanghe (Yellow River) delta and their response to deltaic river channel shifts over the last 200 years. *J. Asian Earth Sci.* **2015**, *108*, 68–80. [CrossRef]
83. Elliott, T. Interdistributary bay sequences and their genesis. *Sedimentology* **1974**, *21*, 611–622. [CrossRef]
84. Tye, R.S.; Kosters, E.C. Styles of interdistributary basin sedimentation: Mississippi delta plain, Louisiana. *Trans. Gulf Coast Assoc. Geol. Soc.* **1986**, *36*, 575–588.
85. Coleman, J.M. Dynamic changes and processes in the Mississippi River delta. *Geol. Soc. Am. Bull.* **1988**, *100*, 999–1015. [CrossRef]
86. Overduin, P.P.; Strzelecki, M.C.; Grigoriev, M.N.; Couture, N.; Lantuit, H.; St-Hilaire-Gravel, D.; Günther, F.; Wetterich, S. Coastal changes in the Arctic. *Geol. Soc. Lond. Spec. Publ.* **2014**, *388*, 103–129. [CrossRef]
87. Héquette, A.; Ruz, M.H. Spit and barrier islands migration in the southeastern Canadian Beafourt Sea. *J. Coast. Res.* **1991**, *7*, 677–698.
88. Gibbs, A.E.; Richmond, B.M. *National Assessment of Shoreline Change: Historical Shoreline Change along the North Coast of Alaska, US–Canadian Border to Icy Cape*; US Department of the Interior, US Geological Survey: Washington, DC, USA, 2015; 110p.
89. Xue, P.; Chen, C.; Beardsley, R.C.; Limeburner, R. Observing system simulation experiments with ensemble Kalman filters in Nantucket Sound, Massachusetts. *J. Geophys. Res. Oceans* **2011**, *116*, C01011. [CrossRef]
90. Piller, W.E.; Pervesler, P.; Golebiowski, R.; Kleemann, K.; Mansour, A.; Rupp, C. The northern Bay of Safaga (Red Sea, Egypt): An actuopalaeontological approach. 1. Topography and bottom facies. *Beitr. Paläont. Österr* **1989**, *15*, 103–147.

91. Marriner, N.; Goiran, J.P.; Morhange, C. Alexander the Great's tombolos at Tyre and Alexandria, eastern Mediterranean. *Geomorphology* **2008**, *100*, 377–400. [CrossRef]
92. Héquette, A.; Ruz, M.H.; Hill, P.R. The effects of the Holocene sea level rise on the evolution of the southeastern coast of the Canadian Beaufort Sea. *J. Coast. Res.* **1995**, *11*, 494–507.
93. Ward, R.D. Sedimentary response of Arctic coastal wetlands to sea level rise. *Geomorphology* **2020**, *370*, 107400. [CrossRef]
94. Da Lio, C.; Tosi, L. Vulnerability to relative sea-level rise in the Po River delta (Italy). *Estuar. Coast. Shelf Sci.* **2019**, *228*, 106379. [CrossRef]
95. Hammar-Klose, E.S.; Pendleton, E.A.; Thieler, E.R.; Williams, S.J.; Norton, G.A. *Coastal Vulnerability Assessment of Cape Cod National Seashore (CACO) to Sea-Level Rise*; Open File Report, 02-233; US Geological Survey: Reston, VA, USA, 2003; 18p.
96. Orford, J.D.; Carter, R.W.G.; Jennings, S.C. Control domains and morphological phases in gravel-dominated coastal barriers of Nova Scotia. *J. Coast. Res.* **1996**, *12*, 589–604.
97. Nienhuis, J.H.; Ashton, A.D.; Nardin, W.; Fagherazzi, S.; Giosan, L. Alongshore sediment bypassing as a control on river mouth morphodynamics. *J. Geophys. Res. Earth Surf.* **2016**, *121*, 664–683. [CrossRef]
98. Bezzi, A.; Casagrande, G.; Martinucci, D.; Pillon, S.; Del Grande, C.; Fontolan, G. Modern sedimentary facies in a progradational barrier-spit system: Goro lagoon, Po River delta, Italy. *Estuar. Coast. Shelf Sci.* **2019**, *227*, 106323. [CrossRef]
99. Maicu, F.; De Pascalis, F.; Ferrarin, C.; Umgiesser, G. Hydrodynamics of the Po river-delta-sea system. *J. Geophys. Res. Oceans* **2018**, *123*, 6349–6372. [CrossRef]
100. Bezzi, A.; Pillon, S.; Popesso, C.; Casagrande, G.; Da Lio, C.; Martinucci, D.; Tosi, L.; Fontolan, G. From rapid coastal collapse to slow sedimentary recovery: The morphological ups and downs of the modern Po River delta. *Estuar. Coast. Shelf Sci.* **2021**, *260*, 107499. [CrossRef]
101. Penland, S.; Boyd, R.; Suter, J.R. Transgressive depositional systems of the Mississippi Delta Plain: A model for barrier shoreline and shelf sand development. *J. Sediment. Petrol.* **1988**, *58*, 932–949.
102. Corbau, C.; Simeoni, U.; Zoccarato, C.; Mantovani, G.; Teatini, P. Coupling land use evolution and subsidence in the Po River delta, Italy: Revising the past occurrence and prospecting the future management challenges. *Sci. Total Environ.* **2019**, *654*, 1196–1208. [CrossRef]
103. Black, K.P.; Andrews, C.J. Sandy shoreline response to offshore obstacles Part 1: Salient and tombolo geometry and shape. *J. Coast. Res.* **2001**, *29*, 82–93.
104. Davies, J.L. *Geographical Variation in Coastal Development*; Longman: London, UK; New York, NY, USA, 1980; 212p.
105. Dally, W.R.; Pope, J. *Detached Breakwaters for Shore Protection*; Technical Report CERC-86-1; U.S. Army Corps of Engineers, Coastal Engineering Research Center, Waterways Experiment Station: Vicksburg, MS, USA, 1986; 62p.
106. Van Rijn, L.C. Coastal erosion and control. *Ocean Coast. Manag.* **2011**, *54*, 867–887. [CrossRef]
107. Alcántara-Carrió, J.; Sasaki, D.; Mahiques, M.; Taborda, R.; Souza, L.A.P. Sedimentary constraints on the development of a narrow deep strait (São Sebastião Channel, SE Brazil). *Geo-Mar. Lett.* **2017**, *37*, 475–488. [CrossRef]
108. Gourlay, M.R. Beach processes in the vicinity of offshore breakwaters. In Proceedings of the 5th Australasian Conference on Coastal and Ocean Engineering, Perth, Australia, 25–27 November 1981; pp. 132–137.
109. Sterr, H. Comparative studies of coastal erosion in the FRG. *J. Coast. Res.* **1990**, *Sp.I. 9*, 821–837.
110. Hofstede, J. Coastal flood defense and coastal protection along the Baltic Sea coast of Schleswig-Holstein. *Die Küste Arch. Res. Technol. North Sea Balt. Coast* **2008**, *74*, 170–178.
111. Giese, G.S.; Borrelli, M.; Mague, S.T. Tidal inlet evolution and impacts of anthropogenic alteration: An example from Nauset Beach and Pleasant Bay, Cape Cod, Massachusetts. *Northeastern Nat.* **2020**, *27* (Suppl. S10), 1–21. [CrossRef]
112. Stéphan, P.; Suanez, S.S.; Fichaut, B. Long-term morphodynamic evolution of the Sillon de Talbert gravel barrier (Brittany, France). *Shore Beach* **2012**, *80*, 19–36.
113. Vespremeanu-Stroe, A.; Preoteasa, L. Morphology and the cyclic evolution of Danube delta spits. In *Sand and Gravel Spits*; Randazzo, G., Jackson, D., Cooper, A., Eds.; Coastal Research Library 12; Springer: Cham, Switzerland, 2015; pp. 327–339.
114. Nuyts, S.; O'Shea, M.; Murphy, J. Monitoring the Morphodynamic Cannibalization of the Rossbeigh Coastal Barrier and Dune System over a 19-Year Period (2001–2019). *J. Mar. Sci. Eng.* **2020**, *8*, 421. [CrossRef]
115. Aubrey, D.G.; Speer, P.E. Updrift migration of tidal Inlets. *J. Geol.* **1984**, *92*, 531–545. [CrossRef]
116. Peterson, C.D.; Murillo-Jiménez, J.M.; Stock, E.; Price, D.M.; Hostetler, S.W.; Percy, D. Origins of late-Pleistocene coastal dune sheets, Magdalena and Guerrero Negro, from continental shelf low-stand supply (70–20 ka), under conditions of southeast littoral-and eolian-sand transport, in Baja California Sur, Mexico. *Aeolian Res.* **2017**, *28*, 13–28. [CrossRef]
117. Murillo de Nava, J.M.; Gorsline, D.S. Holocene and modern dune morphology for the Magdalena coastal plain and islands, Baja California Sur, Mexico. *J. Coast. Res.* **2000**, *16*, 915–925.
118. Bruun, P.; Mehta, A.; Jonsson, I.G. *Stability of Tidal Inlets: Theory and Engineering*; Developments in Geotechnical Engineering; Elsevier: Amsterdam, The Netherlands, 1978; Volume 23, 510p.
119. Nienhuis, J.H.; Ashton, A.D. Mechanics and rates of tidal inlet migration: Modeling and application to natural examples. *J. Geophys. Res. Earth Surf.* **2016**, *121*, 2118–2139. [CrossRef]
120. Lynch-Blosse, M.A.; Kumar, N. Evolution of downdrift-offset tidal inlets: A model based on the Brigantine Inlet system of New Jersey. *J. Geol.* **1976**, *84*, 165–178. [CrossRef]

121. Qi, Y.; Yu, Q.; Gao, S.; Li, Z.; Fang, X.; Guo, Y. Morphological evolution of river mouth spits: Wave effects and self-organization patterns. *Estuar. Coast. Shelf Sci.* **2021**, *262*, 107567. [CrossRef]
122. Chi, S.; Zhang, C.; Wang, P.; Shi, J.; Li, F.; Li, Y.; Wang, P.; Zheng, J.; Sun, J.; Nguyen, V.T. Morphological evolution of paired sand spits at the Fudu river mouth: Wave effects and anthropogenic factors. *Mar. Geol.* **2023**, *456*, 106991. [CrossRef]
123. Leatherman, S.P. Reworking of glacial outwash sediments along outer Cape Cod: Development of Provincetown spit. In *Glaciated Coasts*; FitzGerald, D.M., Roshan, P.S., Eds.; Academic: London, UK, 1987; pp. 447–464.
124. Leatherman, S.P.; Zaremba, R.E. Dynamics of a northern barrier beach: Nauset Spit, Cape Cod, Massachusetts. *Geol. Soc. Am. Bull.* **1986**, *97*, 116–124. [CrossRef]
125. Giese, G.S.; Adams, M.B.; Rogers, S.S.; Dingman, S.L.; Borrelli, M.; Smith, T.L. Coastal Sediment Transport on outer Cape Cod Massachusetts: Observations and Theory. In Proceedings of the Coastal Sediments, Miami, FL, USA, 2–6 May 2011; Rosati, J.D., Wang, P., Roberts, T.M., Eds.; World Scientific Pub Co Inc.: Miami, FL, USA, 2011; pp. 2353–2365.
126. Borrelli, M.; Giese, G.S.; Mague, S.T.; Smith, T.L.; Mittermayer, A.; Legare, B.J.; Solazzo, D. *Potential Impacts to the Nauset Barrier from the Proposed Dredging and Disposal in Nauset Harbor. A Technical Report prepared for the Town of Eastham*; Tech Rep: 19-CL07; Center for Coastal Studies: Provincetown, MA, USA, 2019; 19p.
127. Giese, G.S.; Aubrey, D.G.; Liu, J.T. *Development, Characteristics, and Effects of the New Chatham Harbor Inlet*; Woods Hole Oceanographic Institution: Woods Hole, MS, USA, 1989; 33p.
128. FitzGerald, D.M.; Pendleton, E. Inlet formation and evolution of the sediment bypassing system: New Inlet, Cape Cod, Massachusetts. *J. Coast. Res.* **2002**, *36*, 290–299. [CrossRef]
129. Borrelli, M.; Oakley, B.A.; Giese, G.S.; Boothroyd, J.C. Inlet formation as a result off hydraulic inefficiency leading to further inlet instability. In Proceedings of the Coastal Sediments, Miami, FL, USA, 2–6 May 2011; pp. 519–532.
130. Berman, G. *Longshore Sediment Transport Cape Cod Massachusetts*; Woods Hole Oceanographic Institute & Cape Cod Cooperative Extension: Woods Hole, MS, USA, 2011; 47p.
131. NASA. Coastline Change. NASA Earth Observatory. 2020. Available online: https://earthobservatory.nasa.gov/world-of-change/CapeCod (accessed on 11 April 2023).
132. Giese, G.S.; Mague, S.T.; Rogers, S.S. *A Geomorphological Analysis of Nauset Beach/Pleasant Bay/Chatham Harbor for the Purpose of Estimating Future Configurations and Conditions*; Prepared for the Pleasant Bay Resource Management Alliance; Pleasant Bay Alliance: Harwich, MA, USA, 2009; 32p.
133. Oldale, R.N.; Friedman, J.D.; Williams, R.S. *Changes in Coastal Morphology of Monomoy Island, Cape Cod, Massachusetts U.S. Geological Survey Prof. Paper 750*; U.S. Geological Survey: Reston, VA, USA, 1971; Chapter B; pp. 101–107.
134. Bird, E.C.F. (Ed.) Heidelberg. In *Encyclopedia of the World's Coastal Landforms*; Springer: Dordrecht, The Netherlands, 2010; pp. 641–644.
135. Serizawa, M.; Uda, T.; Miyahara, S. Prediction of formation of recurved sand spit using BG model. In Proceedings of the 36th Coastal Engineering Conference, Baltimore, MD, USA, 30 July–3 August 2018; Volume 36, p. 24.
136. Liu, H. Dynamic changes of coastal morphology following the 2011 Tohoku tsunami. In Proceedings of the 7th International Conference on Asian and Pacific Coasts, Bali, Indonesia, 24–26 September 2013; pp. 594–601.
137. Ruiz, F.; Rodríguez-Vidal, J.; Abad, M.; Cáceres, L.M.; Carretero, M.I.; Pozo, M.; Rodríguez-Llanes, J.M.; Gómez-Toscano, F.; Izquierdo, T.; Font, E.; et al. Sedimentological and geomorphological imprints of Holocene tsunamis in southwestern Spain: An approach to establish the recurrence period. *Geomorphology* **2013**, *203*, 97–104. [CrossRef]
138. Wright, L.D.; Short, A.D. Morphodynamic variability of surf zones and beaches: A synthesis. *Mar. Geol.* **1984**, *56*, 93–118. [CrossRef]
139. Zăinescu, F.I.; Vespremeanu-Stroe, A.; Tătui, F. The formation and closure of the Big Breach of Sacalin spit associated with extreme shoreline retreat and shoreface erosion. *Earth Surf. Process. Landf.* **2019**, *44*, 2268–2284. [CrossRef]
140. Koiwa, N.; Takahashi, M.; Sugisawa, S.; Ito, A.; Matsumoto, H.A.; Tanavud, C.; Goto, K. Barrier spit recovery following the 2004 Indian Ocean tsunami at Pakarang Cape, southwest Thailand. *Geomorphology* **2018**, *306*, 314–324. [CrossRef]
141. Panin, N. The Danube Delta. Geomorphology and Holocene Evolution: A Synthesis. *Geomorphol. Relief Process. Environ.* **2003**, *9*, 247–262. [CrossRef]
142. Preoteasa, L.; Vespremeanu-Stroe, A.; Hanganu, D.; Katona, O.; Timar-Gabor, A. Coastal changes from open coast to present lagoon system in Histria region (Danube Delta). *J. Coast. Res.* **2013**, *65* (Suppl. S1), 564–569. [CrossRef]
143. Frihy, O.; Lawrence, D. Evolution of the modern Nile delta promontories: Development of accretional features during shoreline retreat. *Environ. Geol.* **2004**, *46*, 914–931. [CrossRef]
144. Kelman, D. Strait of Canso transmission line. In *Archaeological Screening & Reconnaissance Antigonish & Inverness Counties, Nova Scotia*; Final Report; Kelman Heritage Consulting: West LaHave, NS, Canada, 2015; 29p.
145. Elliott, E.L. Sandspit of the Otago coast. *N. Z. Geogr.* **1958**, *14*, 65–74. [CrossRef]
146. Canning, P.; Fox, D.; Pontee, N. The benefits of managing spit evolution: A case study in the Exe Estuary, UK. In *Innovative Coastal Zone Management: Sustainable Engineering for a Dynamic Coast*; Cooper, N.J., Ed.; ICE Publishing: London, UK, 2012; pp. 625–635.
147. Karunarathna, H.; Reeve, D.E.; Fox, D.; Box, S.; Pontee, N.; Chadwick, A.; Lawrence, J. Appraising spit dynamics and estuary responses: A coastal management study from the Exe Estuary, UK. In Proceedings of the 31st Coastal Engineering Conference, Hamburg, Germany, 31 August–5 September 2008; World Scientific Publishing Company: Singapore, 2009; pp. 4202–4213. [CrossRef]

148. Johnston, T.W. Sediment Supply, Sediment Transport and Long-Term Shoreline Evolution on "Open" and "Closed" Cellular Coasts: Co. Wexford and Co. Donegal, Ireland. Ph.D. Thesis, The New University of Ulster, Coleraine, Northern Ireland, UK, 1984; 348p.
149. Ruz, M.H. Impact des amenagernents sur l'evolution du littoral de Wexford, sud-est de l´Irlande. *Norois* **1987**, *34*, 261–273. [CrossRef]
150. Anthony, E.J. Patterns of sand spit development and their management implications on deltaic, drift-Aligned coasts: The cases of the Senegal and Volta River delta spits, West Africa. In *Sand and Gravel Spits*; Randazzo, G., Jackson, D., Cooper, A., Eds.; Coastal Research Library 12; Springer: Cham, Switzerland, 2015; pp. 21–36. [CrossRef]
151. Orejarena Rondón, A.F.; Afanador Franco, F.; Ramos de la Hoz, I.; Conde Frías, M.; Restrepo López, J.C. Evolución morfológica de la espiga de Galerazamba, Caribe colombiano. *Bol. Cientif. CIOH* **2015**, *33*, 123–144. [CrossRef]
152. Silva, M.S.; Guedes, C.C.F.; Silva, G.A.M.; Ribeiro, G.P. Active mechanisms controlling morphodynamics of a coastal barrier: Ilha Comprida, Brazil. *Ocean Coast. Res.* **2021**, *69*, 21004. [CrossRef]
153. Lawson, S.K.; Tanaka, H.; Udo, K.; Hiep, N.T.; Tinh, N.X. Morphodynamics and evolution of estuarine sandspits along the bight of Benin coast, West Africa. *Water* **2021**, *13*, 2977. [CrossRef]
154. Restrepo, J.D.; Kjerfve, B.; Correa, I.D.; González, J. Morphodynamics of a high discharge tropical delta, San Juan River, Pacific coast of Colombia. *Mar. Geol.* **2002**, *192*, 355–381. [CrossRef]
155. Villate, D.A.; Portz, L.; Manzolli, R.P.; Alcántara-Carrió, J. Human disturbances of shoreline morphodynamics and dune ecosystem at the Puerto Velero spit (Colombian Caribbean). *J. Coast. Res.* **2020**, *95*, 711–716. [CrossRef]
156. Castillo, M.; Muñoz-Salinas, E.; Sanderson, D.C.W.; Cresswell, A. Landscape evolution of Punta Arena sand spit (SE Baja California Peninsula, NW Mexico): Implications of ENSO on landscape erosion rates. *Catena* **2020**, *193*, 104601. [CrossRef]
157. Nahon, A.; Idier, D.; Senechal, N.; Féniès, H.; Mallet, C.; Mugica, J. Imprints of wave climate and mean sea level variations in the dynamics of a coastal spit over the last 250 years: Cap Ferret, SW France. *Earth Surf. Process. Landf.* **2019**, *44*, 2112–2125. [CrossRef]
158. Orviku, K.; Jaagus, J.; Kont, A.; Ratas, U.; Rivis, R. Increasing activity of coastal processes associated with climate change in Estonia. *J. Coast. Res.* **2003**, *19*, 364–375.
159. Łabuz, T.A. Environmental impacts—Coastal erosion and coastline changes. In *Second Assessment of Climate Change for the Baltic Sea Basin*; The BACC II Team, Ed.; Springer: Cham, Switzerland, 2015; pp. 381–396.
160. Iskander, M.M. Stability of the Northern coast of Egypt under the effect of urbanization and climate change. *Water Sci.* **2021**, *35*, 1–10. [CrossRef]
161. Ogorodov, S.A.; Baranskaya, A.V.; Belova, N.G.; Kamalov, A.M.; Kuznetsov, D.E.; Overduin, P.P.; Shabanova, N.N.; Vergun, A.P. Coastal dynamics of the Pechora and Kara Seas under changing climatic conditions and human disturbances. *Geogr. Environ. Sustain.* **2016**, *9*, 53–73. [CrossRef]
162. Nordstrom, K.F.; Jackson, N.L. Removing shore protection structures to facilitate migration of landforms and habitats on the bayside of a barrier spit. *Geomorphology* **2013**, *199*, 179–191. [CrossRef]
163. Correa, I.D.; Alcántara-Carrió, J.; González, R.D.A. Historical and recent shore erosion along the Colombian Caribbean coast. *J. Coast. Res.* **2005**, *Sp.I. 49*, 52–57.
164. Orford, J. Alternative interpretations of man-induced shoreline changes in Rosslare Bay, southeast Ireland. *Trans. Inst. Br. Geogr.* **1988**, *13*, 65. [CrossRef]
165. Zweers, S. A Study of the Erosion Problem along Rosslare Strand. Master's Thesis, Delft University of Technology, Delft, The Netherlands, 2008; 149p.
166. Simeoni, U.; Corbau, C. A review of the Delta Po evolution (Italy) related to climatic changes and human impacts. *Geomorphology* **2009**, *107*, 64–71. [CrossRef]
167. Stefani, M. The Po River delta region: Depositional evolution, climate change and human Intervention through the last 5000 years. In *Landscapes and Landforms of Italy*; World Geomorphological Landscapes, Soldati, M., Marchetti, M., Eds.; Springer: Cham, Switzerland, 2017; pp. 193–202. [CrossRef]
168. Corbau, C.; Zambello, E.; Nardin, W.; Simeoni, U. Secular diachronic analysis of coastal marshes and lagoons evolution: Study case of the Po River delta (Italy). *Estuar. Coast. Shelf Sci.* **2022**, *268*, 107781. [CrossRef]
169. Vincenzi, A. Coastline Changes in Veneto Region (Italy) from 2012 to 2018 by Means of Multitemporal Orthophotos. Master's Thesis, University of Padua, Padua, Italy, 2022; 118p.
170. Kosyan, R.D.; Krylenko, M.V. Modern state and dynamics of the Sea of Azov coasts. *Estuar. Coast. Shelf Sci.* **2019**, *224*, 314–323. [CrossRef]
171. Hayashi, K.; Hashimoto, K.; Yagisawa, K.; Kobayashi, N. Beach morphologies at Notsukezaki sand spit, Japan. In Proceedings of the 32nd Coastal Engineering Conference, Shanghai, China, 30 June–5 July 2010; p. 2.
172. Itori, S.; Yagisawa, K.; Sasaki, T.; Yanaguchi, R.; Kobayashi, N. Storm-induced erosion on Notsuzezaki sand spit. In Proceedings of the 36th Coastal Engineering Conference, Baltimore, MD, USA, 30 July–3 August 2018; Volume 1, p. 79. [CrossRef]
173. Zenkovich, V.P. *The Shores of the Black and Azov Seas*; Geographical Press: Moscow, Russia, 1958; 374p. (In Russian)
174. Pontee, N.I.; Townend, I.H.; Chesher, T.; McLaren, P. To Breach or not to Breach? Spit dynamics and coastal management. In Proceedings of the Coastal Engineering 2002 Conference: Solving Coastal Conundrums, Cardiff, UK, 7–12 July 2002; Smith, J.K., Ed.; World Scientific Publishing Co. Pte. Ltd.: Cardiff, Wales, UK, 2003; pp. 3799–3811.

175. Stéphan, P.; Suanez, S.; Fichaut, B.; Autret, R.; Blaise, E.; Houron, J.; Ammann, J.; Grandjean, P. Monitoring the medium-term retreat of a gravel spit barrier and management strategies, Sillon de Talbert (North Brittany, France). *Ocean Coast. Manag.* **2018**, *158*, 64–82. ISSN 0964-5691. [CrossRef]
176. Vitousek, S.; Buscombe, D.; Vos, K.; Barnard, P.L.; Ritchie, A.C.; Warrick, J.A. The future of coastal monitoring through satellite remote sensing. *Cambridge Prism. Coast. Futur.* **2023**, *1*, e10. [CrossRef]

remote sensing

MDPI

Article

Holocene Erosional Processes in a Highly Exposed Intertidal Sandstone Reef Inferred from Remote Sensing Data

Nicolás Ferrer [1,*], Kella Santana [2], Javier Martín [1], José Valdazo [2] and Oscar Bergasa [2]

1 Instituto de Oceanografía y Cambio Global, University of Las Palmas de Gran Canaria (IOCAG-ULPGC), Parque Científico Tecnológico Marino de Taliarte, s/n., 35214 Telde, Spain; javier.martin@ulpgc.es

2 Elittoral Estudios de Ingeniería Costera y Oceanográfica SLNE, Edificio Polivalente II, Parque Científico Tecnológico, Campus Universitario de Tafira, 35017 Las Palmas, Spain; ksantana@elittoral.es (K.S.); jvaldazo@elittoral.es (J.V.); obergasa@elittoral.es (O.B.)

* Correspondence: nicolas.fvg@ulpgc.es

Abstract: An intertidal sandstone reef, named *barra* de Las Canteras, protects the western coast of Las Palmas de Gran Canaria city (Canary Islands, Spain). The beach-reef system of Las Canteras constitutes one of the most valuable coastal geomorphological sites in the archipelago. Stratigraphic studies have identified the formation of the reef in the Last Interglacial (MIS 5e) in a coastal sedimentary paleo-environment. The rock structure is highly exposed to the Atlantic swell and consists mainly of a sandstone beachrock with a medium resistance to erosional processes. However, the historical and current erosion rates and the original extent of the reef are not known to date. This paper explores the geomorphological structure of the reef by combining a topo-bathymetric analysis (obtained by differential GPS, multibeam echosounder and hyperspectral sensor) and the analysis of geomorphological features on high-resolution images, obtained with a hyperspectral camera mounted on a UAV. The results provide a comprehensive, high-resolution image of the subaerial and submerged morphology of the reef. The structure reflects the distribution of erosional fronts and the existence of collapsing submarine blockfields and nearshore, uneroded, remnant reliefs. Detailed analysis of these features allows to estimate the probable original extent of the sandstone reef and to relate the erosional retreat processes to the sea-level dynamics during the Holocene.

Keywords: sandstone reef; beachrock; coastal erosion; Las Canteras; Canary Islands

Citation: Ferrer, N.; Santana, K.; Martín, J.; Valdazo, J.; Bergasa, O. Holocene Erosional Processes in a Highly Exposed Intertidal Sandstone Reef Inferred from Remote Sensing Data. *Remote Sens.* **2023**, *15*, 2968. https://doi.org/10.3390/rs15122968

Academic Editor: Paolo Ciavola

Received: 10 April 2023
Revised: 29 May 2023
Accepted: 5 June 2023
Published: 7 June 2023

1. Introduction

Sandstone reefs are important paleoenvironmental markers and natural protectors against erosion in tropical coastal regions [1]. Despite their ecological importance, they have received much less scientific attention than other similar formations, such as coral reefs [2]. The intertidal sandstone reefs of northeastern Brazil are probably the best known and most studied in the world [1,3,4]. In the State of Pernanmuco, intertidal beachrocks crops out parallel to the coast for tens of kilometers. They are mostly composed of parallel stratified layers of calcarenites with 4–5° seaward dips, to which a Holocene age is attributed [4,5]. Their unusual development in this part of the world has motivated their valuation as important geoheritage and geotourism elements [1,6,7]. A remarkable subtropical sandstone reef also occurs in the Canary Islands. The sandstone reef of Las Canteras is up to six meters thick and has subhorizontal parallel laminations that dip gently seaward, so it has been identified as a beachrock formation [8,9]. The sandy grains are strongly compacted by carbonate cement and contain abundant remains of marine organisms, mainly mollusks, algae and foraminifera [10]. Due to its stratigraphic similarity with other sandstone outcrops and marine conglomerates of the eastern Canaries, it has traditionally been considered part of a larger sedimentary unit called Terraza Baja de Las Palmas [11] or Rasa Jandiense [12], dated to the Last Interglacial MIS 5e.

In many coasts, natural reefs are interposed offshore, safeguarding the sedimentary stability of the sandy beaches [13,14]. Rocky and erosional coastal environments support sedimentary deposits and beaches of very diverse characteristics, controlled by complex geological structures and their erosion-deposition dynamics [15]. According to Gallop et al. [16], the existence of rock controls, rather than being infrequent, is the rule on sandy beaches on coasts worldwide. Several studies have pointed out the determinant influence of hard rock structures, such as reefs, headlands or underlying outcrops, on the morphodynamics of sandy beaches [16–20]. They act as protectors against coastal erosion, generating strong energy gradients between exposed and protected sectors [18,21]. Inherited geological structures introduce complexity into wave-breaking patterns, currents and sedimentary dynamics, cross-shore and alongshore, causing mismatches between theoretical morphodynamic models and the actual beach morphologies observed in the environments [22].

A complete understanding of the morphodynamic behavior of rock-controlled coastal systems requires a simultaneous examination of subaerial and subaqueous geomorphic transformations. Topo-bathymetric studies have become frequent, especially in the monitoring of rapid response sedimentary systems, such as beaches [20]. The monitoring of the submerged surfaces presents a greater technical challenge than the monitoring of land surfaces. Traditional techniques have used boat-borne echosounders [23,24], but high costs often do not allow for high-recurrence surveys and they are not operational in shallow intertidal areas. The need for more versatile and less expensive alternative techniques has led to the development of bathymetric procedures, such as video monitoring [20,25] or hyperspectral sensing [26,27]. The results of these new techniques have generally proved to be reliable and represent new tools for research.

Reefs can dissipate much of the incident wave energy on coasts [28,29], which triggers erosion on reef structures, especially during storm events. The progressive erosion of tropical and subtropical sandstone reefs can produce a chain of very negative socio-ecological consequences. From the loss of the ecosystem services linked to the natural protection of the reefs, a drastic alteration of the hydrodynamic regime and a serious impact on beach morphology and marine ecosystems can be expected. Despite the important geodynamic implications, especially for the configuration and geomorphological stability of many sedimentary coasts, the erosion of highly exposed rocky reefs is a poorly investigated phenomenon to date. From the knowledge of the morphodynamics of rocky shores, we know that coastal rock platforms are dynamic and eroded by mechanical and abrasive wave action, as well as by other physical, chemical and biological weathering processes [30,31]. Shore platforms are primarily eroded in two ways: by a progressive micro-erosional lowering [32–35] and by sporadic, larger block detachments usually deposited on the upper platform and reworked by storm waves [36–38]. Moreover, although the movement of non-cohesive sediments can occur at greater depths, erosion of solid rock by waves is strongly concentrated in the break zone, near the still waters and intertidal zone [39–42].

The planning of conservation actions to avoid the future degradation of relevant natural elements such as sandstone reefs, requires a better understanding of the magnitude and forms of past and present erosional processes. In order to contribute to this knowledge, the aim of this work was to obtain, through a combination of different instruments and remote sensing techniques, a high-resolution topo-bathymetric model of the largest sandstone reef of the Canary Islands (the reef of Las Canteras), which provides new insights into its possible original configuration and spatio-temporal evolution. The article contains the area description and the acquisition methodology of a high-resolution topo-bathymetric model of the reef of Las Canteras, combining GPS, echosounder and hyperspectral sensing. The results include a comprehensive analysis of morphometric features, geomorphological units and active processes on the reef. The discussion finally provides an evolutionary and contextualized interpretation of the reef of Las Canteras based on the observations.

2. Study Area

The sandstone reef of Las Canteras is located in the city of Las Palmas de Gran Canaria (Canary Islands, Spain), at a subtropical latitude of the eastern Atlantic Ocean (~28°N and ~15°W), off the coast of Africa (Figure 1a). The reef is popularly known as *la barra* ('the bar') and is divided into six sections [43]. *La barra grande* is the largest fragment and the most relevant in respect to coastal hydrodynamics and sediment transport [44]. It runs parallel to the beach at a distance from the shore of about 200 m, has an approximate length of 850 m and an average intertidal width of 50 m (Figure 1b). *La barra* protects Las Canteras beach from the North Atlantic swell, which in this sector has an average significant height of 1.5 m and a peak period of more than 10 s. The storm waves reach significant heights of more than 4 m [21,45]. The morphodynamics of the different sectors of Las Canteras beach and its sedimentary stability is highly conditioned by the natural protection of *la barra* [21,45–50]. In the most protected coastal sector, normal wave overtopping barely exceeds the reef at low tide, generating an intermediate semi-enclosed coastal lagoon with high-marine biodiversity [51,52]. Likewise, during storm surges, *la barra* prevents the cross-shore transport of sediments offshore, converting the intermediate shallow seabed into a sand reservoir for the beach.

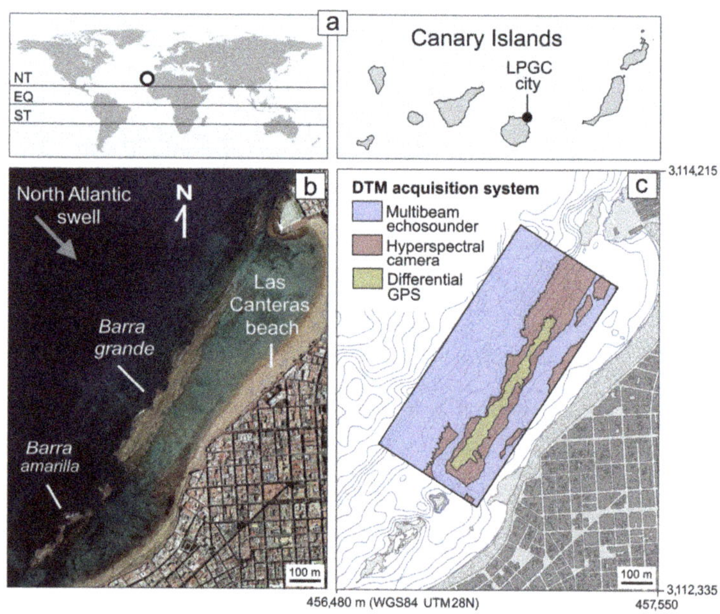

Figure 1. (**a**) Geographical location of *la barra* de Las Canteras, in the Canary Islands (LPGC stands for the city of Las Palmas de Gran Canaria). NT, Northern Tropic, EQ, Ecuator, ST, Southern Tropic. (**b**) Distribution of the main fragments of the reef (*barra grande* and *barra amarilla*) on digital orthophoto at low tide (IDE Canarias, GRAFCAN). (**c**) Zonation of the technological systems used for the acquisition of the digital topo-bathymetric model of *la barra grande*.

It can be assumed that the stratified sandstone reef of *la barra* has average levels of resistance to coastal erosion. However, as no empirical data exist, we do not know the patterns, magnitudes and rates of erosion to which it is subjected to in the short, medium and long term. The degradation of the sandstone reef of Las Canteras may lead to the degradation of the beach it protects, which in turn is one of the most important tourist elements of Las Palmas de Gran Canaria city, that is, of the main economic and demographic center of the Canary Islands.

3. Materials and Methods

3.1. GPS Topography

The topographic surfaces of the upper part of *la barra*, from 0 m above mean sea level (m asl), were obtained in 2019, by carrying on foot, at low tide, a differential GPS configured in kinematic mode to receive altimetry data every 50 cm (Figure 1c). The topographic instrument is a Topcon HiperV model, which receives positions from the GPS and GLONASS satellite constellations, and received real-time kinematic (RTK) corrections, via internet (mobile telephony), through NTRIP protocol from the GRAFCAN Permanent Stations service (https://www.grafcan.es/red-de-estaciones, accessed on 15 January 2019). The device manufacturer defines a spatial accuracy for RTK in kinematic mode of 10 mm + 1 ppm in the horizontal dimension and 15 mm + 1 ppm in the vertical dimension. More than 70,000 altimetry values with xyz coordinates of the upper part of *la barra grande* were obtained from the GPS topography.

3.2. Multibeam Bathymetry

Seabed surfaces around *la barra*, below −1.5 m asl, were obtained in 2019 (simultaneous to GPS topography) by on-board multibeam echosounder, at 3 knots maximum speed at high tide (Figure 1c). The system consists of a NORBIT iWBMS echosounder that emits 256 fan-shaped beams covering an angle of 160°. The direct measurement of sound speed was calculated using data collected by a Valeport Swift-plus profiler. The system was completed by integrated Applanix Wavemaster inertial motion sensors and an Applanix navigation system using Trimble technology. An eight-core, high-capacity computer with HYSWEEP software, synchronized and integrated the data received from all system components: multibeam echosounder, positioning system, heading, motion, tide and sound speed sensors. Post-processing by Patch test calibration, velocity profiling and denoising was performed using Hypack's MBMAX software. After post-processing steps, a regular grid of xyz bathymetric points of the underwater surrounding environment was obtained every 50 cm.

3.3. Hyperspectral Altimetry

High-resolution aerial images were obtained in 2020 using a RESONON Pika L hyperspectral camera attached to a drone with a gimbal system and irradiance sensor (Figure 1c). The camera has a spectral range of 400–1000 nm and a maximum spectral resolution of 2.1 nm. The flights were performed at an altitude of 120 m, following zig-zag trajectories with a spacing of 25 m. Radiometric and geometric correction was performed with SpectrononPro software, using calibration files to convert digital values. The orthophotos were mosaicked using the OrfeoToolbox. The specular reflections of the water were corrected using a *deglinting* algorithm [53] and the anomalous effects generated by the wave foam were corrected using an *inpainting* algorithm [54]. The subaerial part of the image was obtained at a final resolution of 10 cm, while a resolution of 30 cm in the submerged part was adopted to reduce the high-computational expense of solving the equations of the coastal water inversion involved in obtaining the bathymetry.

Bathymetry was obtained using radiative-transfer modelling of seawater in coastal environments. The inherent radiative effects of water, such as absorption and backscattering and the effects of light reflection at the bottom, were modelled in each of the hyperspectral bands, considering the different benthic covers. The model follows the semi-analytical form of the radiative transfer equations (RTE) [55], where the modelled reflectivity in a channel is obtained from the sum of the above factors. Calculating the surface reflectivity involves systems of non-linear equations that are impossible to solve analytically, so we used the Levenberg–Marquardt optimization algorithm [56]. It makes it possible to find the most suitable result by iteratively minimizing the error in respect to the reflectivity obtained by the sensor, applying cost function and minimal epsilon error. This algorithm provides very adequate results for this complex problem and uses moderate computational resources compared to other algorithms. The resulting terrain model had a horizontal resolution of 1 m.

3.4. Detection of Active Processes

The detection of active erosional processes from 1954 to 2022 was carried through an exhaustive examination of historical aerial images of the extensive photo library of the Spatial Data Infrastructure (SDI) of the Canary Islands (GRAFCAN S.A., Canary Islands Government, https://www.idecanarias.es/, accessed on 10 February 2023). It contains 22 photograms between 1954 and 2000, with an almost biannual frequency, and 18 orthophotos between 2000 and 2022, with an almost annual frequency. Conventional aerial photographs from the 1950s to the 1990s have scales between 1:5000 and 1:25,000, and were georeferenced in the WGS84 UTM28N coordinate system. The digital orthophotos from the 2000s have resolutions varying between 12.5 and 50 cm/pixel and are available for GIS viewing via the Web Map Service. The detection of active processes was completed with field surveys at the outer edge of the reef.

3.5. Data Integration and Interpretation

GPS altimetry, composed of more than 70 thousand topographic point coordinates, and echosounder altimetry, composed of more than 1 million bathymetric point coordinates, were integrated into ArcGIS software under a WGS84 UTM28N reference system. The point cloud was used to construct an interpolation grid based on a triangle irregular network (TIN) and its subsequent conversion into a 50 cm resolution raster digital terrain model (DTM). Intertidal areas with no GPS or echosounder data were covered by the 1 m resolution hyperspectral sensor-derived DTM. The resulting digital model covers an area of interest of 380,000 m^2 including *la barra grande* and its nearby underwater environment up to 200 m away (Figure 2).

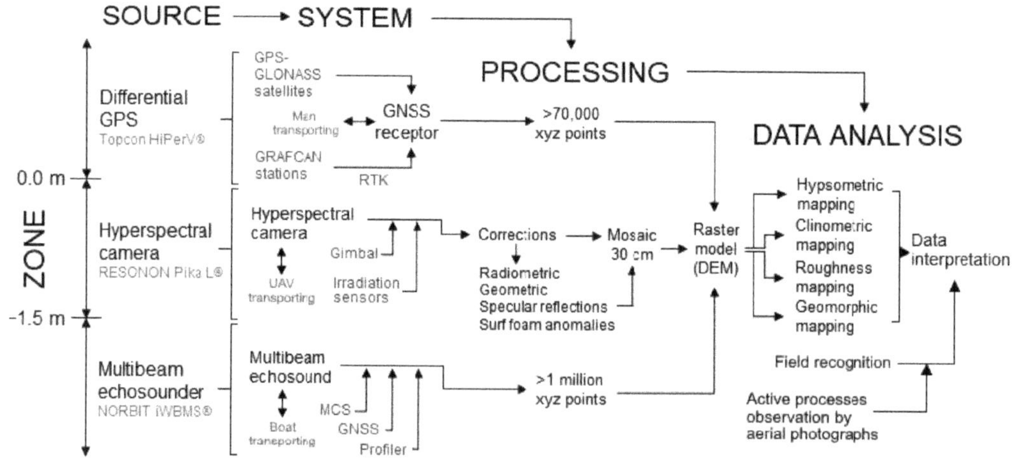

Figure 2. Workflow for the acquisition of topo-bathymetric data, integration into a digital topo-bathymetric model and the geomorphological interpretation of the rocky structure of *la barra grande* (Las Palmas de Gran Canaria, Canary Islands).

The concluding analysis of the digital topo-bathymetric model included two levels (Figure 2). At the first level, the morphometric structure was analyzed through GIS geoprocessing algorithms in order to examine the spatial distribution of altimetric ranges, terrain slopes and surface roughness. Terrain irregularity was calculated using the Vector Ruggedness Measure (VRM) [57,58]. This method measures terrain irregularity as the dispersion of the orthogonal vectors orientation of the grid cells within a raster neighborhood. At the second level of analysis, geomorphological features were recognized combining the morphometric variables with the active processes detected on aerial photographs and the field recognition. Finally, geomorphological units were established to interpret evolutionary processes in the reef.

4. Results

4.1. Morphometric Structure

The morphometric structure entails a strictly morphological description of the surface features based on measured variables. The morphometric of *la barra* has been determined by analyzing the distribution of terrain altimetry, slopes and roughness.

The hypsometric structure (Figure 3a) of *la barra grande* comprises intertidal and subtidal sections. The surface of the main sandstone bank lies entirely in the regional tidal range of the Canary Islands, between −1.5 m asl at maximum low-spring tide, and 1.5 m asl at maximum high-spring tide. However, most of this area, approximately 60,730 m^2, is in the middle to lower intertidal zone, between −1 and 0 m asl. Only a few inner sectors of the sandstone surface, totaling a small extension of 2665 m^2, are developed above 0 m asl, and only in two small topographic outcrops of 12 m^2 each, which are located further inland, the elevations exceed 0.5 m asl. One of them reaches 0.68 m asl, which is the maximum altitude of the topo-bathymetric model. In between, there is a relatively extensive intertidal rocky surface of 14,425 m^2, whose elevations of between −0.5 and 0 m asl determine that it is slightly depressed in respect to the two previous surface bands.

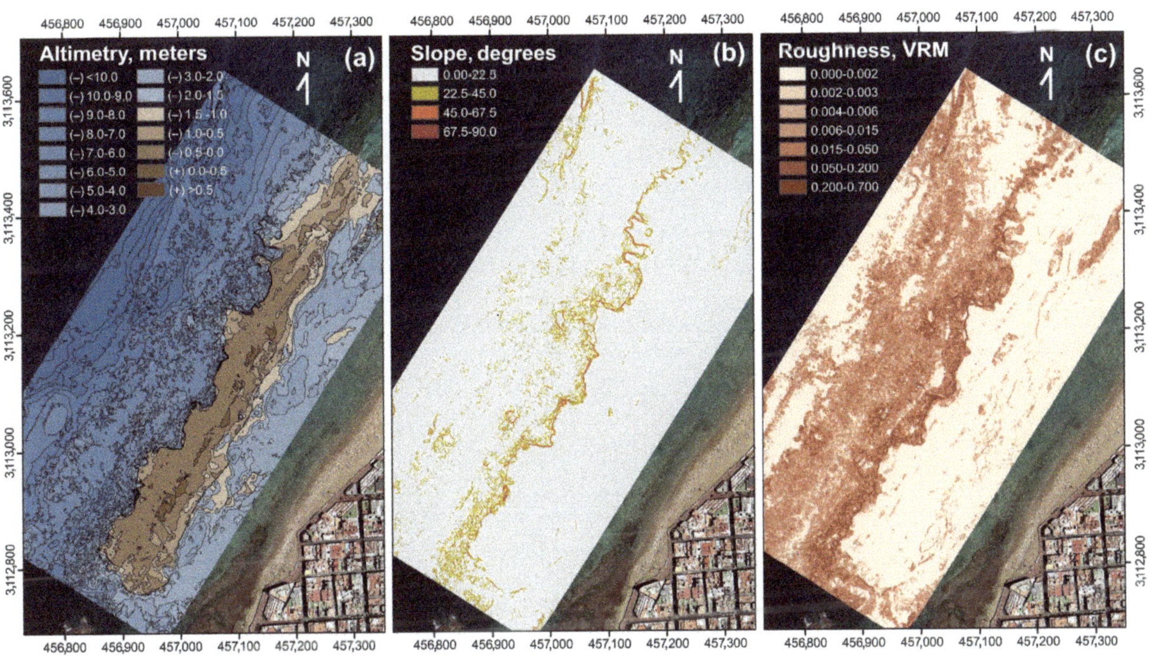

Figure 3. Morphometric variables derived from the digital topo-bathymetric model of *la barra grande* de Las Canteras (maps in metric units of the WGS84-UTM28N system) (**a**) distribution of altimetry and depths (meters); (**b**) distribution of terrain slopes (degrees); (**c**) distribution of terrain roughness (VRM = 0, completely smooth; VRM = 0.7, very rough).

At the seaward outer edge of the reef, the surface changes from intertidal depths of 0 to −1 m, to mean subtidal depths of −5 m. From the subtidal depths, towards offshore, a shallow underwater surface develops, reaching on average a depth of 12 m at a distance of 200 m from the outer edge of the reef. Above it, at least three small subtidal surfaces are clearly distinguishable with prominences of 2 to 3 m in relation to their underwater surroundings (Figure 4). The first is 45 m from the outer nearest edge of the reef, has a top surface at a mean depth of −3.5 m and is surrounded by seabed surfaces of −5 to −6 m depth. The second is 50 m from the nearest outer edge, has a top surface at a mean depth of −2.8 m and is surrounded by submarine surfaces of −6 m depth. Additionally, the third is 100 m from the nearest reef edge, has a top surface at a mean depth of −4.5 m and is surrounded by average depths of −7 m.

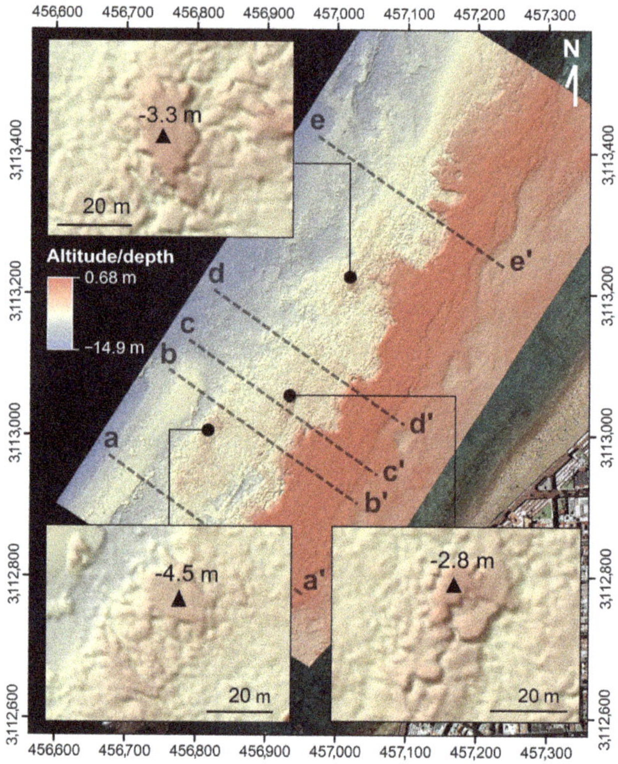

Figure 4. Topographic outcrops on the surrounding offshore seabeds, visible in the digital topo-bathymetric high-resolution model, and the surface roughness profiles of Figure 5 in dashed lines and letters (units in meters of the WGS84-UTM28N system).

The clinometric structure of the topo-bathymetric model of *la barra grande* shows significant zone differentiation (Figure 3b). Above −0.5 m asl, the surfaces are practically flat, although slightly trending NW (i.e., towards offshore) across planes with an average slope of less than 2 degrees. From the intertidal level of −0.5 m asl, the reef slope increases seaward. In some sectors, from −0.5 m asl, planes of variable cross-shore width between 25 and 50 m are observable, with slopes of up to 5 degrees to NW gradually penetrating the lower intertidal and subtidal zone of the water column and connecting the upper surface of the reef with the outer shallow subtidal seabed through small escarps of 1 to 2 m. In the sectors where these gently sloping surfaces are not developed, the almost flat surface of the reef top ends abruptly in the outer edge, in vertical 5–6 m cliffs which fall directly to the subtidal seabed. From here seaward, the outer seabed develops as a wide, low-slope surface, lying 1 to 2 degrees to the NW. In short, the topo-bathymetric model is dominated by flat or slightly sloping surfaces to the NW, except a line of steep slopes in the form of vertical escarpments of 1 to 6 m in height developed along the entire outer seaward-edge of the reef.

According to the observation of the topo-bathymetric model at a scale of 5 m, the surface irregularity, estimated using the VRM indicator, also marks significant spatial zonation in the surfaces of *la barra grande* and its submarine surroundings (Figure 3c). The upper intertidal surface of the reef is essentially smooth at this scale, with values very close to 0.0, that is the minimum roughness value of the VRM index. Additionally, essentially smooth are the inner subtidal seabeds, which lie between the sandstone reef and the present-day beach, at an average depth of −3 m. However, surface roughness increases dramatically on the outer subtidal-marine bottoms, starting from maximum values in the area adjacent to the outer edge of *la barra* and progressively decreasing (Figure 5). Maximum values occur in the areas adjacent to the outer reef escarpments, where values of up to 0.7 are reached (being 1.0 the theoretical maximum roughness value of the VRM index). The irregularity of the seabed decreases progressively towards the outside, as we move away from the edge of *la barra*, until, at a maximum distance of 150 m and −10 m depth, the high values of irregularity at an observation scale of 5 m practically disappear.

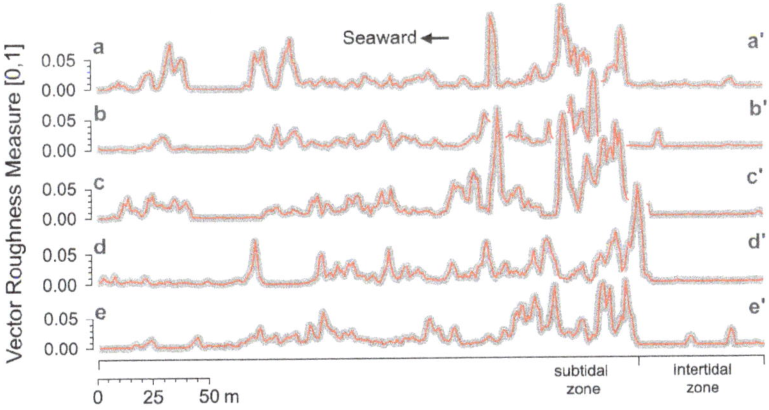

Figure 5. Vector Roughness Measure profiles showing the essentially smooth character of the intertidal zone, the very high surface irregularity of the subtidal zone adjacent to the outer seaward reef edge and the decreasing roughness towards offshore. The profile locations for letters a–e and a′–e′ can be found in Figure 4.

4.2. Geomorphological Units

Geomorphological units are morphometrically similar features linked to a common forming processes. The morphometric results and subsequent field reconnaissance has made it possible to distinguish seven main geomorphological units in *la barra grande* of Las Canteras (Figure 6). The morphogenetic character of these units provides a spatio-temporal evolutionary interpretation, in which erosion processes play a primordial role.

Firstly, the reef currently shows a top platform of 50 m average width (U1, Figure 6), with abundant development of microerosional landforms (rills, potholes, ridges, etc.), as a product of the intense daily physical, chemical and biological weathering in the middle and upper intertidal zone (Figure 7). However, according to the roughness values, this platform is essentially smooth at the metric scale of the topo-bathymetric model. It develops from the lower limit of the middle intertidal zone to the upper intertidal zone at its inner landward edge. A drop of approximately 1 m between the inner and outer edge of the top platform determines the existence of a gentle slope towards the NW of between 1 and 2 degrees. However, it also declines in a NNE direction of less than 1 degree, determining the gradual lowering of *la barra grande* to subtidal depths at its northern end. An inner fringe of this top platform is topographically discordant with the rest of the surface, being sunk between 0.5 and 1 m from its natural level (Figure 7). This is due to the extraction, throughout the 18th and 19th century, of the calcareous sandstones of *la barra* for different urban uses in the city of Las Palmas de Gran Canaria (U2, Figure 6). However, three preserved remnants of the mining activities (U3, Figure 6) show that the elevation of the inner edge was naturally at about 0.5 m asl, allowing the original pre-mining surface to be projected and reconstructed.

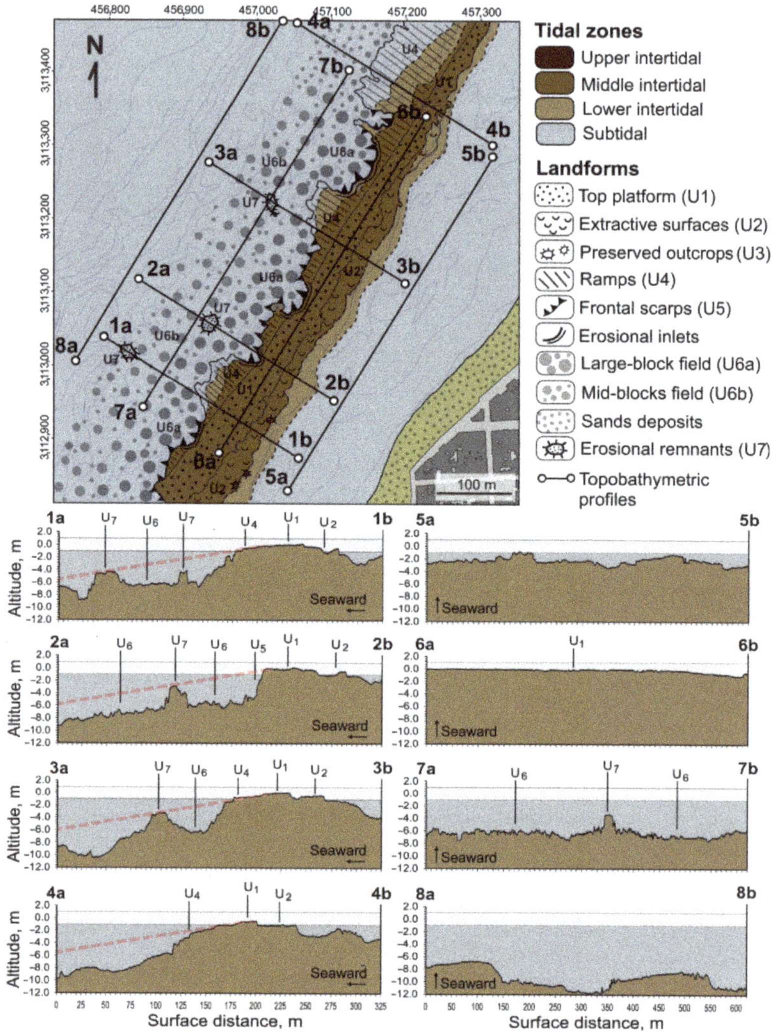

Figure 6. Geomorphological map with transverse (1a–1b to 4a–4b) and longitudinal (5 to 8) topo-bathymetric profiles (units in meters of the WGS84-UTM28N system). Geomorphological units: U1; intertidal top platform; U2, extractive depression surfaces; U3, preserved rocky outcrops; U4, intertidal ramps; U5, frontal escarpments; U6, submarine blockfields; U7, erosional remnants. The dotted lines represent the seaward projection of the intertidal ramps (U4) with 2–3 degrees of inclination, showing their theoretical connection with the erosional remnants (U7).

The outer edge of the top platform is planimetrically sinuous. The headlands are intertidal ramps (U4, Figure 6) of variable widths of between 25 and 50 m, with average slopes of approximately 5 degrees to the NW. On its surface, there are abundant parallel erosional rills visible in the field as a result of the mechanical action of the waves run-up at low tide (Figure 7). The intertidal ramps gradually penetrate the water table, ending in small subtidal escarpments (Figures 7 and 8). Between headlands, the inlets are incipient erosional corridors as a result of the retreat of the intertidal ramps by wave action. Instead of ramps, we observe receding concavities in the inlets (Figure 7), where the outer edge of the top platform ends in vertical cliffs of greater height. Ultimately, the outer edge of the reef is steep, but the height of the frontal escarpments (U5, Figure 6) varies from 2 to 6 m depending on the intensity or magnitude of marine erosional retreat, which is deduced from the degree of planview penetration in the concavities (erosion) or the degree of prominence of the intertidal ramp-like headlands (preservation).

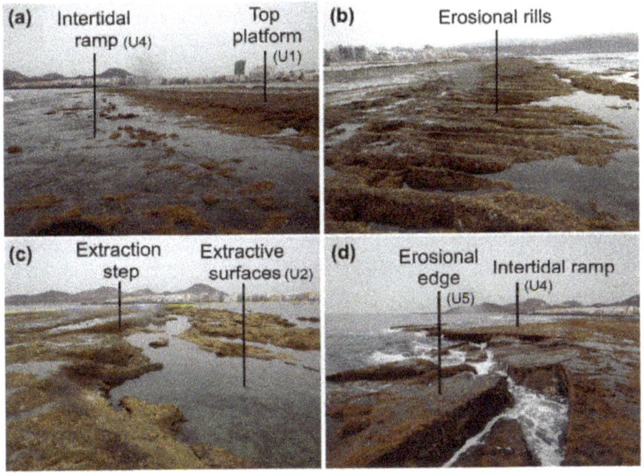

Figure 7. Photographs of intertidal geomorphological units: (**a**,**b**) intertidal top platform (U1, Figure 6) and ramps (U4, Figure 6) with erosional micromorphologies of weathering and parallel rill erosion; (**c**) extractive depressions (U2, Figure 6); (**d**) erosional edge and top of frontal escarpments (U5, Figure 6). Photographs are north-facing, except for (**b**), which is south-facing.

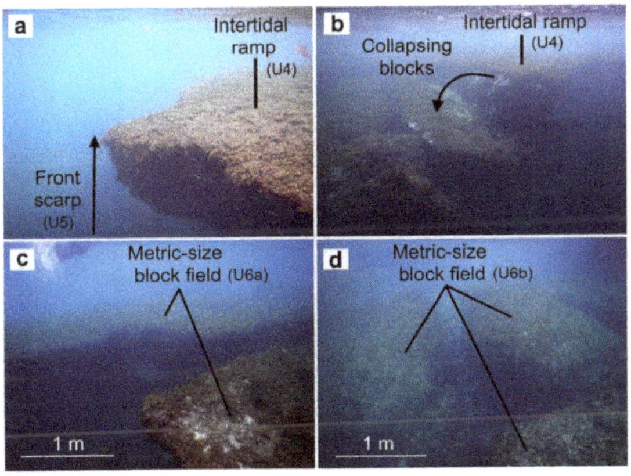

Figure 8. Photographs of subtidal geomorphological units: (**a**) outer section of an intertidal ramp (U2, Figure 6) ending in a frontal escarpment (U4, Figure 6); (**b**) collapsing blocks on the frontal walls of the intertidal ramp; (**c**,**d**) metric-sized blocks on the seabed adjacent to the outer reef edge (U6a, Figure 6). Scale of the photographs (**c**,**d**) is approximate.

At subtidal depths towards open water, the submarine platform is shallow with a gentle slope to offshore and very high surface roughness at the scale of the digital topo-bathymetric model. The analysis of the hyperspectral images together with the field inspections allowed to verify that this roughness corresponds to the presence of an extensive field of large sandstone blocks (U6). It extends up to a distance of approximately 150 m and up to −10 m in depth from the outer edge of the intertidal top platform (Figure 8). As can be deduced from their spatial distribution and composition, the blocks are the product of detachments from the outer reef front (Figure 8). In other words, they result from the successive collapses and rockfalls that have taken place in the sandstone wall during the retreat caused by mechanical wave erosion.

The size of the rock fragments in the blockfield decreases with distance from the outer edge of the intertidal platform (Figure 5). A first submarine zone of width of 50 m (U6a, Figure 6) of more modern and larger rockfalls, runs parallel to the outer edge, and can be distinguished from a second submarine zone of 100 m width (U6b, Figure 6) of older and smaller collapsing blocks. The detection of seabed features in the high-resolution hyperspectral images allowed to establish a statistical estimate of the size of the blocks in the shallow submarine blockfield adjacent to the edge of the intertidal platform (U6a) (Figure 9). According to these results, about 60% of the blocks have surface sizes between 5 and 25 m^2; almost 20% have sizes between 25 and 125 m^2; and some blocks are larger than 125 m^2.

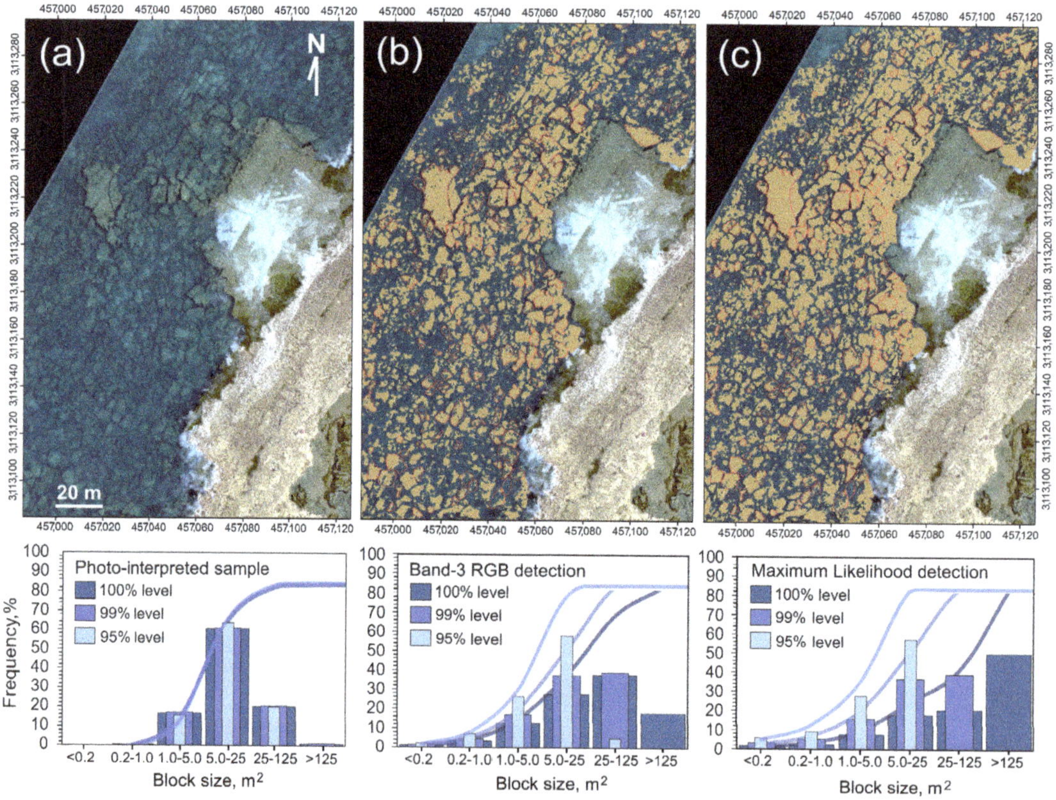

Figure 9. Blockfield on the near seabeds adjacent to the outer reef edge (WGS84-UTM28N units in meters) (**a**) High-resolution hyperspectral RGB image. (**b**) Blocks detection by light slicing in the green RGB band (brown) compared to the recognition by direct photo-interpretation at 590 random points (red contours). (**c**) Block detection by maximum likelihood algorithm from 822 RGB signature samples (brown) compared to direct photo-interpretation at 590 random points (red contours). The histograms show the distribution of the surface size of the blocks according to the three detection methods, considering 100%, 99% and 95% of the size distribution obtained.

In the middle of the blockfield originated by the marine erosional dismantling of the reef, the topo-bathymetric model showed the existence of topographic outcrops (Figure 4) interpretable as erosional remnants of *la barra* de Las Canteras (U7, Figure 6). The morphometric structure of these three rocky remnants allows to assimilate them with non-eroded surfaces of *la barra*, which would have been separated from the main calcareous sandstone mass during its dismantling and erosional retreat over time. Therefore, the erosional remnants can be considered as key geomorphological indicators of the original extension of *la barra* and of the total longitudinal magnitude of its erosional retreat. The first is located 100 m from the outer edge of *la barra* (X.456825, Y.3113020, in UTM28N) and has a flat surface of 270 m^2; the second is located 50 m from the outer edge (X.457015, Y.3113220) and has a flat, more fragmented surface of 220 m^2; and the third is located ~40 m away (X.456935, Y.3113060) and has a flat surface of 130 m^2 (Figure 4).

4.3. Active Erosional Processes

Examination of the extensive series of aerial photograms and historical orthoimages available for this area in the SDI-Canarias (GRAFCAN S.A., Canary Islands Government), revealed the occurrence of at least five decametric detachments on the sidewalls of *la barra* in the last 68 years, from 1954 to 2022 (Figure 10). They add up to a rock loss of 698.5 m^2, with 43.1 m of cumulative retreat along 116 m. Between 1963 and 1975, a rockfall event occurred (UTM X.456990, Y.3113010), in which 150 m^2 of rock was detached, leading to a maximum linear retreat of 8.8 m over a north-facing ramp sidewall. The frame from 1974 is the first in which this rockfall was clearly visible. Between 1987 and 1998, two rockfalls also occurred on north-facing ramp sidewalls. The larger of these (UTM X.457105, Y.3113240), measuring 214 m^2, had a maximum linear retreat of 9.2 m; and the smaller (UTM X.4567042, Y.3113075), measuring 115 m^2, had a maximum linear retreat of 9.3 m. The frames from 1989 and 1996 are, respectively, the first in the series in which they were clearly visible. The last two rockfalls occurred on west-facing erosional walls. The first (UTM X.457150, Y.3113270) occurred between 2004 and 2007 and affected 70 m^2 of rock mass with a maximum retreat of 6.8 m; and the second (UTM X.457010, Y.3113020) occurred between 2017 and 2019, and affected 145 m^2 of rock mass, leading to a linear retreat of 9.0 m. These collapses were first observed in the frames from 2007 and 2019, respectively.

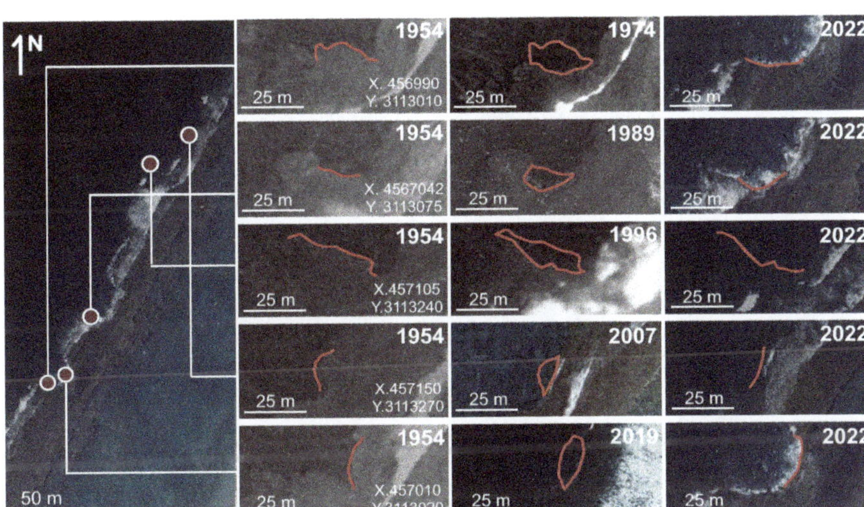

Figure 10. Rock areas eroded over the last 68 years (1954–2022) on the sidewalls of *la barra*, identified from the 22 historical aerial photographs and 18 digital orthophotos of the SDI-Canarias (GRAFCAN S.A., Canary Islands Government). The central column displays the first frames of the historical series in which the rock detachment is clearly visible. The red lines show the contours of the rocky edges. X,Y coordinates are metric units of the WGS84-UTM28N system.

5. Discussion

According to the geomorphological indicators found in the topo-bathymetric digital model and in the verifications carried out in situ, the current configuration of *la barra* must be interpreted as a rocky structure partially dismantled by natural and anthropic erosive processes. The interpretation of the results also leads to an estimation of the probable magnitude, patterns and rates of this disintegration over time.

While the natural erosional processes have mainly taken place at the front of the rocky structure, due to marine erosion, the erosional processes of human origin linked to the mining activities that took place centuries ago, are evident on its inner margin. According to Pérez-Torrado et al. [59], the extracted stone was mainly used as filters to remove water impurities thanks to the purification capacity of this sandstone rock. The reconstruction of the surface, according to high-resolution topo-bathymetry, had allowed to calculate the volume of calcarenite extraction at approximately 14,290 m^3; a value very close to the 15,000 m^3 estimated in previous field measurements [60].

The geomorphological configuration of the outer edge of the reef and the adjacent seabed provides new data to determine the spatio-temporal magnitude and mechanisms of marine erosion acting on *la barra*. The height of the bordering escarpment along the entire outer reef correlates with its planview morphology: the rocky headlands end in smaller escarpments than the rocky inlets, which form steep concavities where the highest falls are found. This difference in height-morphology is indicative of differential erosion processes and contrasting degrees of current retreat along the erosional frontal wall of the reef. In addition, the existence of an extensive submarine blockfield, extending up to 150 m from the outer edge of *la barra* (Figure 3c), more strongly supports the course of an extensive process of rock leveling due to prolonged marine erosion.

In this sector of the coastline of the island of Gran Canaria known as Confital Bay, the island shelf is approximately 4 km wide and has maximum depths below −100 m [61]. The blockfield of *la barra* de Las Canteras is mainly developed on a nearshore seabed of −5 to −10 m depth. It therefore extends over the inner sector of the island shelf that has been subjected to repeated flooding and emersion over geological time, due to Quaternary marine sea-level oscillations [62,63]. The sandstone blocks covering this surface are chaotically arranged and are of medium and large size, especially those covering the seabed adjacent to the intertidal rocky reef (Figures 5, 8 and 9).

Direct observation of historical rockfalls on the reef front, places the origin of the submarine blockfield by analogy to the present day in the disintegration of *la barra* by prolonged wave action, and the successive and numerous collapses of the outer wall. Examination of the historical series of aerial images available reinforces this position (Figure 10). The present day observations are consistent with the geomorphological indicators found in the topo-bathymetric digital model and shows that the erosion of *la barra* is mainly produced by large episodic rockfalls. Most of the detached blocks distributed on the closest outer seabeds are between 5 and 125 m^2 (Figure 9), while the five detachments detected in the historical period, since the middle of the 20th century, involve the uprooting of surfaces of more than 100 m^2, 20 m longitudinal wall lengths, and retreats of up to 9 m (Figure 10). They have occurred at north-facing points of ramps sidewalls or at west-facing points of inner walls of the erosive concavities. The frequency of this events over the last 68 years allows us to estimate an approximate return period of almost 14 years. However, this calculation may be underestimating the actual mean recurrence of smaller rockfalls, as observations are limited in time and constrained by the resolution and georeferencing quality of aerial imagery. Considering the total intertidal surface above −1 m depth, except for terrains degraded by extractions, the total detachment of 698.5 m^2 represents a loss over the last 68 years of 1.6% of the reef; an annual a rate of 0.02%/year.

The results indicate that rockfalls triggered by storm-wave impact are the primary erosion mode on *la barra*. Chemical and biological weathering processes are expected to act secondarily at a micro-scale, weakening the rocks and preparing them for mechanical wave erosion. The presence of small sandbanks in the nearshore blockfield enables wave abrasion and may explain the development of rill-erosion morphologies on the intertidal reef ramps (Figure 7b).

The wave exposure and the pattern of discontinuities in the calcareous sandstone strata must determine the spatial distribution of the erosion in *la barra*, as is usual in these types of rocky coasts [38,64,65]. The distribution of historical rockfalls is consistent with the morphologies of the erosional front, as they occurred preferentially in pre-existing concavities (Figure 10). The concavities must be interpreted as incipient cross-shore erosional corridors that will ultimately fragment the reef longitudinally. Up to five in-progress erosional corridors can be observed today on *la barra grande* (Figures 6 and 10). Considering the strong influence of discontinuities on the patterns of coastal erosion in sedimentary rocks [38,64,65], the formation of these corridors may reflect the structural control of a set of discontinuities of NW–SE direction, which is also consistent with the dominant swell in this coastal sector. In turn, the northern orientation of some of the rockfalls suggests another direction of retreat, according to a possible NNE–SSW pattern of structural discontinuities, parallel to the planview layout of the reef and orthogonal to the NW–SE set of discontinuities. The topo-bathymetric model shows alignments consistent with this NNE–SSW retreat pattern in the disposition of the erosional submarine remnants (Figure 4).

The slight general NW tilt of the surface of *la barra* de Las Canteras is the topographic reflection of a seaward-dipping stratification, resulting from its paleo-coastal sedimentary origin [8,9,59]. These slopes, which are approximately 5° in the intertidal ramps, would have an essentially structural nature as a result of the cementation of an ancient beach formation (beachrock). The offshore projection of the slopes of the intertidal ramps in the topo-bathymetric digital model, links coherently with the series of submarine pinnacles detected at distances of between 50 and 100 m from the outer edge of *la barra*. They have flat surfaces of more than 100 m^2, also slightly tilted to the NW. This reinforces the interpretation of these submarine pinnacles as preserved erosional remnants of *la barra* (as 'sea stacks' on rocky cliffed coasts), which indicate former positions of the erosional front (see topo-bathymetric profiles 1, 2 and 3 in Figure 6). Together with the blockfield identified, the existence of these preserved remnants allows projection of the original extension of *la barra* up to an average distance of 150 m from the current outer edge. Bearing in mind that, at present, the average intertidal width of *la barra grande* is between 50 m and 100 m, depending on sectors, we currently witness a retreating of about 1/3 to 2/3 of the original rocky structure.

Stratigraphic studies have traditionally suggested that *la barra* was formed in the Last Interglacial period (MIS5e), approximately 110 ka ago, in an intertidal or subtidal coastal sedimentary environment [10]. Further cementation and consolidation of *la barra* into a calcarenite rock would probably have occurred during the marine regression. In the Last Glacial Period, the sea level dropped to the edge of the island shelf to more than −100 m depth [66]. Paleontological evidences indicate that the sea level remained up to two meters above present-day sea level in much of the Middle and Upper Holocene in the low-latitude Northeastern Atlantic region [67–69]. Considering that the erosional extension observed in the high-resolution topo-bathymetry is mostly between −5 and −10 m depth, the erosive capacity of waves on the subtidal sandstones would have been very limited since the positioning of the sea level at or above present-day levels 6500 years ago. As the most likely hypothesis, we would date the erosion of most of the subtidal structure of *la barra* in an earlier period, between 9000 and 6500 years B.P. (Lower and Middle Holocene), when the rising sea level transited from depths of −10 m to the present level. Since then (6500 years ago to the present), with different degrees of effectiveness depending on the metric variability of the Holocene mean sea levels [69], the erosion of the current intertidal rocks would have begun, observable as an active process at present (Figure 10). However, this interpretation should be cautiously considered, as recent studies dated similar beachrocks outcrops at Las Canteras beach to the Holocene transgression [9,70], which would place the formation and erosion processes of *la barra* much closer together in time.

6. Conclusions

Sandstone reefs are highly effective protective barriers against coastal erosion and safeguard the sedimentary stability of many sandy beaches. For this reason, *la barra* de Las Canteras is a natural coastal sandstone reef of great social and ecological value for the Canary Islands. Due to the apparent resistance of the rocks to the force of the waves and the relative slowness of the erosive processes, insufficient efforts have been made so far to know the magnitude, rate and form of the past and present erosion in *la barra*.

By means of high-resolution topo-bathymetry, obtained by combining different remote sensing technologies (echosounder, GPS and hyperspectral sensor), this work has provided new data to interpret features of its original configuration and its spatio-temporal evolution.

According to the geomorphological evidence provided in this paper, the original width of that sedimentary formation, which today is between ~50 and ~100 m (depending on the degree of erosion by sectors), could have been up to 200 m. The main hypothesis is that the erosional retreat of the sandstone reef occurred during the Holocene sea level rise. The erosion of the subtidal rocky structure of *la barra*, converted into a field of rockfall blocks below −5 m asl, together with the erosional semi-preserved remnants visible in the present topo-bathymetry, could be attributed to the transient eustatic rise from the Lower to the Middle Holocene. The erosion of the intertidal rocky masses (i.e., the intertidal platforms and ramps on whose walls active erosions are observed at present) would have occurred since the positioning of the sea at the present level, or up to 2 m above it in some time phases, about 6500 years ago.

The underwater and subaerial geomorphology of the reef shows how erosion follows preferential lines conditioned by wave exposure and, most probably, the geometry of an orthogonal fracture system. Differential erosion on the transverse fractures leads to the formation of concavities, the development of transverse erosional corridors (NW-SE) and, ultimately, the longitudinal fragmentation of *la barra grande*. Therefore, the erosive disintegration of the reef is marked by a process of fragmentation into pieces, as can be seen today in the different fragments of *la barra* along Las Canteras beach (*barra grande* and *barra amarilla*). At the same time, the likely exploitation of the longitudinal fracture lines produces a parallel retreat vector (NNE–SSW) which is concentrated on the north-facing walls of the intertidal rocky ramps. This can be seen in the present-day rockfalls and as is coherently attested to by the marked NNE–SSW dispositions of the subtidal erosional remnants found in the topo-bathymetric model.

At present, the erosion of *la barra* is a slow process marked by episodic collapses with an average frequency of almost 14 years, with up to 100 m^2 of rock mass losses and sudden retreats of 10 m on the outer edge of the reef. Over the past 68 years, 1.6% of the reef's preserved surface has been eroded; a low rate of 0.02%/year. However, it can be expected that the development of erosional corridors and the gradual loss of volume in *la barra grande* may lead to significant modifications of the waves and sea currents regime in the near future, even before its complete disintegration. It is therefore necessary to further investigate the erosional processes of *la barra* de Las Canteras, combining geomorphological monitoring techniques with hydrodynamic modelling, in order to project likely future scenarios and predict their ecological, cultural and economic consequences.

Author Contributions: Conceptualization, N.F.; Methodology, N.F. and K.S.; Software, K.S. and J.M.; Validation, K.S. and J.M.; Formal analysis, N.F.; Investigation, N.F.; Resources, J.V.; Data curation, N.F. and J.M.; Writing—original draft, N.F.; Writing—review & editing, K.S., J.M., J.V. and O.B.; Project administration, O.B.; Funding acquisition, O.B. All authors have read and agreed to the published version of the manuscript.

Funding: This research is a result of the project 'Evaluación de la capacidad de usos de La Barra de Las Canteras, considerando especies catalogadas y hábitats de la Red Natura 2000', co-funded by the Regional Ministry of Territorial Policy, Sustainability and Security (Government of the Canary Islands) and the European Regional Development Fund (ERDF) program 'Canarias Avanza con Europa'. Nicolás Ferrer is a beneficiary of the Postdoctoral contract programme 'Catalina Ruiz,

2021' (APCR2021010018/6431001) co-funded by the Canary Islands Government and the European Social Fund.

Data Availability Statement: Not applicable.

Conflicts of Interest: The authors declare no conflict of interest.

References

1. Guimarães, T.; Mariano, G.; Barreto, A.; Sá, A.A. Beachrocks of Southern Coastal Zone of the State of Pernambuco (Northeastern Brazil): Geological resistance with history. *Geoheritage* **2017**, *9*, 111–119. [CrossRef]
2. Soares, M.D.O.; Rossi, S.; Martins, F.A.S.; Carneiro, P.B.D.M. The forgotten reefs: Benthic assemblage coverage on a sandstone reef (Tropical South-western Atlantic). *J. Mar. Biol. Assoc. UK* **2017**, *97*, 1585–1592. [CrossRef]
3. Darwin, C. On a remarkable bar of sandstone off Pernambuco, on the coast of Brazil. *Lond. Edinb. Dublin Philos. Mag. J. Sci.* **1841**, *19*, 257–260. [CrossRef]
4. Mabesoone, J.M. Origin and age of the sandstone reefs of Pernambuco (Northeastern Brazil). *J. Sedim. Res.* **1964**, *34*, 715–726. [CrossRef]
5. Chaves, N.S.; Sial, A.N. Mixed oceanic and freshwater depositional conditions for beachrocks of Northeast Brazil: Evidence from carbon and oxygen isotopes. *Int. Geol. Rev.* **1998**, *40*, 748–754. [CrossRef]
6. Pereira, L.S.; de Carvalho, D.M.; da Cunha, L.S. Methodology for the semi-quantitative evaluation of geoheritage applied to coastal geotourism in João Pessoa (Paraíba, Northest Brazil). *Geoheritage* **2019**, *11*, 1941–1953. [CrossRef]
7. Da Silva, M.L.N.; do Nascimento, M.A.L. Ecosystem services and typology of urban geodiversity: Qualitative assessment in Natal town, Brazilian Northeast. *Geoheritage* **2020**, *12*, 57. [CrossRef]
8. Alonso, I. North coast: Las Canteras beach. In *Tendencias Actuales en Geomorfología Litoral*; Hernández-Calvento, L., Alonso, I., Mangas, J., Yanes, A., Eds.; University of Las Palmas de Gran Canaria: Las Palmas de Gran Canaria, Spain, 2005; pp. 219–238.
9. Menéndez, I.; Herrera-Holguín, C.; Mangas, J. Upper Quaternary coastal palaeoenvironments and palaeosea levels in Las Canteras beach, Gran Canaria (Canary Islands, Spain). *Mar. Geol.* **2020**, *429*, 106322. [CrossRef]
10. Pérez-Torrado, F.J.; Mangas, J. Origin and geological evolution of the Las Canteras bar (Las Palmas de Gran Canaria). *Vector Plus* **1994**, *1*, 4–13.
11. Balcells, R.; Barrera-Morate, J.L.; Ruiz, M.T. Geological map of sheet 1101-I-II, scale 1:25,000. (Las Palmas de Gran Canaria). In *Memoir*; ITGE: Madrid, Spain, 1990; p. 131.
12. Meco, J.; Petit-Maire, N.; Fontugne, M.; Shimmield, G.; Ramos, A.J. The Quaternary deposits in Lanzarote and Fuerteventura (Eastern Canary Islands, Spain): An overview. In *Climates of the Past. Proceeding CLIP Project*; Meco, J., Petit-Maire, N., Eds.; Universidad de Las Palmas de Gran Canaria (ULPGC): Las Palmas, Spain, 1997; pp. 123–136.
13. Eversole, D.; Fletcher, C.H. Longshore sediment transport rates on a reef-fronted beach: Field data and empirical models Kaanapali Beach, Hawaii. *J. Coast. Res.* **2003**, *19*, 649–663.
14. Bosserelle, C.; Gallop, S.L.; Haigh, I.D.; Pattiaratchi, C.B. The influence of reef topography on storm-driven sand flux. *J. Mar. Sci. Eng.* **2021**, *9*, 272. [CrossRef]
15. Trenhaile, A. Rocky coasts—their role as depositional environments. *Earth Sci. Rev.* **2016**, *159*, 1–13. [CrossRef]
16. Gallop, S.L.; Kennedy, D.M.; Loureiro, C.; Naylor, L.A.; Muñoz-Pérez, J.J.; Jackson, D.W.; Fellowes, T.E. Geologically controlled sandy beaches: Their geomorphology, morphodynamics and classification. *Sci. Environ.* **2020**, *731*, 139123. [CrossRef]
17. Muñoz-Pérez, J.J.; Tejedor, L.; Medina, R. Equilibrium beach profile model for reef-protected beaches. *J. Coast. Res.* **1999**, *15*, 950–957.
18. Gallop, S.L.; Bosserelle, C.; Eliot, I.; Pattiaratchi, C.B. The influence of limestone reefs on storm erosion and recovery of a perched beach. *Cont. Shelf Res.* **2012**, *47*, 16–27. [CrossRef]
19. Gallop, S.L.; Bosserelle, C.; Eliot, I.; Pattiaratchi, C.B. The influence of coastal reefs on spatial variability in seasonal sand fluxes. *Mar. Geol.* **2013**, *344*, 132–143. [CrossRef]
20. Rodríguez-Padilla, I.; Castelle, B.; Marieu, V.; Morichon, D. Video-Based Nearshore Bathymetric Inversion on a Geologically Constrained Mesotidal Beach during Storm Events. *Remote Sens.* **2022**, *14*, 3850. [CrossRef]
21. Alonso, I.; Vilas, F. The influence of boundary conditions on beach zonation. In *Coastal Dynamics*; ASCE: Reston, VA, USA, 1994; pp. 417–431.
22. Jackson, D.W.T.; Cooper, J.A.G.; Del Rio, L. Geological control of beach morphodynamic state. *Mar. Geol.* **2005**, *216*, 297–314. [CrossRef]
23. Dugan, J.P.; Morris, W.D.; Vierra, K.C.; Piotrowski, C.C.; Farruggia, G.J.; Campion, D.C. Jetski-based nearshore bathymetric and current survey system. *J. Coast. Res.* **2001**, *17*, 900–908.
24. Maleika, W.; Palczynski, M.; Frejlichowski, D. Interpolation methods and the accuracy of bathymetric seabed models based on multibeam echosounder data. In Proceedings of the Intelligent Information and Database Systems: 4th Asian Conference, ACIIDS 2012; Proceedings, Part III 4, Kaohsiung, Taiwan, 19–21 March 2012; Springer: Berlin/Heidelberg, Germany, 2012; pp. 466–475.
25. Stockdon, H.F.; Holman, R.A. Estimation of wave phase speed and nearshore bathymetry from video imagery. *J. Geophys. Res. Oceans* **2000**, *105*, 22015–22033. [CrossRef]

26. Lee, Z.; Carder, K.L.; Mobley, C.D.; Steward, R.G.; Patch, J.S. Hyperspectral remote sensing for shallow waters: 2. Deriving bottom depths and water properties by optimization. *Appl. Opt.* **1999**, *38*, 3831–3843. [CrossRef]
27. Lyzenga, D.R.; Malinas, N.P.; Tanis, F.J. Multispectral bathymetry using a simple physically based algorithm. *IEEE Trans. Geosci. Remote Sens.* **2006**, *44*, 2251–2259. [CrossRef]
28. Ferrario, F.; Beck, M.W.; Storlazzi, C.D.; Micheli, F.; Shepard, C.C.; Airoldi, L. The effectiveness of coral reefs for coastal hazard risk reduction and adaptation. *Nat. Commun.* **2014**, *5*, 1–9. [CrossRef]
29. Gallop, S.L.; Young, I.R.; Ranasinghe, R.; Durrant, T.H.; Haigh, I.D. The large-scale influence of the Great Barrier Reef matrix on wave attenuation. *Coral Reefs* **2014**, *33*, 1167–1178. [CrossRef]
30. Robinson, L.A. Erosive processes on the shore platform of northeast Yorkshire, England. *Mar. Geol.* **1977**, *23*, 339–361. [CrossRef]
31. Bird, E.C. *Coastal Geomorphology: An Introduction*; John Wiley & Sons: Hoboken, NJ, USA, 2008.
32. Gill, E.D.; Lang, J.G. Micro-erosion meter measurements of rock wear on the Otway coast of southeast Australia. *Mar. Geol.* **1983**, *52*, 141–156. [CrossRef]
33. Stephenson, W.J.; Kirk, R.M.; Hemmingsen, S.A.; Hemmingsen, M.A. Decadal scale micro erosion rates on shore platforms. *Geomorphology* **2010**, *114*, 22–29. [CrossRef]
34. Stephenson, W.J.; Kirk, R.M.; Kennedy, D.M.; Finlayson, B.L.; Chen, Z. Long term shore platform surface lowering rates: Revisiting Gill and Lang after 32 years. *Mar. Geol.* **2012**, *299*, 90–95. [CrossRef]
35. Trenhaile, A.S.; Porter, N.J. Shore platform downwearing in eastern Canada; A 9–14 year micro-erosion meter record. *Geomorphology* **2018**, *311*, 90–102. [CrossRef]
36. Stephenson, W.J.; Naylor, L.A. Geological controls on boulder production in a rock coast setting: Insights from South Wales, UK. *Mar. Geol.* **2011**, *283*, 12–24. [CrossRef]
37. Naylor, L.A.; Stephenson, W.J.; Smith, H.C.; Way, O.; Mendelssohn, J.; Cowley, A. Geomorphological control on boulder transport and coastal erosion before, during and after an extreme extra-tropical cyclone. *Earth Surf. Process. Landf.* **2016**, *41*, 685–700. [CrossRef]
38. Buchanan, D.H.; Naylor, L.A.; Hurst, M.D.; Stephenson, W.J. Erosion of rocky shore platforms by block detachment from layered stratigraphy. *Earth Surf. Process. Landf.* **2020**, *45*, 1028–1037. [CrossRef]
39. Trenhaile, A.S. *The Geomorphology of Rock Coasts*; Clarendon Press: Oxford, UK, 1987; Volume 1168.
40. Trenhaile, A.S. Modeling the development of wave-cut shore platforms. *Mar. Geol.* **2000**, *166*, 163–178. [CrossRef]
41. Walkden, M.J.A.; Hall, J.W. A predictive mesoscale model of the erosion and profile development of soft rock shores. *Coast. Eng.* **2005**, *52*, 535–563. [CrossRef]
42. Thébaudeau, B.; Trenhaile, A.S.; Edwards, R.J. Modelling the development of rocky shoreline profiles along the northern coast of Ireland. *Geomorphology* **2013**, *203*, 66–78. [CrossRef]
43. Alonso, I. Sedimentary Processes at Las Canteras Beach (Gran Canaria). Ph.D. Thesis, University of Las Palmas de Gran Canaria, Las Palmas de Gran Canaria, Spain, 1993.
44. Arriola-Velásquez, A.; Tejera, A.; Guerra, J.G.; Alonso, I.; Alonso, H.; Arnedo, M.A.; Martel, P. Spatio-temporal variability of natural radioactivity as tracer of beach sedimentary dynamics. *Estuar. Coast. Shelf Sci.* **2019**, *231*, 106476. [CrossRef]
45. Alonso, I.; Vilas, F. Sedimentary variability at Las Canteras beach (Gran Canaria). *Geogaceta* **1996**, *20*, 428–430.
46. Martínez-Martínez, J.; Gordo Rojas, M.C.; Santana, J.M.; Jiménez, J.A.; Veloso Quevedo, J.J. Sedimentary dynamics at Las Canteras beach (Las Palmas de Gran Canaria). *Rev. Obras Públ.* **1988**, *2*, 145–152.
47. Martínez-Martínez, J.; Álvarez Espejo, R.; Bilbao, I.A.; Rodríguez Cabrera, R. Analysis of sedimentary processes on the Las Canteras beach (Las Palmas, Spain) for its planning and management. *Eng. Geol.* **1990**, *29*, 377–386. [CrossRef]
48. Alonso, I. Spatial beach morphodynamics. An example from Canary Islands, Spain. *Litoral* **1994**, *94*, 169–183.
49. Alonso, I. Medición del transporte de sedimentos por olas y corrientes. *Vector Plus Sci. Cult. Misc.* **1999**, *14*, 13–23.
50. Sánchez García, M.J.; Casamayor Font, M.; Alonso Bilbao, I.; Rodriguez, S. Long-term evolution of the coastline at Las Canteras beach (Canary Islands). *Geotemas* **2015**, *15*, 85–88.
51. Tuya, F.; Asensio, M.; Bosch, N.E.; García, A.; Navarro, A. Partitioning multiple diversity dimensions of nearshore fish assemblages within a coastal seascape. *Hydrobiologia* **2019**, *834*, 87–102. [CrossRef]
52. Tuya, F.; Asensio, M.; Navarro, A. "Urbanite" rays and sharks: Presence, habitat use and population structure in an urban semi-enclosed lagoon. *Reg. Stud. Mar. Sci.* **2020**, *37*, 101342. [CrossRef]
53. Martin, J.; Eugenio, F.; Marcello, J.; Medina, A. Automatic sun glint removal of multispectral high-resolution WorldView-2 imagery for retrieving coastal shallow water parameters. *Remote Sens.* **2016**, *8*, 37. [CrossRef]
54. Guillemot, C.; Le Meur, O. Image inpainting: Overview and recent advances. *IEEE Signal Process. Mag.* **2013**, *31*, 127–144. [CrossRef]
55. Zhonping, L. Hyperspectral remote sensing for shadow waters: 2. Deriving bottom depths and water properties by optimization. *Appl. Opt.* **1999**, *38*, 18.
56. Gavin, H.P. *The Levenberg-Marquardt Method for Nonlinear Least Squares Curve-fitting Problems*; Department of Civil and Environmental Engineering, Duke University Durham: Durham, NC, USA, 2013.
57. Hobson, R.D. Surface Roughness in Topography: A Quantitative Approach. In *Spatial Analysis in Geomorphology*; Chorley, R.J., Ed.; Harper & Row: London, UK, 1972; pp. 221–245.

58. Sappington, J.M.; Longshore, K.M.; Thomson, D.B. Quantifying Landscape Ruggedness for Animal Habitat Analysis: A case Study Using Bighorn Sheep in the Mojave Desert. *J. Wildl. Manag.* **2007**, *71*, 1419–1426. [CrossRef]
59. Pérez-Torrado, F.J.; Cabrera, M.C.; Hansen Machín, A.; Rodriguez-Gonzalez, A. *Geolodía 13 Gran Canaria: An Urban Geological Walk*; Las Palmas de Gran Canaria: Las Palmas, Spain, 2013.
60. Otero Paz, I. *La Historia de Las Palmas de Gran Canaria, Enterrada en la Playa de Las Canteras*; elDiario.es: Las Palmas de Gran Canaria, Spain, 2016. Available online: https://www.eldiario.es/canariasahora/cultura/palmas-gran-canaria-enterrada-canteras_1_4042039.html (accessed on 20 April 2023).
61. Instituto Español de Oceanografía. *Topobathymetric Map of the Canary Islands 1:700,000 (UTM 28N Ellipsoid WGS84 Projection)*; Instituto Español de Oceanografía, Ministerio de Educación y Ciencia: Madrid, Spain, 2006; ISBN 84-958-77-30-9.
62. Meco, J.; Guillou, H.; Carracedo, J.C.; Lomoschitz, A.; Ramos, A.J.G.; Rodríguez-Yánez, J.J. The maximum warmings of the Pleistocene world climate recorded in the Canary Islands. *Palaeogeogr. Palaeoclimatol. Palaeoecol.* **2002**, *185*, 197–210. [CrossRef]
63. Zazo, C.; Goy, J.L.; Hillaire-Marcel, C.; Gillot, P.Y.; Soler, V.; Gonzalez, J.Á.; Ghaleb, B. Raised marine sequences of Lanzarote and Fuerteventura revisited-a reappraisal of relative sea-level changes and vertical movements in the eastern Canary Islands during the Quaternary. *Quat. Sci. Rev.* **2002**, *21*, 2019–2046. [CrossRef]
64. Kennedy, D.M.; Dickson, M.E. Lithological control on the elevation of shore platforms in a microtidal setting. *Earth Surf. Process. Landf. J. Br. Geomorphol. Res. Group* **2006**, *31*, 1575–1584. [CrossRef]
65. Naylor, L.A.; Stephenson, W.J. On the role of discontinuities in mediating shore platform erosion. *Geomorphology* **2010**, *114*, 89–100. [CrossRef]
66. Rijsdijk, K.F.; Hengl, T.; Norder, S.J.; Otto, R.; Emerson, B.C.; Ávila, S.P.; López, H.; van Loon, E.; Tjørve, E.; Fernández-Palacios, J.M. Quantifying surface-area changes of volcanic islands driven by Pleistocene sea-level cycles: Biogeographical implications for the Macaronesian archipelagos. *J. Biogeogr.* **2014**, *41*, 1242–1254. [CrossRef]
67. Ballouche, A.; Lefevrè, D.; Carruesco, C.; Raynal, J.P.; Texier, J.P. Holocene environments of coastal and continental Morocco. In *Quaternary Climate in Western Mediterranean*; Lopez Vera, F., Ed.; Universidad Autónoma de Madrid: Madrid, Spain, 1986; pp. 517–531.
68. Barusseau, J.P.; Descamps, C.; Giresse, P.; Monteillet, J.; Pazdur, M. Nouvelle définition des niveaux marins le long de la côte nord-mauritienne (Sud du Banc d'Arguin) pendant les cinq derniers millénaires. *Comptes Rendus L'académie Sci. Sér. 2 Mécanique Phys. Chim. Sci. L'univers Sci. Terre* **1989**, *309*, 1019–1024.
69. Meco, J.; Lomoschitz, A.; Rodríguez, Á.; Ramos, A.J.; Betancort, J.F.; Coca, J. Mid and Late Holocene sea level variations in the Canary Islands. *Palaeogeogr. Palaeoclimatol. Palaeoecol.* **2018**, *507*, 214–225. [CrossRef]
70. Herrera-Holguín, A.C.; Menéndez, I.; Mangas, J. Beachrock, paleosuelo y aeolianita holocenas en el arco central de la playa de las canteras (NE de Gran Canaria, España). In Proceedings of the XV Reunión Nacional de Cuaternario Bizkaia Aretoa, Bilbao, Spain, 1–5 July 2019; Libro de Resúmenes; Universidad del País Vasco/Euskal Herriko Unibertsitatea: Leioa, Spain, 2019; pp. 243–246.

remote sensing

MDPI

Article

Contribution of Reverse Dune Migration to Stabilization of a Transgressive Coastal Dune Field at Lagoa do Peixe National Park Dune Field (South of Brazil)

Rogério Portantiolo Manzolli [1,*], Luana Carla Portz [1], Angela Fontán-Bouzas [2,3], Volney Junior Borges Bitencourt [4] and Javier Alcántara-Carrió [1]

1 Department of Geology and Geochemistry, Universidad Autónoma de Madrid, 28049 Madrid, Spain; luana.portz@uam.es (L.C.P.); javier.alcantara@uam.es (J.A.-C.)
2 Department of Geodynamics, Stratigraphy and Paleontology, Faculty of Geological Sciences, Complutense University of Madrid, C/José Antonio Novais, 12, 28040 Madrid, Spain; anfontan@ucm.es
3 CIM-UVIGO GEOMA, University of Vigo, 36310 Vigo, Spain
4 Instituto do Meio Ambiente de Santa Catarina (IMA), Florianópolis 88020-300, Brazil; volneybitencourt@ima.sc.gov.br
* Correspondence: rogerio.manzolli@uam.es

Abstract: Coastal dunes that transgress typically move landward, while their reverse movement is not well understood. The article discusses the study of barchan and barchanoid dunes in the Lagoa do Peixe National Park in the coastal plain of Rio Grande do Sul, Brazil. The aim of the study is to analyze seasonal patterns and long-term trends in the direction and migration rates of these dunes, which can pose a threat to the lagoon if they invade its space. The crest migration of 12 dunes was monitored by satellite images between July 2003 and December 2018, and DGPS topographic surveys were performed on five dunes between 2010 and 2018. The migration rates obtained were combined with an analysis of the meteorological data and calculations of the drift potential for eolian sediment transport. The wind regime in the study area shows a multidirectional pattern, with the predominant wind direction being from the NE, followed by the ENE direction. The wind direction also exhibits a seasonal behavior, with the winds from the first quadrant being dominant during spring and summer months and a gradual increase in winds from the second and third quadrants from the end of summer to winter. The dune crest migration rates in the Lagoa do Peixe National Park show an average of 16.55 $m \cdot yr^{-1}$ towards WSW–W, mainly controlled by the direction of the effective winds. However, intense SSW–WSW winds caused by cold fronts in the past generate the reverse migration of dunes towards ENE–E. The reverse migration of dunes explains the steadiness of the dune fields at CPRGS and is a factor controlling dune stabilization and the geomorphological evolution of transgressive coastal dune fields. The article highlights the importance of monitoring dune movement to understand their responses to natural and anthropogenic stressors and to protect sensitive ecosystems.

Keywords: coastal dune; barchan dune; sediment transport; drift potential; migration rate

Citation: Manzolli, R.P.; Portz, L.C.; Fontán-Bouzas, A.; Bitencourt, V.J.B.; Alcántara-Carrió, J. Contribution of Reverse Dune Migration to Stabilization of a Transgressive Coastal Dune Field at Lagoa do Peixe National Park Dune Field (South of Brazil). *Remote Sens.* **2023**, *15*, 3470. https://doi.org/10.3390/rs15143470

Academic Editors: Ramón Blanco Chao, Germán Flor-Blanco and José Juan de Sanjosé Blasco

Received: 14 April 2023
Revised: 27 June 2023
Accepted: 28 June 2023
Published: 10 July 2023

1. Introduction

Coastal dunes are naturally developed on sandy beaches, from tropical to arctic conditions [1,2]. They constitute large sediment accumulations of which form, size, and orientation vary according to the beach profile, coast orientation, direction and speed of the wind, particle size, and type of vegetation of each location [3–5].

Coastal dunes are dynamic environments that develop and evolve as the result of a complex interaction between sand, wind, vegetation, and external pressures [6]. They form when sediment deposited over the beach dries, and then it is blown by the wind landward [7]. Their occurrence is, therefore, directly related to sand supply and favorable wind patterns [8]. Coastal dunes tend to develop in beaches where there is a large sediment supply, enough wind to move this sediment landward, and a backshore space to accumulate this sediment [9]. The sediment availability depends on both the volume of sediments and its grain size [10,11]. The beach profile also controls the development of dunes; usually, the development of dunes is favored in dissipative beaches better than in reflective ones [6].

The geomorphological evolution and stabilization of coastal dune fields can be primarily related to either climate variations or human interferences, which are able to transform active dunes into stable ones and vice-versa. Changes in vegetation cover, due to both climate variations and human activities, contribute to dune stabilization [12–14]. The model developed by [15] predicts that long droughts with strong winds might result in the reactivation of dunes. Even relatively small changes in climatological parameters can generate changes in both vegetal cover and geomorphological configurations [15–17]. Monitoring the dune movement is essential in this scenario; dune migration rates can provide essential information about the dunes' responses to natural (wind and rain regimes, vegetation cover) and anthropogenic stressors [18].

The monitoring of dune systems is being conducted in several regions, using different techniques after the spatial and temporal scales of the study, such as drones. Thus, aerial photographs and/or satellite images are often used to evaluate changes in the dunes' position and morphology, particularly for long-term studies, i.e., ten years or more [18–26] By contrast, short-term studies and fieldwork measurements with GPS were initially the most usual method for mapping and measurements of the dunes' slip faces [27–29]. However, nowadays, LiDAR and drone systems are more frequently used [30–33].

Most of the dune fields in the coastal plain of Rio Grande do Sul (CPRGS) are currently located in the Middle and South regions. These dune fields have a history of stabilization and vegetation growth, which are related to high precipitation levels, reduced wind speed, and large sediment supply [34], as well as alterations in the regional hydrology that influence groundwater levels [35]. The objective of this study is to analyze the normal and reverse migration of the transgressive dunes, related to the typical annual wind patterns and the occurrence of cold fronts, respectively, and their implications for the stabilization of the dune field of the Lago do Peixe Natural Park.

2. Study Area

The study area is located on the dune field close to the Peixe lagoon (middle region of the CPRGS). This dune field and coastal lagoon form the Lagoa do Peixe National Park, designated as a Ramsar site, but the lagoon is at risk of being invaded by the transgressive barchan and barchanoid dunes.

2.1. Regional Setting and Study Area—Relevance of Dune Fields in Rio Grande do Sul

The relief adjacent to the coast of Rio Grande do Sul varies abruptly. The North region is formed by the high mountains of the Serra Geral escarpment; in the South, the escarpment ends, and a large alluvial coastal plain prevails with a flat and open surface. The CPRGS was reworked during the transgressive and regressive cycles of the Quaternary. Initially, deposits of coalescing alluvial fans developed at the end of the Tertiary because of the transport processes of terrigenous clastic sediments associated with upland environments. Afterwards, these deposits were laterally overlapped by four barrier-lagoon depositional systems, whose formation was controlled by the sea level fluctuations during the Quaternary (Barriers I, II, and III, from the Pleistocene, and Barrier IV, from the Holocene), creating a succession of marine and lagoon terraces [36–38]. Consequently, this alluvial plain has a complex system of sandy barriers that protect a large lagoon system (Patos lagoon and Mirim lake) and a series of other waterbodies, isolated or interconnected with the sea through narrow and shallow channels [36].

The coast of Rio Grande do Sul has several factors that favors the formation and evolution of one of the most extensive systems of the coastal dunes of Brazil, with low roughness topography (coastal plain), appropriate wind regimes, and a large supply of fine quartz sand [39]. In these sectors, transgressive dune fields are formed. These are broad, eolian sand deposits formed by the downwind movement of sand sheets and free dunes over vegetated to semi-vegetated terrain [40]. They range from small sheets to large-scale sand seas, typically bordered by precipitation ridges and often fronted by deflation basins and plains [41]. Barchan dunes and barchanoid ridges are common features along the southern Brazilian coastal dunes [34,42–44]. In some sectors with larger dune fields (such as the middle of RS), transverse dunes can also be identified. In the marginal portions of the dune fields, places with less sand and a more abundant presence of vegetation, there are predominant parabolic sand dunes [45]. In fact, in Rio Grande do Sul, the dunes are located along the entire coast, being drastically reduced in sectors where urbanization is more developed [46]; in such cases, only frontal dunes are formed [45,47–50].

The Lagoa do Peixe National Park (LPNP) is in the Middle region of the CPRGS, between the Patos lagoon and the Atlantic Ocean (Figure 1). The Park includes several ecosystems that are representative of the region, such as beaches, dunes, salt marshes, saline swamps, and the lagoon.

The Peixe lagoon is 35 km long, 1 m wide, and approximately 30 cm deep. The origin of the Peixe lagoon is related to the formation of a marine and eolian sediment barrier during the Holocene transgression. In fact, it corresponds to the lagoon portion of the Barrier-Lagoon System IV. Moreover, there is a direct connection with the sea in the southern sector of the lagoon; it happens through an artificial opening of the barrier during winter months that persists until the beginning of summer, when the prevailing winds deposit marine sediments, blocking the opening [51].

The dunes cover approximately 45% of the LPNP area [52], forming a continuous band alongshore. They are mainly composed of unconsolidated quartz sand with grain size (mean sediment grain size 0.215 mm) [53]. The dunes are very prominent, and they are better represented at the northern area, where the higher dunes with heights over 15 m and perpendicularly orientated in relation to the NE wind direction are located [51]. The width of the dune field varies alongshore of the national park, reaching a maximum of 5.15 km in the north and a minimum of 0.70 km in the south. The speed and direction of dune migration on the east side of the LPNP are currently among the discussion topics regarding management practices. Given the low depth of the lagoon, changes in these patterns could greatly affect the input of sediment to the lagoon system, causing high-impact environmental changes.

2.2. Climate Control of Dune Fields in Rio Grande do Sul

The regional winds blowing in the study area are connected to the atmospheric flow over Rio Grande do Sul. This flow is generated by the interaction among the Atlantic Subtropical Anticyclone, the intermittent movements of polar masses, and the barometric depression of northeastern Argentina [54]. The variations on this atmospheric system result in wind seasonal patterns. During spring and summer, the weather in the coastal plain is usually warm and windy, particularly with winds from NE and E; during fall–winter, the area is dominated by cold fronts coming from SW–NE, oftentimes regularly [55]. The winds intensity varies along the coast, with a positive speed gradient from North to South. In response to the prevailing winds, the free dunes migrate towards SW [29,47,54,56].

The dimensions of the coastal dune fields in CPRGS are associated with wind patterns but also with the rainfall regime of each sector. The dune fields at the northern littoral are narrower (1300 to 1400 m wide) due to the local higher precipitation, lower wind drift potential (DP), and smaller sand supply. Southwards, the dune fields increase their width (reaching 6900 m) as the precipitation decreases, and the wind DP and the sand supply increases in the area [34].

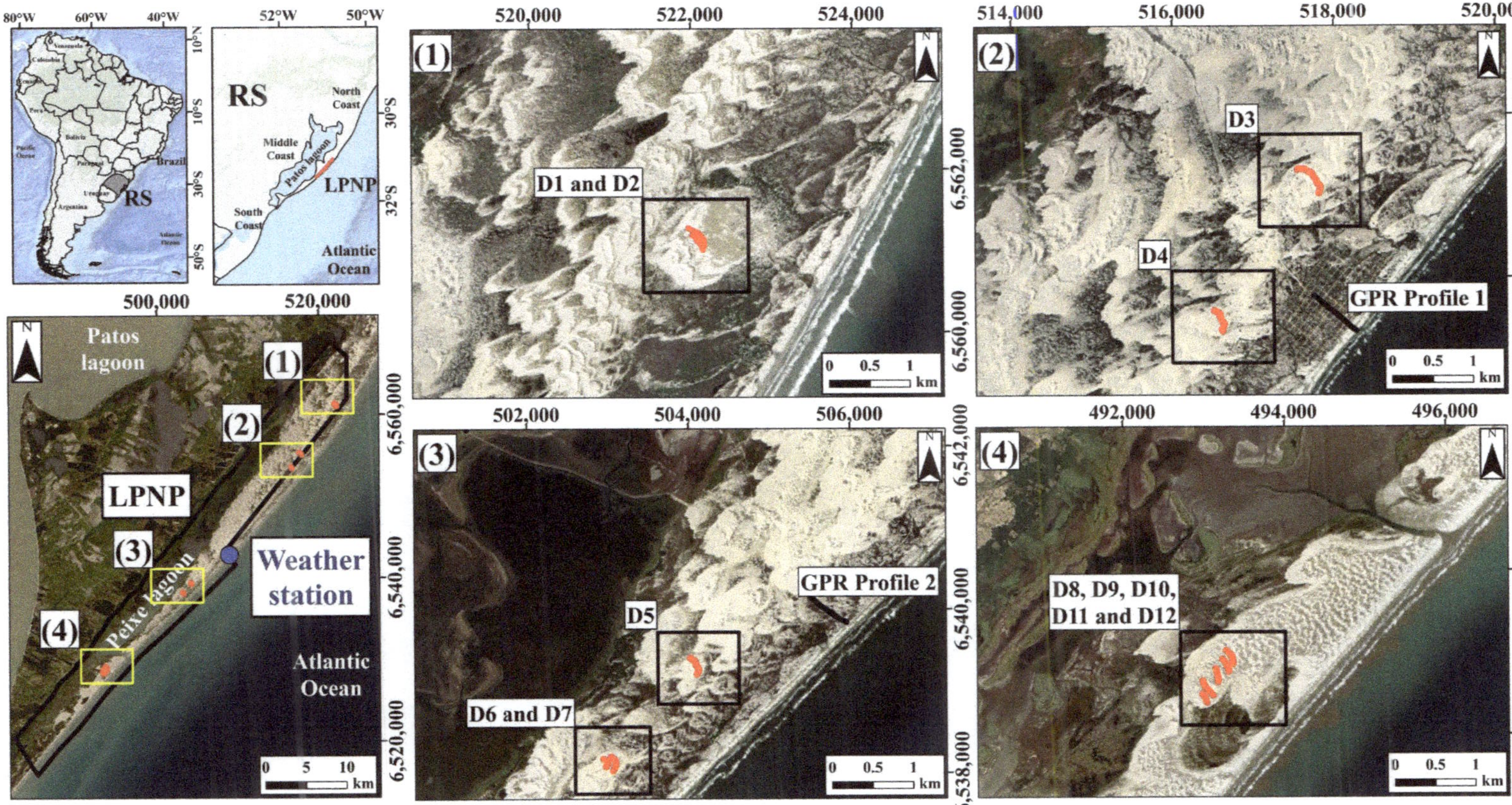

Figure 1. Location of the study area on the middle sector of the coastal plain of Rio Grande do Sul (South of Brazil). On the left image, the black line is delimiting the Lagoa do Peixe National Park, and the blue point is the location of the weather station. In the right images, for sectors 1 to 4, the red lines indicate the crests of the monitored dunes (Image base: ArcMap® 10.8).

2.3. Coastal Barrier Stratigraphy

According to several studies, the stratigraphy of the study area shows thick eolian deposits [57,58]. The GPR surveys conducted at the Peixe lagoon exhibited up to 15 m of thickness [59,60]. This is a result of both the coastline orientation relative to the NE winds [61] and the high sediment supply rate resulting from a jam in the longshore transport [62]. The silting of back-barrier lagoons by the transgression of dune sands provides a platform for barrier translation during the post-glacial marine transgression (PMT). Probably, such processes were operating since before the sea level maximum of the mid-Holocene (~6 ka ago) [57,58]. Thus, the eolian component plays a crucial role in the development of this barrier system during the mid- to late Holocene [57,58,61,63]. Overall, the middle coast of Rio Grande do Sul shows a transgressive stratigraphy with recent evidence of progradation in a few stretches [61,63]. Figures 2 and 3 shows a 500 m long processed (A) and interpreted (B) 80-MHz GPR profile perpendicular to the coast (in the vicinity of the Peixe lagoon) [60]. This interpreted stratigraphic cross section is supported by sedimentological and geochronological data [58].

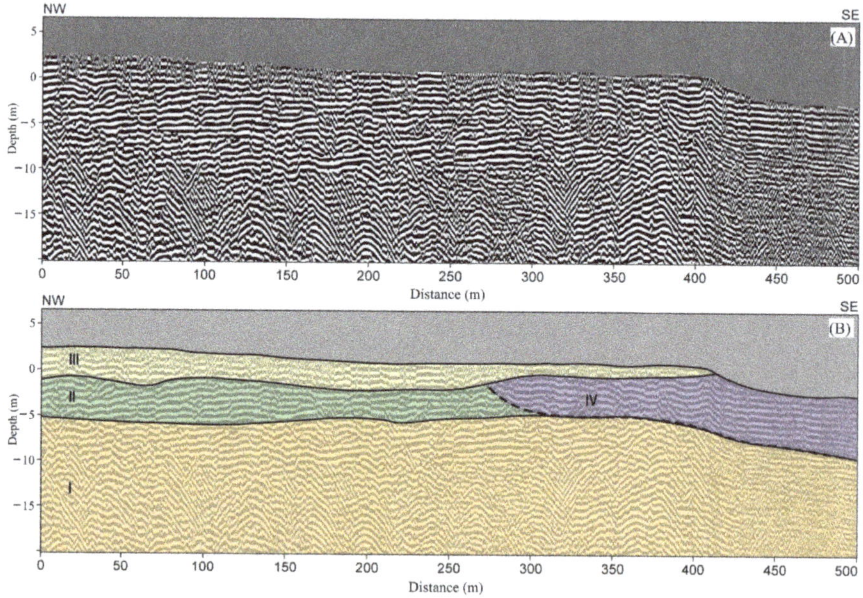

Figure 2. 2-D GPR Profile 1 from Balneário Mostardense, (**A**) processed and (**B**) interpreted. Key: I—undifferentiated deposits, corresponding to the Pleistocene substrate, II—lagoonal, III—eolian, and IV—beach (including post-beach, tidal flat, and foreshore). The dotted line marks a reversal in reflection patterns. See location in Figure 1. (Adapted from [60]).

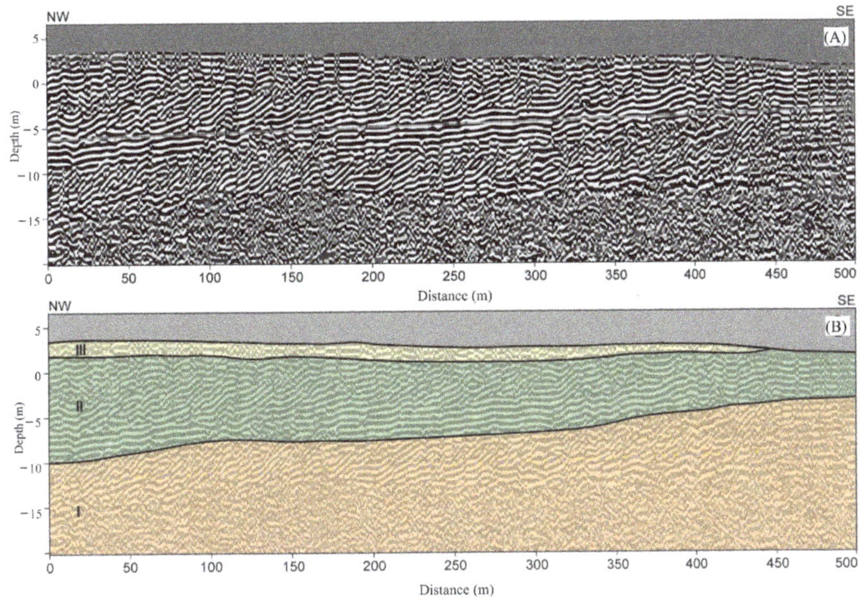

Figure 3. 2-D GPR Profile 2 from North Lagoa do Peixe, (**A**) processed and (**B**) interpreted Key: I—pre-Holocene substrate; II—lagoon margin clinoforms, and III—eolian capping. Landward to NW. See location in Figure 1. (Adapted of [60]).

3. Materials and Methods

3.1. Meteorological Data Analysis

The wind (speed and direction) and precipitation data for the study period were obtained from the Mostardas-INMET A834 meteorological station (altitude: 4 m) and provided by the Brazilian National Meteorological Institute—INMET (see location in Figure 1). This station has been operating since 13 March 2008, to 2018.

The wind data were analyzed and the rose diagram was plotted by the software GRAFER® 8.0, considering five speed classes (0–3; 3–6; 6–9; 9–12, and >12 $m \cdot s^{-1}$) and the sixteen main directions of wind (N, NEN, NE, ENE, E, ESE, SE, SES, S, SWS, SW, WSW, W, WNW, NW, and NWN). They were five defined study periods for wind data analysis, in accordance with the dates of the topographic surveys: P1 (12 December 2010–7 December 2014), P2 (8 December 2014–1 April 2015), P3 (2 April 2015–20 April 2016), P4 (21 April 2016–9 December 2016), and P5 (10 December 2016–31 December 2018), with TP being (12 December 2010–31 December 2018) the total study period for the meteorological data.

For a better understanding and discussion of the results, the wind directions were subdivided into 4 quadrants: Q1 (N, NNE, NE, ENE); Q2 (E, ESSE, SE, SSE); Q3 (S, SSW, SW, WSW); and Q4 (W, WNW, NW, NNW). Furthermore, the wind directions were also classified based on their relation to the input or output of sediments in the dune system. Wind directions towards the land include NE, ENE, E, ESE, SE, SSE, S, and SSW, while wind directions towards the sea are NNE, N, NNW, WNW, W, WSW, and SW. This segmentation allows for a more precise analysis of the specific influences of these wind directions on the input or removal of sediment s in the dune field.

Monthly average rainfall and directional rainfall histograms were computed from the database of the same meteorological station. In addition, the directional histograms for the above periods were plotted considering wind direction, which allowed for identifying wind directions.

3.2. Computation of Potential Eolian Sand Transport

The first set of wind roses, for each period of dune migration analysis P1 to P5 and the total period TP, were plotted after the determination of drift potential (DP), real drift potential (RDP), and directional drift potential (DDP) according to Fryberger and Dean [64]. In these wind roses, the sand drift (SD) for each wind direction is expressed in vector units (V.U.).

Furthermore, the sand roses were computed considering the equation of Bagnold [65] for the prediction of bed load eolian sand transport, i.e., considering sediment characteristics and surface roughness too, instead of only wind characteristics as in Fryberger and Dean [64], according to Alcántara-Carrió and Alonso [66].

The empirical equation of Bagnold [65] for the prediction of eolian sand transport was utilized to calculate the potential eolian sand transport, because previous calibrations with sediment traps in the study area showed that this equation is the one that provides the best agreement with the sand transport in the area [48]. This equation considers the average grain size of sediment, the average sediment density, and the saturated air density, as well as the wind speed and direction.

The wind data were transformed to a 10 m height according to Bagnold [65]; however, the meteorological station was placed at a 4 m height. The wind values during precipitation events were considered zero, i.e., without the ability to cause eolian sand transport. The threshold wind speed used to define a transport event was defined based on the equation by Bagnold [65] to isolate the periods where there was potential for transport and exclude the remaining data. Thus, the minimum wind speed for the initiation of motion was defined as 5 $m{\cdot}s^{-1}$, which is the minimum velocity required for sediment saltation. Therefore, only wind speeds above this velocity were considered for the sediment transport calculations. The value obtained by the equation expresses the relative amount of sand potentially transported by the wind during the time that wind was blowing. The results were compared with the previous studies in the study area and region.

3.3. Topographic Surveys

The detailed morphology and topographic evolution of five dunes from the central sectors 2 and 3 were monitored by DGPS. In sector 2, D3 and D4 correspond to barchanoid dunes; in sector 3, D5 corresponds to a barchanoid dune, while D6 and D7 are two barchan dunes (Figure 1). Six topographic surveys were carried out in December 2010 (only dunes D3 and D4), December 2014, April 2015, April 2016, December 2016, and December 2018. These surveys were performed with DGPS equipment in cinematic mode with real-time correction (RTK—real time kinematic), using a post-processed GNSS. The data were acquired using a Topcon RTK-S86T GNSS unit with the GLONASS option (datum: WGS84), having both a planar metric precision and a planar altimetric precision smaller than 1 cm.

The data processing was performed in a geographic information system using ArcMap® 10.8 software. To generate the digital terrain models (DTM), the inverse distance weighted (IDW) interpolation method was used. This method was selected for its capacity to incorporate geostatistical analyses. These analyses showed a good correlation between the measured and interpolated data, making it possible to estimate the quality of the elevation points predicted in terms of a variance estimate. The DTMs obtained through the IDW interpolation presented a very low error indicated by the distribution, even considering the differences in the number of points sampled in different years.

3.4. Remote Sensing

The satellite images were also analyzed to determine the migration of the dune crest, considering a set of 12 dunes from the four sectors of the LPNP. Thus, this remote sensing analysis was performed for the same dunes D3 to D7 from central sectors 2 and 3, monitored by DGPS topographic measurements, and seven additional dune crests: D1 and D2 from sector 1 and D8 to D12 from sector 4, in the northern and southern margin of the study area, respectively (Figure 1).

Google Earth images were employed from the available dates (July 2003; July 2005; October 2005; December 2014; December 2018). The initial images up to the year 2014 are considered interval 1, and between 2014 and 2018, they are considered interval 2. All the remote sensing products were reprojected to the UTM projection (zone 22S) and WGS84 datum. A minimum of 10 control points per image were used in the georeferencing process, and the mean squared error was less than 1.0 m. The satellite images were also georeferenced using the same control points. The use of images from Google Earth is a technique already established in several publications [67–73]. The dune slip faces and crests were used to map the dunes by photointerpretation. These morphologies correspond to light or dark linear ridges in the images, depending on the relationship between the lighting configuration (subsolar azimuth) and the trends and types of dunes [74]. Each dune was analyzed and vectorized considering these characteristics. These features were mapped using a line segment (vector data).

3.5. Determination of Dune Migration Rates

Digital terrain models (DTMs) were created based on the topographical survey data. The methodology developed by Xia and Dong [75] was utilized to determine the dune migration rates. Thus, the dune crest lines were traced over the surfaces as the lines with the highest elevation and with the greatest slip face slopes. Then, these lines were divided into points where the vectorial distance to each point was measured. The vectorial average of the points was used to obtain the average movement of the crests, as well as the migration direction. All the processing and calculations were performed in ArcMap® 10.8.

4. Results

4.1. Climate

4.1.1. Wind

The wind regime shows a multidirectional pattern with the predominance of the NE wind direction, being the secondary direction from the ENE. The landward winds are more dominant than the seaward winds, and therefore, there is an eolian sediment input blowing from the beaches to the dune fields (Figure 4 and Table 1).

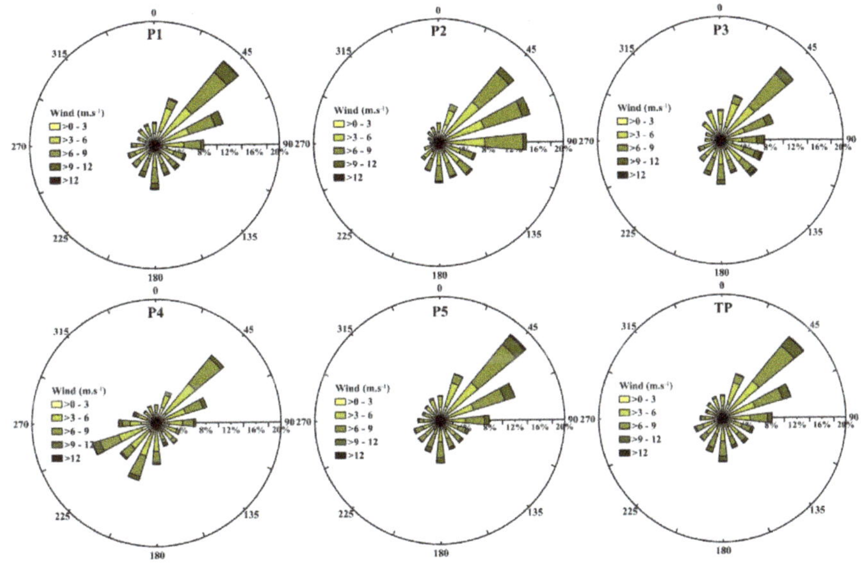

Figure 4. Wind roses by speed classes (m·s^{-1}) for the periods P1, P2, P3, P4, P5, and TP. See methodology for the temporal limits of the periods.

Table 1. The wind frequency data (%) for each direction and study period.

	Wind Frequency																	
	N	NNE	NE	ENE	E	ESE	SE	SSE	S	SSW	SW	WSW	W	WNW	NW	NNW	Landwa	Seaward
P1	3.7	7.9	17.5	11.5	7.9	5.3	5.2	5.2	7.3	5.5	4.7	4.5	3.8	2.9	3.4	3.6	65.4	34.6
P2	3.2	6.5	15.7	15.3	14.1	6.2	7.0	6.1	6.6	4.9	3.7	2.5	1.5	1.8	2.4	2.7	75.8	24.2
P3	4.4	7.0	15.4	10.5	8.6	7.2	7.4	5.6	7.2	5.9	4.1	3.5	2.6	2.6	3.6	4.6	67.6	32.4
P4	2.9	5.3	14.1	8.5	6.4	4.0	4.4	4.0	6.8	9.8	7.6	10.7	6.0	4.0	2.6	2.9	57.9	42.1
P5	4.0	7.9	17.9	12.6	7.9	4.9	4.8	5.0	6.9	5.4	4.6	4.5	3.6	3.0	3.2	3.8	65.4	34.6
TP	3.8	7.5	17.0	11.6	8.1	5.4	5.4	5.1	7.1	5.8	4.8	4.8	3.7	2.9	3.3	3.7	65.5	34.5

For all periods, the landward wind directions were predominant (NE; ENE; E; ESE; SE; SSE; S; SSW), with the lowest percentage identified in P4 with 57.90% and the highest frequency of occurrence in P2 with 75.8%. The frequency of occurrence for the predominant NE wind direction ranged between 17.90% in P5 and 14.10% in P4, considering that the latter period does not include the summer season, between the months of January to March. Furthermore, this period was the only one that presented the secondary direction in WSW, with a frequency of 10.73%; in the other study periods, the secondary direction was ENE, oscillating between 15.34% and 10.47% for P2 and P3, respectively. (Figure 4 and Table 1).

The wind direction shows seasonal behavior, with the predominance of winds coming from the first quadrant (i.e., N, NNE, NE, and ENE) during the spring (October, November, December) and summer months (January, February, March). The frequencies of the winds from the third and fourth quadrants are low, with the exception of the winds from the South in October. From the end of the summer, in March, there is a gradual increase in winds from the second and third quadrants and a decrease in the frequency of winds from the first quadrant, mainly from the NE. By contrast, with the beginning of the winter, in June, the predominance of the wind's provenance is from the third quadrant Q3, with the highest frequencies from the WWS (Figure 5).

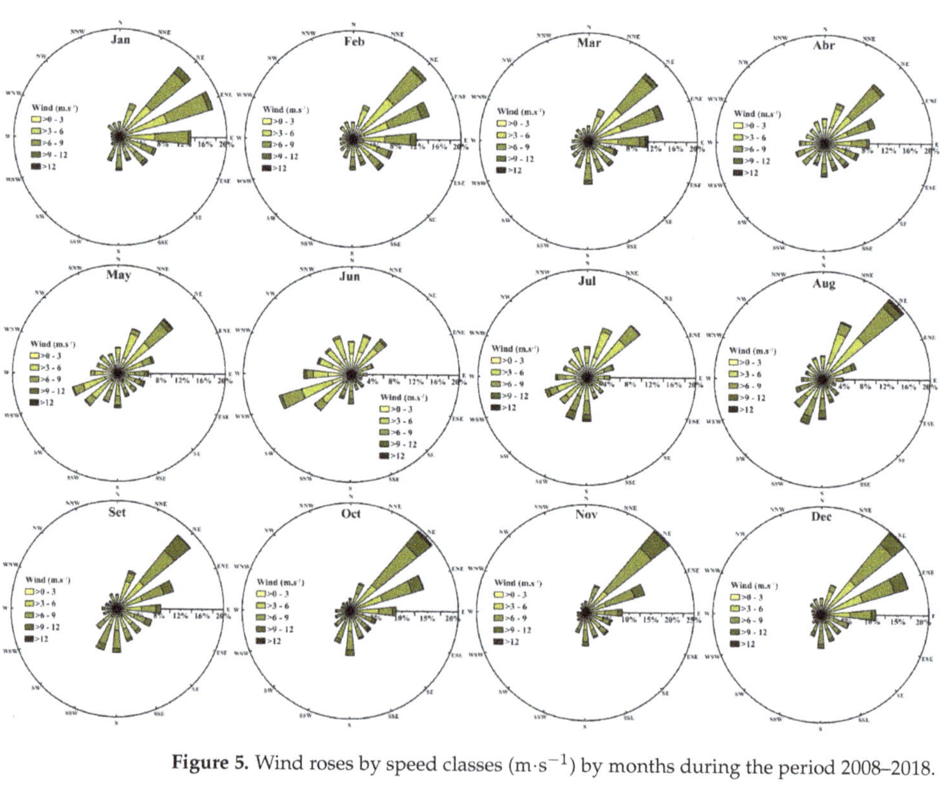

Figure 5. Wind roses by speed classes ($m{\cdot}s^{-1}$) by months during the period 2008–2018.

The wind speed intensity also varies with the seasons, with the highest speeds from the NE between August and December, the highest speeds from the WWS direction in May and June, and the highest speeds from the S and SSW are in July and August.

4.1.2. Rainfall

The frequency and volume of the rainfall during the study period showed great variability (Figure 6). The total annual precipitation values ranged between 966.2 (2011) and 1757.2 mm (2015), with an average of 1268.2 mm. The rainiest year was 2015, with 17% of the total precipitation of the whole study period, while the other years presented percentages between 9.5 and 12.7%.

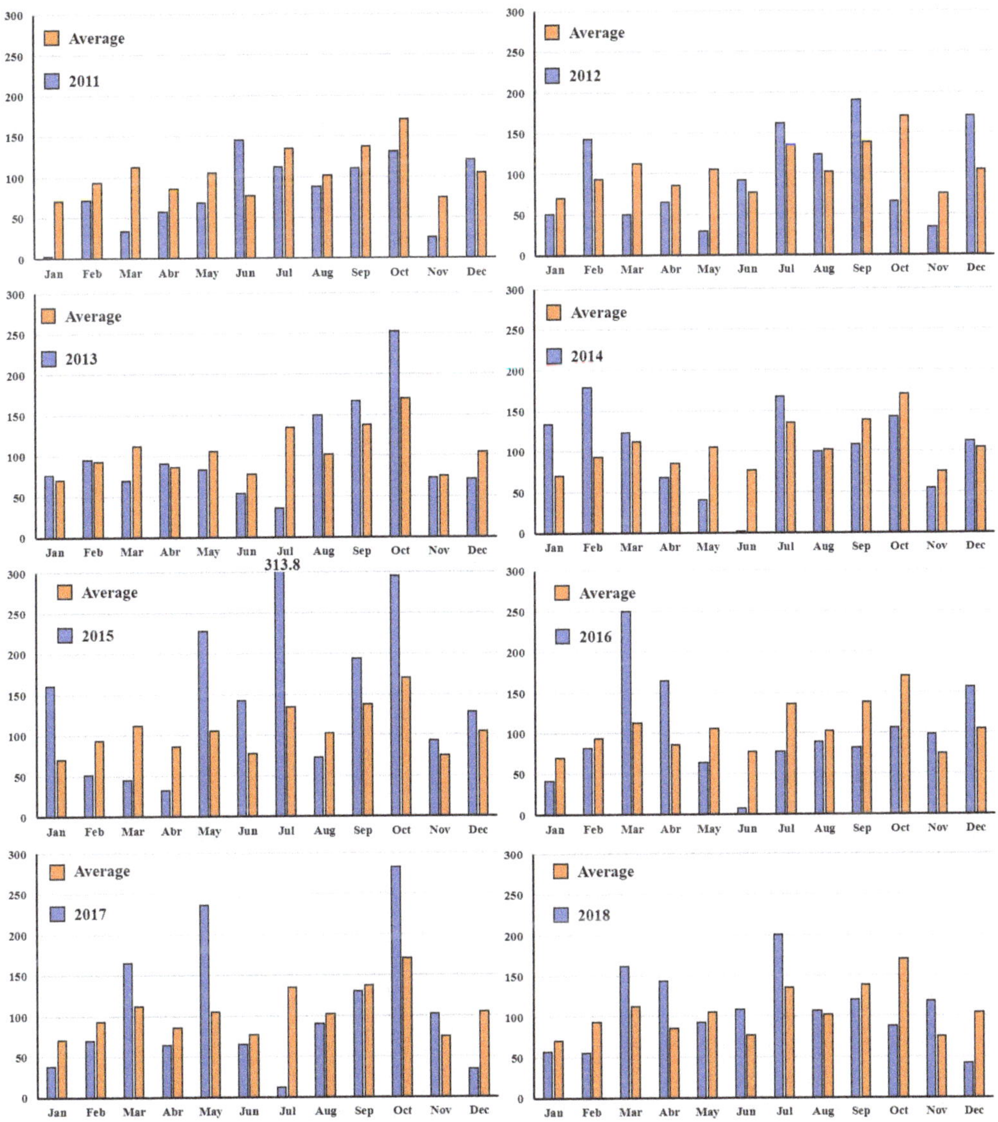

Figure 6. Tha rainfall data by month for the years between 2011 and 2018 (blue bars), compared to the average for the entire analyzed period from 2008–2018 (orange bars).

The highest precipitation volumes were associated with the winter and early spring period (June–September) and the lowest averages with the fall (March–June). The volume of precipitation showed a seasonal behavior, presenting few anomalous values to the monthly average for the entire period. In this interval of analysis, October was the month that presented the highest average precipitation (170.5 mm), with maximum values in 2015 (295.6 mm) and minimum values in 2012 (65.8 mm). The month with the highest volume of precipitation was July 2015 (313.8 mm), well above the average of 135.0 mm for this month (Figure 6). The frequency of hours with rainfall, after the wind approach direction, shows that NE winds present the highest frequency of rain, but for period P3, they were from the second quadrant (Figure 7).

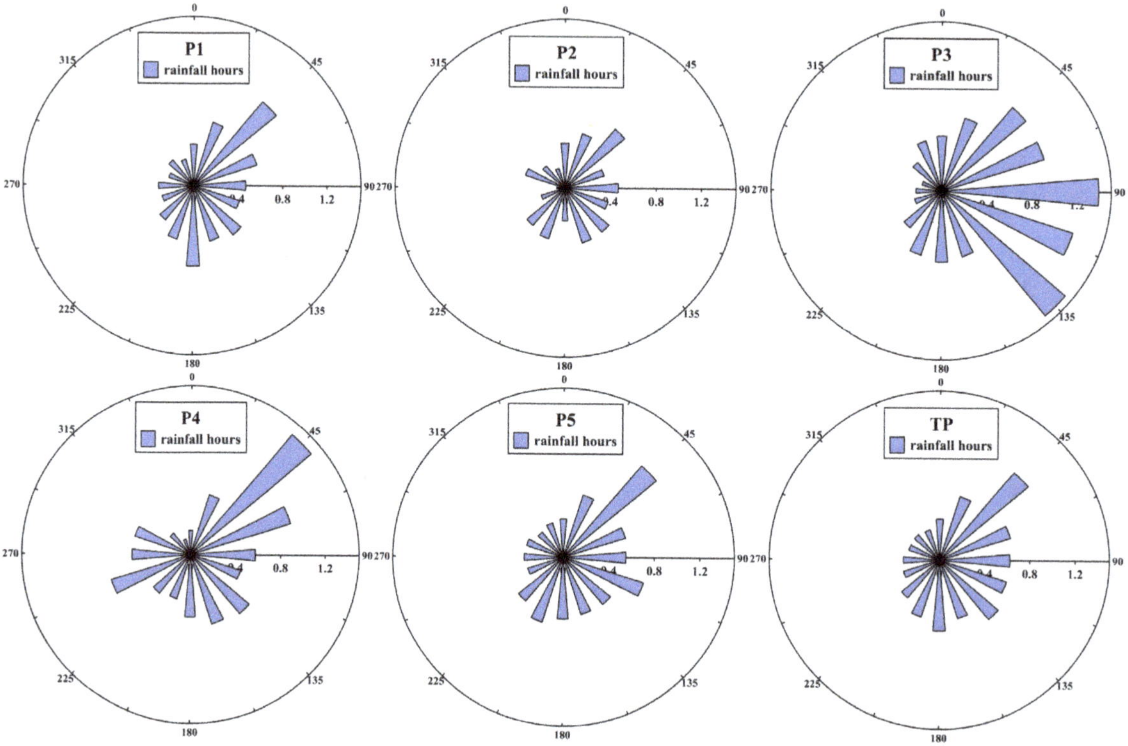

Figure 7. The frequency of rainfall hours by wind direction for the analyzed periods for each analyzed period (P1 to P5) and the total period (TP).

4.2. Eolian Sediment Drift Potential (Fryberger and Dean' Method)

Sand roses for each period of dune migration analysis show that the sand drift potential values are characterized by important variability. The most active period in terms of SD was P5, with values of 98.1 v.u., which is demarcating a high-energy environment. The lowest values (68.8 v.u.) were observed for period P2 (Figure 8).

The RDD for periods P1 to P3 and P5, as well as for the total study period (TP), was toward the W/WNW (Figure 8). However, the wind pattern in the fall–winter period P4 presented a significant change, with the predominance of SSW and WSW winds, and, therefore, the RDD was towards the NNE. The RDP values ranged from 16.8 (P4) to 59.8 (P3).

The wind directional variability measured through the RDP/DP ratio ranged from 0.22 (P4) to 0.65 (P2). The RDP/DP values near one indicated a unidirectional drift potential, and values near zero indicated a multidirectional drift potential. Only period P4 can be considered low according to Fryberger and Deans' Index, indicating a high directional variability of the winds in this period. The other periods are considered intermediate. The relatively high value of 0.65 v.u. for P2 suggests that winds in this period blow predominantly from the NE/ENE quadrant, moving the sand towards the SW.

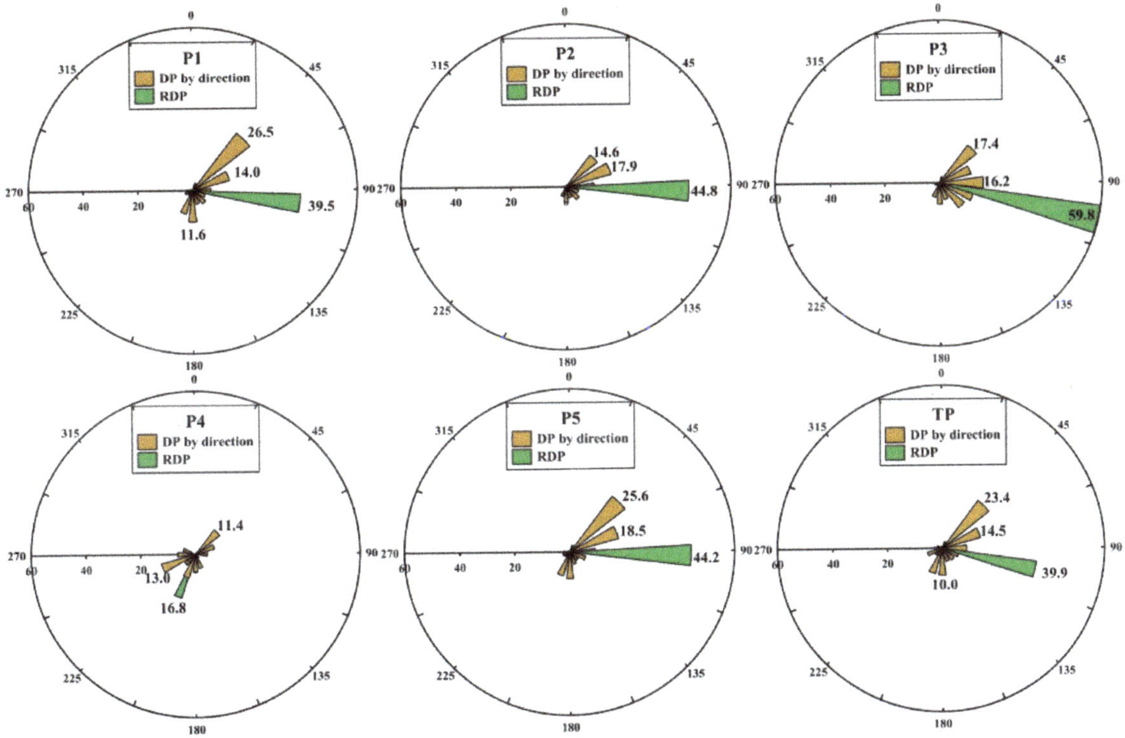

Figure 8. Potential sand transport roses for each analyzed period (P1 to P5) and the total period (TP), determined after the classical method of Fryberger and Dean (1979).

4.3. Prediction of Eolian Sediment Transport (Bagnold' Equation)

The sediment transport during the dune monitoring periods, predicted by Bagnold's equation (1941) and considering the sediment transported during rainfall events as null, shows a significant variability with a resultant direction towards W for all periods, except for P4. Thus, a very characteristic pattern is observed in periods P1–P3 and P5, where there was almost no transport associated with seaward (N–SW) winds, and most of the transport was associated with NNE/NE winds. The period P4 shows the resultant transport towards NNE/NE, i.e., an inversion in the resulting transport component. Finally, the resultant for the full period (TP) was also towards the W. The estimated values of sediment transported for the total period (TP) was 300.639 Kg·m^{-1}, the NNE/NE wind being responsible for the largest volume (maximum of 207.428) of estimated sediment transport (Figure 9).

4.4. Dune's Morphology and Migration

The morphological features of the dune system of the Lagoa do Peixe National Park present a complexity regarding the dunes dimensions. Several factors influence these features, from the sediment supply (interrupted in some sectors by anthropic activities—sector 2); the presence of vegetation, both natural and introduced; the antecedent topography of the dune system (wetlands, wind deflation zones, and blowouts); and the seasonal wind direction variation.

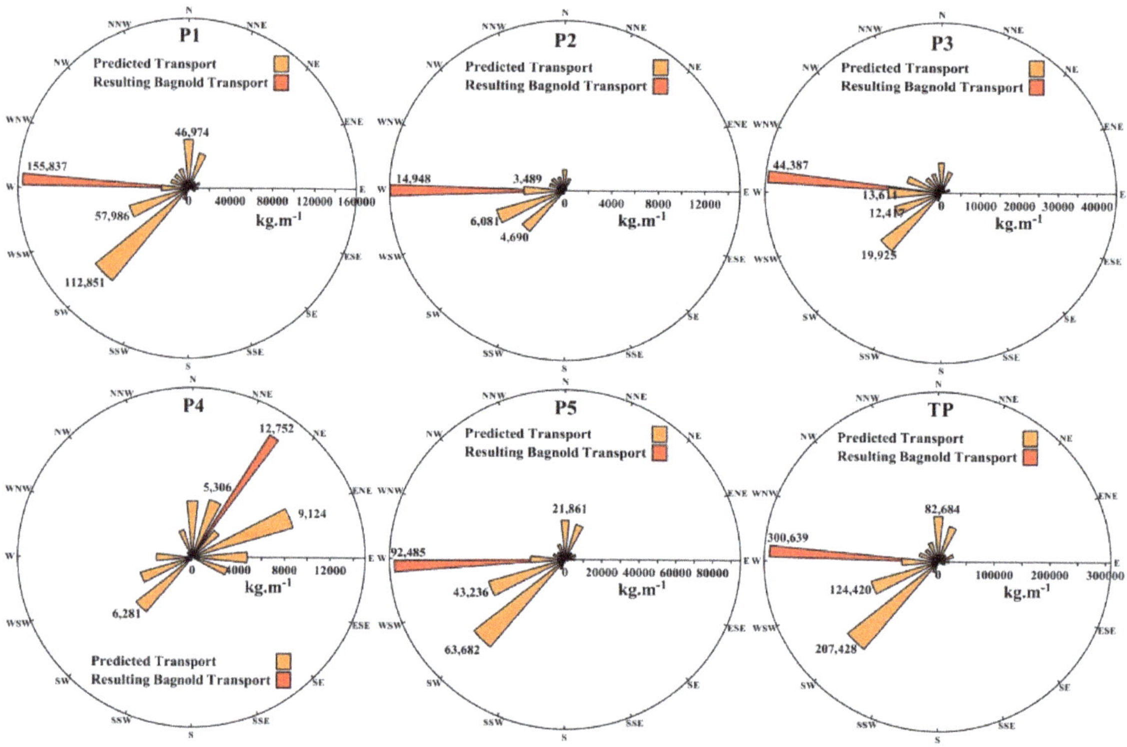

Figure 9. Modified Sand rose diagrams for the periods P1, P2, P3, P4, P5, and TP considering Bagnold' equation (1941). For each period, the prediction of the bed load sediment transport (Kg·m^{-1}) per direction is in orange and the resulting sediment transport in red.

The dune system of the LPNP is very dynamic with the continuous movement of sands, which provides constant variations in the dune dimensions. Table 2 shows the main characteristics of each dune at the beginning and at the end of the analyzed period. The maximum dune height was obtained only from the dunes monitored by DGPS. Three of these were classified as barchanoid ridges (D3, D4 and D5) and two as isolated barchan dunes (D6 and D7). As an example, the digital elevation model of dune D3 in April 2016 is shown (Figure 10).

Table 2. Morphological characteristics of the dunes at the beginning and at the end of the analyzed period. The types of the dunes were classified as isolated barchan (IB) and barchanoid ridges (BR).

		Sector 1		Sector 2		Sector 3			Sector 4				
		D1	D2	D3	D4	D5	D6	D7	D8	D9	D10	D11	D12
Dunes features	Type	IB	IB	BR	BR	BR	IB	IB	IB	IB	IB	IB	IB
	Initial distance from the beach (m)	1459	1453	1260	1237	810	771	675	901	774	741	727	721
	Final distance from the beach (m)	1555	1650	1439	1262	915	782	870	928	917	925	861	843
	Initial width(m)	44	51	116	75	89	30	83	82	92	117	171	95
	Final width(m)	60	54	72	85	68	69	53	130	75	71	113	66
	Initial crest length(m)	325	240	453	255	297	114	216	146	184	304	268	92
	Final crest length(m)	288	162	426	239	143	150	93	196	173	168	223	118
	Maximum initial height (m)	---	---	7.1	5.0	5.5	6.5	7.5	---	---	---	---	---
	Maximum final height (m)	---	---	6.7	6.7	5.3	7.5	5.0	---	---	---	---	---

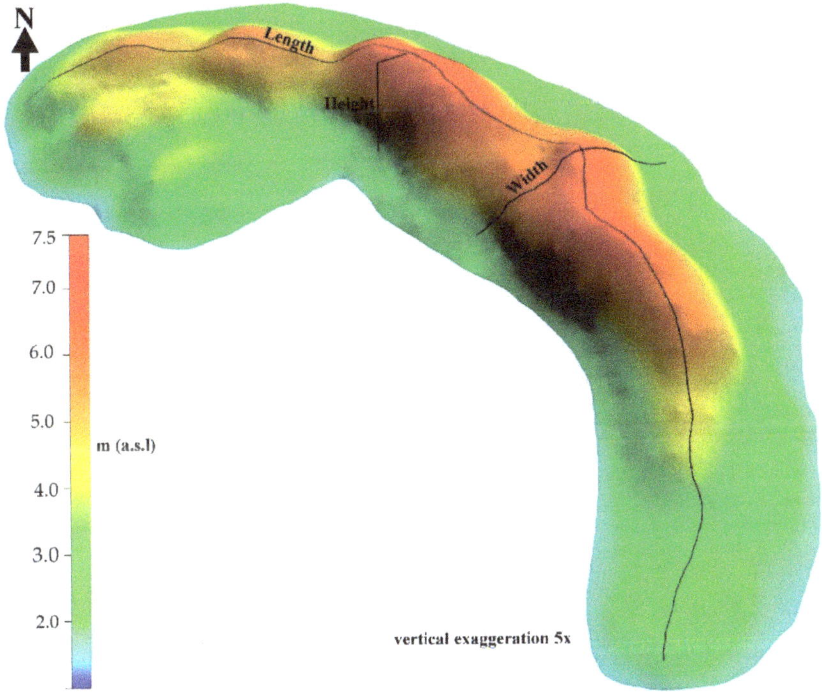

Figure 10. Example of the digital elevation model to obtain the morphological characteristics (dune D3 in April 2016). (Scale in meters above sea level; vertical exaggeration 5×).

In general, the barchanoid ridges and isolated barchan dunes showed inland migration during the sampling period, as evidenced by the increasing distances from the beach. The isolated barchan dunes showed a tendency to become narrower and shorter, whereas the barchanoid ridges showed a tendency to become narrower but with varying crest lengths (Table 2).

In sector 1, in the northern area of the PNNL, the dunes showed an increase in width and a decrease in crest length, whereas in sector 2 in the northern-central area, dune D3 reduced its width and D7 increased, with both dunes slightly decreasing their crest length, and dune D4 showed the highest growth in height, with an increment of 34%.

However, in sector 3 in the southern-central area, dune D5 (BC) was one of the dunes that showed the greatest reduction (≈50%) in crest length, while keeping its height almost stable. Furthermore, in sector 3, the two IB dunes, D6 and D7, presented the greatest discrepancies in their initial and final morphology. While dune D6 showed an increase in all parameters (width, crest length, height), dune D7 was the opposite, with decreases in all parameters. The largest increase in dune width was in D6, and the largest decrease in crest length was measured in D7. In Sector 4 in the southern area, there was a different behavior among the dunes, in which D8, located further north in this sector, showed an increase in width and crest length but, however, a smaller distance from the coast. On the other hand, the D10 dune showed the greatest reduction in crest width and length in this sector (Table 2).

The dune migration rates in sector 1 showed an important variation among the analyzed intervals of satellite images (Figure 11 and Table 3). The highest rate for dune migration was identified in the northern region of the PNNL for dune D2, with rates of 2.18 m/month (2005–2014/Interval 1), followed by 1.67 m/month (2014–2018/Interval 2). The resulting migration direction was 231° for dune crest D1 and 259° for dune crest D2.

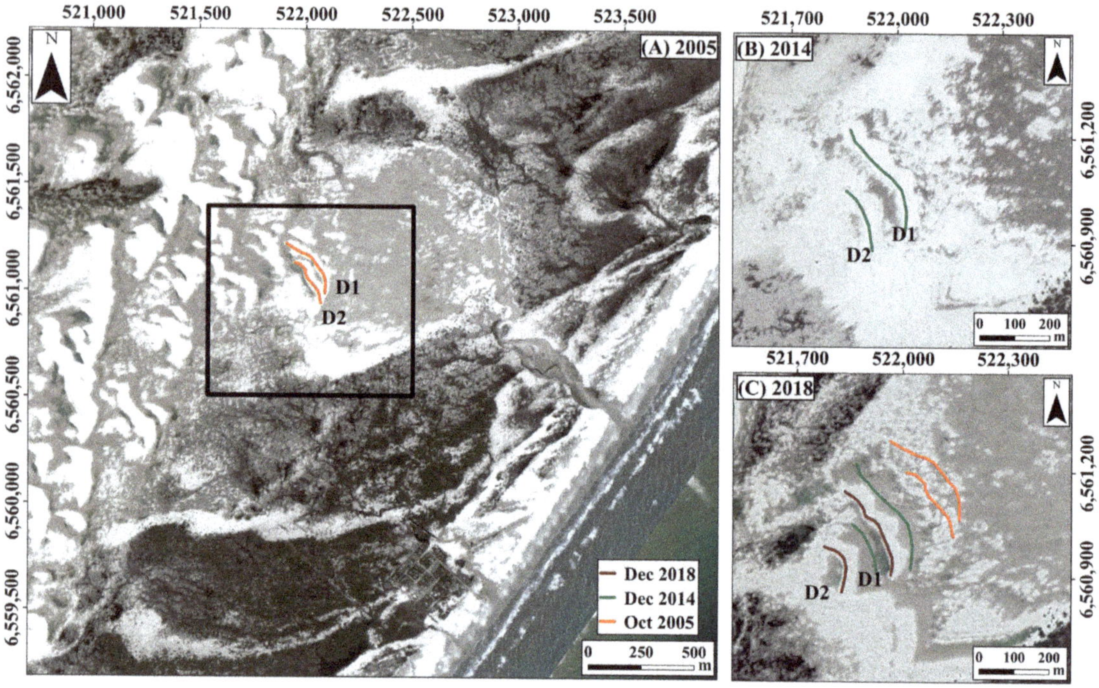

Figure 11. The migration of dunes D1 and D2 in the northern sector 1. The initial location of the dune crests in 2005 (**A**), the location of both dune crests in 2014 (**B**), and the image of 2018 with the overlay of the dune crest migration from 2005 to 2018 (**C**).

Table 3. The migration of the dune crests between interval satellite image: migration rate (m·month) and direction (°N).

	Sector 1				Sector 2				Sector 3					Sector 4									
	D1		D2		D3		D4		D5		D6	D7	D8		D9		D10		D11		D12	
Interval 1	1.67	231°	2.18	259°	1.87	263°	1.1	241°	1.31	232°			1.21	269°	1.4	270°	1.44	274°	1.88	275°	1.44	276°
Interval 2	1.38	236°	1.67	267°	1.22	284°	0.67	301°					1.00	260°	1.63	276°	1.59	262°	0.75	280°	0.73	283°

In the north central sector, both dunes (D3 and D4) presented similar initial distances from the coastline (1260 and 1237 m). However, the two dunes presented different sizes with a reasonable contrast between the initial crest widths (116 and 75 m). Similarly, the initial lengths of the crest were also different (426 and 255 m) (Table 2). During the period analyzed, the behaviors of the dunes' geomorphological changes were opposite; while D3 reduced its dimensions, D4 increased its dimensions, even with respect to the crest height (Table 2).

On the other hand, the direction and rates of migration were relatively similar, as can be seen in Figure 12 and Table 3.

However, if we analyze separately the directions and migration rates (Table 4), we will see that the crests move with some difference, especially in the periods P2 and P4. This fact may be associated with the presence of the urbanization of Mostardas beach (Figure 12).

Further south, in southern-central sector (Figure 13), D5 it was a Barchanoid Ridges (BR), demonstrated a small reduction in crest width (from 89 to 68 m) and a significant reduction in crest length, with a 50% reduction (297 to 143 m), keeping the crest height stable.

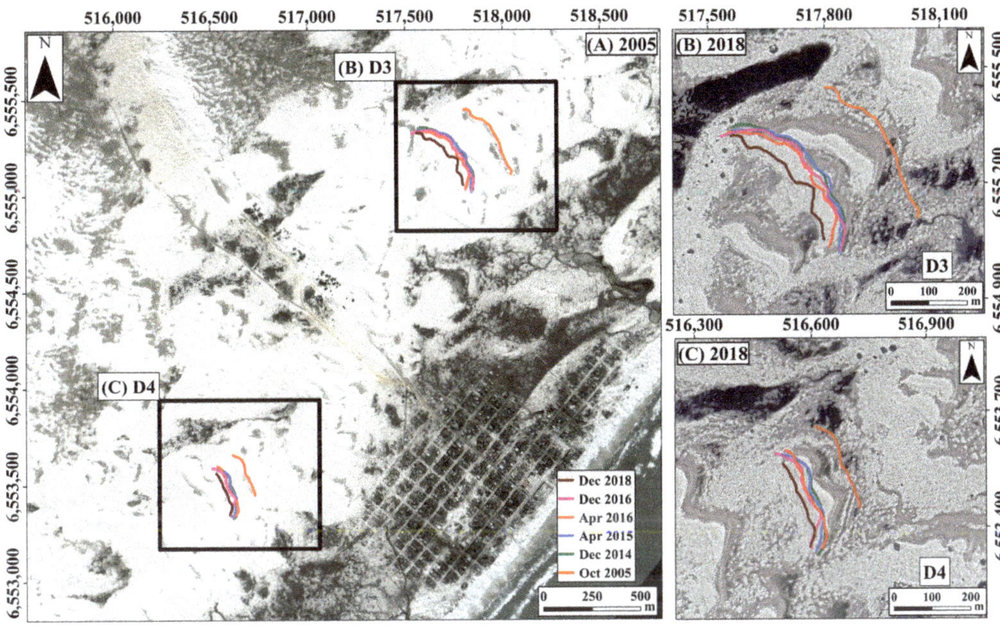

Figure 12. The migration of dunes D3 and D4 in the northern-central sector 2. (**A**) The initial location of the dune crests in 2005, and images of 2018 with the overlay of the crest migration from 2005 to 2018 of dunes D3 (**B**) and D4 (**C**), respectively.

Table 4. The migration of the dune crests between five monitoring periods: migration rate (m·month) and direction (°N).

	Sector 2				Sector 3					
	D3		D4		D5		D6		D7	
P1							0.74	221°	0.81	277°
P2	0.58	207°	0.9	268°	1.76	253°	1.88	256°	1.38	290°
P3	1.36	240°	0.84	255°	1.02	267°	1.08	267°	1.42	282°
P4	1.35	61°	1.34	91°	1.10	78°	2.18	77°	2.54	63°
P5	1.02	241°	1.01	271°	0.96	260°	0.73	272°	1.17	270°

D6 and D7 are very close to each other geographically, with mutual influence on the variations of the morphological characteristics. D7 showed the biggest changes with a drastic reduction in its dimensions. It exhibited an initial distance of 675 m from the beach and a final distance of 870 m. Its initial and final widths were 83 m and 53 m, respectively. The initial ridge length was 216 m, which decreased to 93 m at the end of the sampling period. It had a maximum initial height of 7.5 m and a maximum final height of 5.0 m.

In contrast, D6 showed the greatest increments, more than doubling the crest width (from 30 m to 69 m) and increasing the crest length by 31.5% (from 114 to 150 m), as well as an increase in crest height by 1 m during the period analyzed (Table 2).

In Sector 4, south of the mouth of the Peixe lagoon, the set of 5 IB dunes behaves as an almost homogeneous set with punctual variations between each one. The morphological variations of the dune cluster are subjected to the sedimentary supply from the deflation plains between the dune cluster and the mouth of the Peixe lagoon (Figure 14). In addition, each of the dunes also provides sediment to the adjacent dunes. The direction and migration rates of the dune cluster can be observed in Table 2, and in Figure 14, we can see the design of the dune crests at the different intervals of the satellite images.

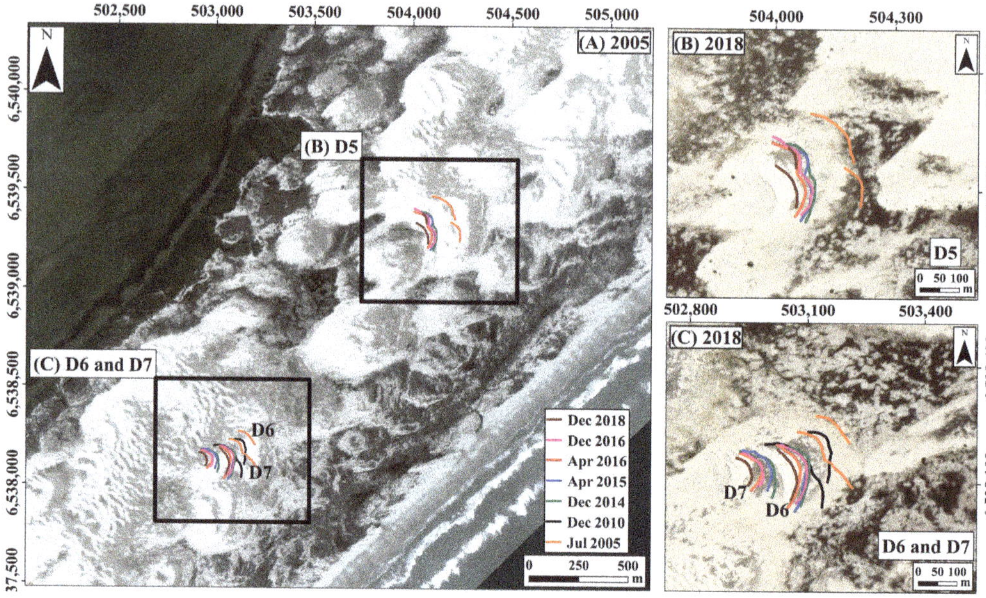

Figure 13. Migration of the dunes D5, D6, and D7 in sector 3. The initial location of the monitored ridges in 2005 (**A**) and the images of 2008 with the overlay of the dune crest D5 (**B**), D6, and D7 (**C**) in seven stages from 2005 to 2018.

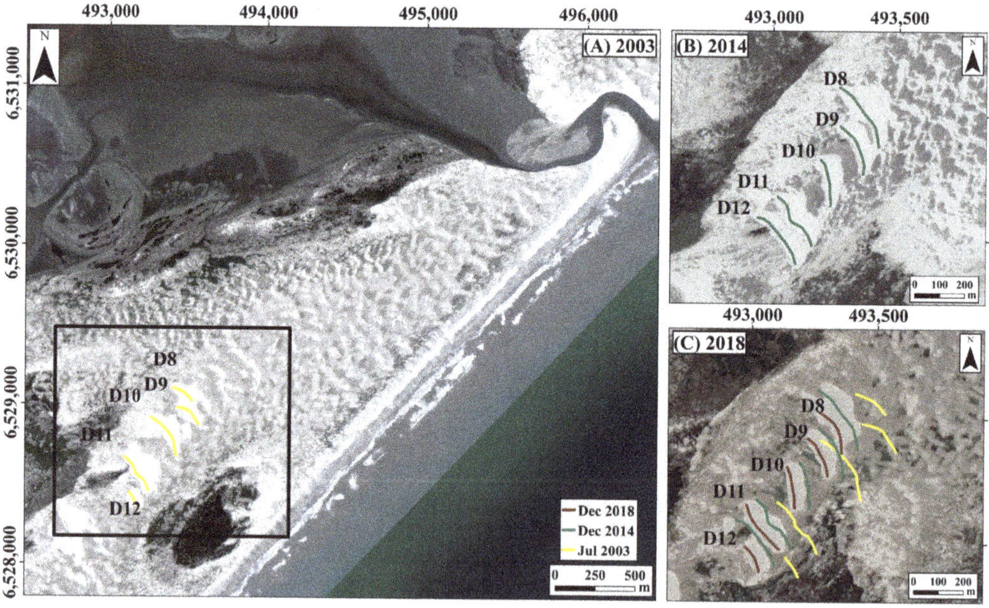

Figure 14. The migration of dunes D8 to D12 in sector 4. The initial location of the dune crests in 2003 (**A**), location in 2014 (**B**), and the image of 2018 with the overlay of the migration of dune crests from 2003 to 2018 (**C**).

The isolated barchan and barchanoid ridges dunes evidenced a complex migration pattern, with no exact parallel migration of their crests. In general, for the total study period, a westward migration was observed, showing the isolated barchanoid dunes larger displacement distances and migration rates than the barchanoid crests, while migration directions were similar, towards WSW–W (Figures 11–14).

5. Discussion

5.1. Normal Dunes Migration and Seasonal Reversion

Reverse bedforms migration is usual in bidirectional flows, i.e., waves and tides and seasonally reverse winds [76]. The reverse migration of dunes can be deduced from seasonal topographic monitoring [77] and numerical modeling [78].

The coastal plain of Rio Grande do Sul presents a high-energy wind regime (DP > 400 UV), according to the classification of Fryberger and Dean [64]. Regarding the directional variability, the low RDP/DP ratios (Table 5) reflect the obtuse bimodal characteristics of the wind regime [29], with predominant winds coming from the NE (Figure 4). The direction of the potential eolian sediment transport also evidences a high variability. The shoreline orientation of the study area is such that the NE winds transport the sediment parallel to the coast. Nevertheless, considering the direction of all the effective winds, the net eolian sediment transport in an annual scale is toward the WNW–W (Figures 8 and 9).

Field monitoring in the study area facilitated the observation of a reversal in the direction of dune migration. The direction of the migration of the dune crests during the entire study period (December 2010–December 2018) was towards WSW–W, which is consistent with the regional dune dynamics (RDD), morphological configuration, and the overall migration of the dune field, as depicted in Figure 4 and Table 4. This trend was also observed during periods P1, P2, P3, and P5. However, from April to December 2016, the dune crests migrated towards ENE–E, as shown in Figure 4 and Table 4, and the RDD changed to NNE (Figures 8 and 9). Therefore, there was a significant change in both the potential eolian sediment transport and the actual dune migration during period P4, albeit not in the same direction.

The normal dune migration and seasonal reversion are a phenomenon observed in various regions worldwide. Reversing transverse dune migration is associated with bimodal seasonal wind regimes [77–81] and so are relatively rare eolian bedforms [78]. Reversing migration has been described for transverse dunes oriented normal to the shoreline [78], desert linear dunes [82], continental barchan dunes [83], and linear coastal dunes [10].

Similar to the coastal plain of Rio Grande do Sul, reversing dune migration was reported at the Mpekweni site in South Africa [78]. The study used 3D computational fluid dynamic modeling to examine the behavior of the near surface airflow traveling over transverse (reversing) dunes on a beach system. When the wind direction is reversed, the dune morphology is rapidly modified, particularly at the dune crests, where rounding and aerodynamic smoothing takes place. The study provides detailed insights into how 3D airflow behavior is modified according to the incident flow direction of reversing dune ridges and the resulting implications for their topographical modification.

The Itapeva dune field, situated in Brazil, presents inverted dunes characterized by major and minor slip faces oriented in opposing directions. This research emphasizes the significance of regional geomorphology in altering local wind patterns and its consequent influence on dune dynamics. Continued monitoring of the dunes has revealed the occurrence of simultaneous slip faces oriented in different directions within the dune field [84].

The reversion of the dune migration direction during P4 reveals the contribution of the effective SSW–WSW winds for the sediment transport. The presence of cold fronts, more frequently in winter, generates extreme SSW–WSW winds [85,86], and therefore, the cold fronts can be considered responsible for the reversion of the migration direction of the transgressive barchan and barchanoid dunes. As reported by Machado and Calliari [86], between 1948 and 2013, there was an increase in extreme wind events coming from the SW quadrant. Therefore, this increase can result in a decreasing trend of the net migration rates for the transgressive dunes and, consequently, an apparent stabilization. Coastal dune stabilization is related to an increase in climate changes or human activities and the consequent increase in vegetal cover [13,87,88]. On the other hand, Costas and co-workers [89] suggest that vegetation plays a passive role with respect to topography, and external agents: sediment supply, beach width, and wind intensity, are much more effective. However, several environmental factors may lead to periods of dune field stabilization, such as climatic variations in precipitation (including the role of groundwater fluctuations), changes in the wave and wind regime, sea level variations, shoreline erosion, grazing activities, and sediment supply, as well as the influence of exotic flora and fauna [90].This study shows that the increase in reversing migration, due to changes in seasonal wind patterns, i.e., the increase in polar fronts in the study area, is another factor that can contribute to coastal dune stabilization. Conceptual models of the evolution of transgressive dune field systems [91] might also include this reverse migration.

Wind energy is frequently considered one of the main controlling factors over dune activity in areas where there is low human activity [92]. In the Peixe lagoon dune field, given the lack of urbanization, the dunes migrate according to the influence of the prevailing winds.

Rain precipitation is another factor that controls the dune migration, due to humidity avoidance or difficult sediment transport [93]. The different direction between the potential eolian sediment transport and the effective dune migration can be an effect of precipitation associated with cold fronts. Theoretically, the sediment transport should be considered negligible during rainfall events [94]. However, the intensity of SSW–WSW winds suggests that during precipitation events (Figure 7) or after a fast desiccation of sand, due to the absence of grass or other vegetation in the dunes, the eolian sediment transport can be produced, in accordance with the studies of Logie [95] and Sarre [96].

The dune field is wider in the northern sector of the study area, the barchanoid dunes are more abundant and the barchan dunes present higher dimensions. Therefore, a higher availability of sediments for the northern sector can be deduced, in accordance with a higher volume of eolian sediments [11]. The net annual eolian sediment transport is landward and, therefore, the eolian flux of sediments between both sectors is negligible. It implies that this higher availability of sediments in the northern sector must be only related to the input of sediments from the beaches. The shoreline orientation shows a small progradation or cuspate morphology in the northern sector, if compared with the general shoreline orientation in the region (Figure 1). Therefore, it can imply a higher concentration of wave energy in the northern sector of the study area, inducing a more dissipative stage of the beach, which favors a higher input of eolian sediments and the better development of dunes, according to the criteria of Short and Hesp [6]. The little changes in the shoreline orientation control the local variability in the eolian sediment availability and, consequently, the distribution and height of barchan and barchanoid dune types, as well as the location of the Peixe lagoon in the southern sector.

5.2. Dune's Migration Trends for the Coastal Plain of Rio Grande do Sul

Several studies have described the eolian potential drift (RDP, RDD) and migration rates of dunes in the coastal plain of Rio Grande do Sul (Table 4), despite only two of them [29,97] conducting direct field measurements.

Tomazelli [29] analyzed the wind regime and the dune migration rates in the northern region of the CPRGS. In total, the wind data series comprised 13 years, from January 1970 to December 1982. This study suggests a resultant drift direction (RDD) to SW (220°N) due to the large predominance of NE and E winds, associated with the edge of the South Atlantic Anticyclone. The analysis of the drift potential data revealed a large seasonal variation, which is a result of changes in the wind regime of this coastal region. According to the author, during fall and winter (i.e., from April to August), there is a reduction in the drift potential and a significant change in the drift direction, which, due to the predominance of winds coming from W and SW, practically reverses its direction. During spring–summer, drift potential increases (average November value: 203.9 UV) and the drift direction returns to its regular orientation, towards SW. The analysis of aerial photographs and satellite images, for a total period of 27 years, was conducted on different dune morphological types and revealed annual average rates between 10 and 38 m·yr^{-1} (Table 4). Direct field measurements conducted on a barchanoid dune with an average height of 8 m over three years showed an average migration rate of approximately 26 m·yr^{-1} in the SW direction (230°N) [29].

Table 5. The comparative table for real potential drift (RDP), resultant drift direction (RDD), migration rates (m/year) and kind of dunes for several dune fields in the coastal plain of Rio Grande do Sul.

	Region	Period of Study	RDP (UV)	RDD	Migration Rate (m·yr^{-1})	Type of Dune	Author
Northern Coast	Torres	1970–1982	800	WSW	-	-	[29]
	Torres	1970–1982	±23	NW	-	-	[58]
	Torres	2008–2015	4.65	-	-	-	[98]
	Tramandaí	2003–2005	±55	SW	-	-	[58]
	Tramandaí	2008–2015	50.85	-	-	-	[98]
	Imbé	1970–1982	1442	SW	-	-	[29]
	Imbé	1948–2003	±44	SW	-	-	[58]
	Magisterio, Pinhal	1986–1989	-	SW	26	Barchanoid	[29]
	Magisterio, Pinhal	1974–1987	-	SW	14.7	Parabolic	[34]
	Magisterio, Pinhal	1974–1987	-	SW	24	Barchan	[34]
	North/middle Coast	1948–1967	-	SW	14.0	Barchan	[29]
	North/middle Coast	1967–1974	-	SW	11.0–32.0	Barchanoid	[29]
Middle coast	Dunas Altas, Palmares do Sul	1987–1999	-	SW	22.5	Parabolic	[34]
	Dunas Altas, Palmares do Sul	1987–1999	-	SW	28.0	Barchanoid	[34]
	Mostardas	1957–2000	±45	W	-	-	[58]
	Peixe lagoon	2010–2018	99.4	WSW-W	16.55	Barchan/Barchar	Present study
	Rio Grande	1970–1982	409	W	-	-	[29]
Southern coast	Taim	2003	57.57	ENE	0	Barchan	[97]
	Taim	2004	103	N	0	Barchan	[97]
	Taim	2005	105.2	NW	20	Barchan	[97]
	Chuí	2003–2006	±90	NWN	-	-	[34]

Guimarães [97] studied the migration of a barchan dune on the southern sector of CPRGS between 2003 and 2005. During 2003, the direction remained between E–NE, while between 2004 and 2005, it changed to NNE. Significant dune migration was not registered between 2003 and 2004, only variations in volume, orientation, and height. However, during 2005, the dune migrated 20 m to the NW and W. According to the author, there is larger wind direction variability in this sector region when compared to other areas of the CPRGS. The author highlights that the years of 2004 and 2005 were atypical, including the Catarina Hurricane (the only hurricane registered in the state).

Another important study in regional scale was conducted by Martinho [34]. DP and RDP variations were analyzed on the northern and middle sectors of CPRGS. Dune migration analyses were also conducted using aerial photographs (1948, 1974, and 1989) and LANDSAT satellite images (1980, 1987, and 1999, with resolutions of 60, 30, and 15 m, respectively). This study identified dune migration rates ranging from 14.7 to 28 m·yr^{-1}, with a direction between 214 and 234°N (Table 3).

When comparing the real drift potential (RDP) between studies, it is possible to see a large discrepancy of results, even though the same method of Fryberger and Dean [64] was used. On the other hand, the resultant drift direction (RDD) results are very similar. Overall, all authors emphasize an RDP reduction from the southern and middle sectors to the northern sector, the last one represented by the Torres dune field. According to Martinho [34], the winds on the northern sector are less strong when compared to the middle sector of the CPRGS, resulting in smaller values of RDP. The NE winds blow parallel to the coast, with its corridor limited by the Serra Geral escarpments. On the other hand, in the middle sector, the winds are stronger and present larger RDP values; in this sector the winds reach the coast at an oblique angle and have a larger wind corridor due to the gentle topography of the coastal plain. In the middle sector, the RDD ranges from SW to W. This sector presents a larger range of wind directions, even though the winds from NE and S are stronger and more important. The south sector has a larger variability in wind direction, presenting an annual average toward WSW but also periods with eolian sediment transport to the north. As demonstrated by Guimarães [97] and the present study, a short-term analysis demonstrates that the migration rates are not constant over the years and can result in null or reverse migrations because of the seasonal wind variations.

The RDP values must be interpreted as representative of the wind energy for each sector. The efficiency of the eolian sand transport depends also on the local characteristics of the surface over which the wind acts. Therefore, the RDP values are not necessarily equal to the real drift, since local characteristics inherent to the terrain surface (topographic variations, humidity levels, grain sizes, and presence of vegetation) over where the wind blows can affect the amount of sand that is effectively transported [29,64].

Considering the effective winds for the dunes formation and their migration rates, a reduction in wind speed and prevailing direction can result in reduced transport and stabilization of the dune field. The analysis of the Imbé meteorological station shows interdecadal variations on the RDP: it decreased between 1948 and 1955, significantly increased between 1955 and 1964, continually decreased from 1964 to 1988, and a short increase occurred from 1988 to 2003 [58]. Similarly, the historical series of meteorological data show a decrease in the monthly average wind speed and an increase in precipitation since 1961 at the extreme north of the CPRGS [99].

The dune migration rates presented by [29,34], obtained from satellite images between 1948 and 1987, can be considered relevant on a large-scale context of dune field changes. However, these results no longer reflect the current migration rates, given that these fields have suffered significant changes and stabilization processes. By contrast, in 1948, for example, the Torres and Mostardas dune fields were completely active without vegetation or deflation areas.

6. Conclusions

The climate data were analyzed to depict trends in dune behavior. The SW–NE shoreline orientation of the study area generates NE winds, which are the most frequent and carry sediments parallel to the coast. Nevertheless, considering the multidirectional wind directions during the year, the resultant drift direction is WSW–W, which generates a net annual input of sediments landward.

The slightly cuspate shoreline of the northern sector induces a more dissipative stage of the beaches and a higher input of sediments to the dune field. Consequently, there is a higher availability of sediments, evidenced by a higher development of barchanoid and barchan dunes with higher dimensions and a wider extension of the dune field in the northern sector. Altogether, this determines the location of the Peixe lagoon in the southern sector of the study area.

The annual average dune migration rates ranged from xx–xx (Barchan) and xx–xx (Barchanoid). The dune crest migrations in the dune field close to the Peixe lagoon show average rates of 16.55 m·year^{-1} for dunes. The predominant direction of migration for both types of dunes is towards WSW–W. This dune crest migration is mainly controlled by the direction of the effective winds due to the absence of vegetation in the dune field and the lack of urbanization in the national park. These dune migration patterns are in agreement with both the sediment transport calculations and the geomorphological configuration of the transgressive dune field.

The seasonal reversion of dune migration occurs during occasional periods. The seasonal analysis shows the relevance of the effective SSW–WSW winds for the sediment transport, dune migration, and dune field stabilization. The past cold fronts closest to Rio

Grande do Sul generate these intense winds and, consequently, a reversion in the migration direction of the barchan and barchanoid dunes, changing it to migrate towards the NNE. This reversion is coherent with the reversion in the direction of the eolian sediment transport indicated by the previous studies, based on the calculation of wind drift potential. This result warns of the possibility of a reduction in the migration rates of the transgressive dunes, due to both the tendency to increase the occurrence of extreme events and the greater entry of cold fronts in this sector of the RS coast.

Consequently, the identified reverse migration of dunes explains the stabilization of the dune fields at CPRGS. The decrease in both speed and frequency of prevailing winds results in reduced transport and stabilization of the dune field. Therefore, reverse dune migration results in a factor controlling dune stabilization and the geomorphological evolution of transgressive coastal dune fields.

The implications of our research extend beyond the local level and have wider international implications. The coastal dune fields are valuable ecosystems around the world that serve as natural barriers against coastal erosion, protect coastal communities, and provide critical habitats for numerous plant and animal species. Understanding the dynamics of dune behavior and the factors influencing their migration and stability is critical for effective coastal management and adaptation strategies in the face of ongoing climate change. By elucidating the complex interplay between wind patterns, sediment transport, and dune morphology, our findings contribute to a better understanding of coastal systems and may inform decision-making processes for the conservation and sustainable management of these fragile coastal environments.

Author Contributions: Conceptualization, R.P.M. and L.C.P.; methodology, R.P.M., J.A.-C. and L.C.P.; software, R.P.M., V.J.B.B. and L.C.P.; validation, R.P.M.; formal analysis, R.P.M., J.A.-C., V.J.B.B. and L.C.P.; investigation, R.P.M., L.C.P., A.F.-B., V.J.B.B. and J.A.-C.; resources, R.P.M., L.C.P., A.F.-B., V.J.B.B. and J.A.-C.; data curation, R.P.M. and L.C.P.; writing—original draft preparation, R.P.M., L.C.P. and J.A.-C.; writing—review and editing, R.P.M., L.C.P., A.F.-B., V.J.B.B. and J.A.-C.; visualization, R.P.M., L.C.P., A.F.-B., V.J.B.B. and J.A.-C.; supervision, R.P.M. and L.C.P.; project administration, L.C.P.; funding acquisition, L.C.P. All authors have read and agreed to the published version of the manuscript.

Funding: This research was funded by Conselho Nacional de Desenvolvimento Científico e Tecnológico (CNPQ/Brazil), grant number 471729/2013-2.

Data Availability Statement: The data presented in this study are available on request from the corresponding author.

Acknowledgments: This study is a contribution to the research project "Reforestation versus dune environment: a strategy for the management of permanent protection areas (APPS)—Dune field of the Lagoa do Peixe National Park".

Conflicts of Interest: The authors declare no conflict of interest.

References

1. Martinez, L.L.; Psuty, N.P. *Coastal Dunes, Ecology and Conservation. Ecological Studies*; Springer: Berlin/Heidelberg, Germany, 2004; Volume 171.
2. Hesp, P.A.; Martínez, M.L. Disturbance Processes and Dynamics in Coastal Dunes. In *Plant Disturbance Ecology*; Elsevier: Amsterdam, The Netherlands, 2007; pp. 215–247.
3. Davidson-Arnott, R.G.D.; Law, M.N. Seasonal Patterns and Controls on Sediment Supply to Coastal Foredunes. Long Point, Lake Erie. In *Coastal Dunes: Form and Process*; Nordstrom, K., Psuty, N., Carter, R.W.G., Eds.; Wiley: New York, NY, USA, 1990; pp. 117–200.
4. Packham, J.R.; Willis, A.J. *Ecology of Dunes, Salt Marsh and Shingle*; Springer: New York, NY, USA, 1997; ISBN 978-0-412-57980-6.
5. Walker, I.J.; Davidson-Arnott, R.G.D.; Bauer, B.O.; Hesp, P.A.; Delgado-Fernandez, I.; Ollerhead, J.; Smyth, T.A.G. Scale-Dependent Perspectives on the Geomorphology and Evolution of Beach-Dune Systems. *Earth-Sci. Rev.* **2017**, *171*, 220–253. [CrossRef]
6. Short, A.D.; Hesp, P.A. Wave, Beach and Dune Interactions in Southeastern Australia. *Mar. Geol.* **1982**, *48*, 259–284. [CrossRef]
7. Bird, E.C.F. Coasts. In *An Introduction to Systematic Geomorphology. Australian National*; Australian National University Press: Canberra, Australian, 1976.

8. Hesp, P.A. The Beach Backshore and Beyond. In *Handbook of Beach and Shoreface Morphodynamics*; Short, A.D., Ed.; John Wiley and Sons Ltd.: Hoboken, NJ, USA, 1999; pp. 145–270.
9. Goldsmith, V. Coastal Dunes. In *Coastal Sedimentary Environments*; Davis, R.A., Ed.; Springer: Berlin/Heidelberg, Germany, 1978; pp. 171–235.
10. Alcántara-Carrió, J.; Alonso, I. Aeolian Sediment Availability in Coastal Areas Defined from Sedimentary Parameters. Application to a Case Study in Fuerteventura. *Sci. Mar.* **2001**, *65*, 7–20. [CrossRef]
11. Fontán Bouzas, A.; Alcántara-Carrió, J.; Montoya Montes, I.; Barranco Ojeda, A.; Albarracín, S.; Rey Díaz de Rada, J.; Rey Salgado, J. Distribution and Thickness of Sedimentary Facies in the Coastal Dune, Beach and Nearshore Sedimentary System at Maspalomas, Canary Islands. *Geo-Mar. Lett.* **2013**, *33*, 117–127. [CrossRef]
12. Xu, Z.; Mason, J.A.; Lu, H. Vegetated Dune Morphodynamics during Recent Stabilization of the Mu Us Dune Field, North-Central China. *Geomorphology* **2015**, *228*, 486–503. [CrossRef]
13. Anthonsen, K.L.; Clemmensen, L.B.; Jensen, J.H. Evolution of a Dune from Crescentic to Parabolic Form in Response to Short-Term Climatic Changes: Råbjerg Mile, Skagen Odde, Denmark. *Geomorphology* **1996**, *17*, 63–77. [CrossRef]
14. Tsoar, H.; Blumberg, D.G. Formation of Parabolic Dunes from Barchan and Transverse Dunes along Israel's Mediterranean Coast. *Earth Surf. Process. Landf.* **2002**, *27*, 1147–1161. [CrossRef]
15. Yizhaq, H.; Ashkenazy, Y.; Tsoar, H. Why Do Active and Stabilized Dunes Coexist under the Same Climatic Conditions? *Phys. Rev. Lett.* **2007**, *98*, 98–101. [CrossRef]
16. Yizhaq, H.; Ashkenazy, Y.; Tsoar, H. Sand Dune Dynamics and Climate Change: A Modeling Approach. *J. Geophys. Res. Earth Surf.* **2009**, *114*, 1–11. [CrossRef]
17. Lavee, H.; Imeson, A.C.; Sarah, P. The Impact of Climate Change on Geomorphology and Desertification along a Mediterranean-Arid Transect. *Land Degrad. Dev.* **1998**, *9*, 407–422. [CrossRef]
18. Lancaster, N. Eolian Features and Processes. In *Geological Monitoring*; Young, R.N., Ed.; Geological Society of America: Boulder, CO, USA, 2009; pp. 1–25.
19. Bigarella, J.J.; Klein, A.H.D.F.; Menezes, J.T.; Vintém, G. Sub-Tropical Coastal Dunes: Examples from Southern Brazil. *J. Coast. Res.* **2005**, *42*, 113–137.
20. Hernández-Calvento, L.; Jackson, D.W.T.; Medina, R.; Hernández-Cordero, A.I.; Cruz, N.; Requejo, S. Downwind Effects on an Arid Dunefield from an Evolving Urbanised Area. *Aeolian. Res.* **2014**, *15*, 301–309. [CrossRef]
21. Bailey, S.D.; Bristow, C.S. Migration of Parabolic Dunes at Aberffraw, Anglesey, North Wales. *Geomorphology* **2004**, *59*, 165–174. [CrossRef]
22. Madole, R.F. Spatial and Temporal Patterns of Late Quaternary Eolian Deposition, Eastern Colorado, U.S.A. *Quat. Sci. Rev.* **1995**, *14*, 155–177. [CrossRef]
23. Jimenez, J.A.; Maia, L.P.; Serra, J.; Morais, J. Aeolian Dune Migration along the Ceará Coast, North-Eastern Brazil. *Sedimentology* **1999**, *46*, 689–701. [CrossRef]
24. Girardi, J.D.; Davis, D.M. Parabolic Dune Reactivation and Migration at Napeague, NY, USA: Insights from Aerial and GPR Imagery. *Geomorphology* **2010**, *114*, 530–541. [CrossRef]
25. Miot da Silva, G.; Hesp, P.A. Increasing Rainfall, Decreasing Winds, and Historical Changes in Santa Catarina Dunefields, Southern Brazil. *Earth Surf. Process. Landf.* **2013**, *38*, 1036–1045. [CrossRef]
26. García-Romero, L.; Hesp, P.A.; Peña-Alonso, C.; Miot da Silva, G.; Hernández-Calvento, L. Climate as a Control on Foredune Mode in Southern Australia. *Sci. Total Environ.* **2019**, *694*, 133768. [CrossRef]
27. Hansen, E.; DeVries-Zimmerman, S.; van Dijk, D.; Yurk, B. Patterns of Wind Flow and Aeolian Deposition on a Parabolic Dune on the Southeastern Shore of Lake Michigan. *Geomorphology* **2009**, *105*, 147–157. [CrossRef]
28. Bristow, C.S.; Lancaster, N. Movement of a Small Slipfaceless Dome Dune in the Namib Sand Sea, Namibia. *Geomorphology* **2004**, *59*, 189–196. [CrossRef]
29. Tomazelli, L. O Regime Dos Ventos e a Taxa de Migração Das Dunas Eólicas Costeiras Do Rio Grande Do Sul, Brasil. *Pesquisas em Geociências* **1993**, *20*, 18. [CrossRef]
30. de Souza Matos-Carneiro, M.; Ferreira, B.; Gregorio, M.; Pessanha, P.; Lins da Silva, D.; de Oliveira-Vital, S. Datos Espaciales LIDAR En La Caracterización Geomorfológica Del Campo de Dunas Costeras Del Río de Fogo, Río Grande Do Nort—Brasil. *Rev. Geográfica De América Cent.* **2018**, *2*, 315–348. [CrossRef]
31. Pickart, A.J.; Hesp, P.A. Spatio-Temporal Geomorphological and Ecological Evolution of a Transgressive Dunefield System, Northern California, USA. *Glob. Planet. Chang.* **2019**, *172*, 88–103. [CrossRef]
32. Ruessink, B.G.; Arens, S.M.; Kuipers, M.; Donker, J.J.A. Coastal Dune Dynamics in Response to Excavated Foredune Notches. *Aeolian. Res.* **2018**, *31*, 3–17. [CrossRef]
33. Andriolo, U.; Gonçalves, G.; Sobral, P.; Fontán-Bouzas, Á.; Bessa, F. Beach-Dune Morphodynamics and Marine Macro-Litter Abundance: An Integrated Approach with Unmanned Aerial System. *Sci. Total Environ.* **2020**, *749*, 141474. [CrossRef] [PubMed]
34. Martinho, C.T.; Hesp, P.A.; Dillenburg, S.R. Morphological and Temporal Variations of Transgressive Dunefields of the Northern and Mid-Littoral Rio Grande Do Sul Coast, Southern Brazil. *Geomorphology* **2010**, *117*, 14–32. [CrossRef]
35. Seeliger, U.; Cordazzo, C.V.; Oliveira, C.P.L.; Seeliger, M. Long-Term Changes of Coastal Foredunes in the Southwest Atlantic. *J. Coast. Res.* **2000**, *4*, 1068–1072.
36. Villwock, J.A. Geology of the Coastal Province of Rio Grande Do Sul. *Pesquisas* **1984**, *16*, 5–49.

37. Villwock, J.A.; Tomazelli, L.J.; Loss, E.L.; Dehnhardt, E.A.; Bachi, F.A.; Dehnhardt, B.A. Geology of the Rio Grande Do Sul Coastal Province. In *Quaternary of South America and Antarctic Peninsula*; CRC Press: Boca Raton, FL, USA, 1986; Volume 4, pp. 79–97.
38. Villwock, J.A.; Tomazelli, L.J. Geologia Costeira Do Rio Grande Do Sul. *Notas Técnicas* **1995**, *8*, 1–45.
39. Tomazelli, L.J.; Villwock, J.A. Considerações Sobre o Ambiente Praial e a Deriva Litorânea de Sedimentos Ao Longo Do Litoral Norte Do Rio Grande Do Sul. *Pesquisas* **1992**, *19*, 3–12. [CrossRef]
40. Hesp, P.A.; Dillenburg, S.R.; Barboza, E.G.; Tomazelli, L.J.; Ayup-Zouain, R.N.; Esteves, L.S.; Gruber, N.L.S.; Toldo, E.E.; Tabajara, L.L.C.D.A.; Clerot, L.C.P. Beach Ridges, Foredunes or Transgressive Dunefields? Definitions and an Examination of the Torres to Tramandaí Barrier System, Southern Brazil. *An. Acad. Bras. Cienc.* **2005**, *77*, 493–508. [CrossRef]
41. Hesp, P.A.; Thom, B.G. Geomophology and Evolution of Active Transgressive Dunefields. In *Coastal Dunes: Form and Process*; Nordstrom, K.F., Psuty, N.P., Carter, R.W.G., Eds.; Jonh Wiley & Sons Ltd: Hoboken, NJ, USA, 1990; pp. 253–288.
42. Bigarella, J.J.; Klein, A.H.D.; Menezes, J.T.; Vintem, G. Southern Brazilian Coastal Dunes: Movement and Structures. *J. Coast. Res.* **2006**, *39*, 1–15.
43. Tomazelli, L.J. Contribuição Ao Estudo Dos Sistemas Deposicionais Holocênicos Do Nordeste Da Província Costeira Do Rio Grande Do Sul, Com Ênfase No Sistema Eólico. Tese de Doutorado, Universidade Federal do Rio Grande do Sul, Porto Alegre, Brasil, 1990.
44. Giannini, P.C.F. Sistemas Deposicionais Do Quaternário Costeiro Entre Jaguaruna e Imbituba, SC. Ph.D. Thesis, Universidade Estadual de São Paulo: São Paulo, Brazil, 1993.
45. Tomazelli, L.J.; Barboza, E.G.; Rosa, L.M.C.C. Geomorfologia e Potencial de Preservação Dos Campos de Dunas Transgressivos de Cidreira e Itapeva, Litoral Norte Do Rio Grande Do Sul, Brasil. *Pesquisas Geociências* **2008**, *35*, 47–55. [CrossRef]
46. Esteves, L.S. Variabilidade Espacio-Temporal Dos Deslocamentos Da Linha de Costa No Rio Grande Do Sul. Tese de Doutorado, Universidade Federal do Rio Grande do Sul, Porto Alegre, Brasil, 2004.
47. Portz, L.C.; Manzolli, R.P.; Gruber, N.L.S.; Correa, I.C.S. Turismo e Degradação Na Orla Do Rio Grande Do Sul: Conflitos e Gerenciamento. *Desenvolv. E Meio Ambiente* **2010**, *22*, 153–166. [CrossRef]
48. Portz, L.; Manzolli, R.P.; Hermanns, L.; Alcántara Carrió, J. Evaluation of the Efficiency of Dune Reconstruction Techniques in Xangri-Lá (Rio Grande Do Sul, Brazil). *Ocean. Coast. Manag.* **2015**, *104*, 78–89. [CrossRef]
49. Portz, L.; Jardim, J.P.M.; Manzolli, R.P.; Gruber, N.S. Impacts on the Dunes System: Natural Dynamic versus Anthropogenic Interference. *Ambiente Sociedade* **2016**, *19*, 135–154. [CrossRef]
50. Portz, L.; Manzolli, R.P.; Alcántara-Carrió, J.; Rockett, G.C.; Barboza, E.G. Degradation of a Transgressive Coastal Dunefield by Pines Plantation and Strategies for Recuperation (Lagoa Do Peixe National Park, Southern Brazil). *Estuar. Coast. Shelf. Sci.* **2021**, *259*, 107483. [CrossRef]
51. MMA (Ministério do Meio Ambiente) Sitio Ransar. Parque Nacional Da Lagoa Do Peixe. Planejamento Para o Sucesso Da Conservação. Secretaria de Biodiversidade e Florestas. Gerência de Biodiversidade Aquática e Recursos Pesqueiros. Available online: http://www.mma.gov.br/publicacoes/biodiversidade/category/53-biodiversidade-aquatica.html (accessed on 5 November 2019).
52. Moraes, V.L. Uso Do Solo e Conservação Ambiental No Parque Nacional Da Lagoa Do Peixe e Entorno (RS). Master's Dissertation, Universidade Federal do Rio Grande do Sul, Porto Alegre, Brasil, 2009; p. 120.
53. Martinho, C.T.; Dillenburg, S.R.; Hesp, P. Wave Energy and Longshore Sediment Transport Gradients Controlling Barrier Evolution in Rio Grande Do Sul, Brazil. *J. Coast. Res.* **2009**, *252*, 285–293. [CrossRef]
54. Camargo, O.A.; Silva, F.J.L. *Atlas Eólico: Rio Grande Do Sul*; SEMC-Secretaria de Energia Minas e Comunicações: Porto Alegre, Brazil, 2002.
55. NIMER, E. *Climatologia Do Brasil*; IBGE: Rio de Janeiro, Brazil, 1989.
56. Martinho, C.T. Morfodinâmica e Evolução de Campos de Dunas Transgressivos Quaternários Do Litoral Do Rio Grande Do Sul. Doctorate Dissertation, Universidade Federal do Rio Grande do Sul: Porto Alegre, Brasil, 2008; p. 215.
57. da Motta, L.M.; Toldo, E.E.; de Sá, L.E.; de Almeida, B.; Nunes, J.C. Sandy Sediment Budget of the Midcoast of Rio Grande Do Sul, Brazil. *J. Mar. Res.* **2015**, *73*, 49–69. [CrossRef]
58. Martinho, C.T.; Dillenburg, S.R.; Hesp, P.A. Mid to Late Holocene Evolution of Transgressive Dunefields from Rio Grande Do Sul Coast, Southern Brazil. *Mar. Geol.* **2008**, *256*, 49–64. [CrossRef]
59. Bitencourt, V.J.B.; Dillenburg, S.R.; Manzolli, R.P.; Barboza, E.G. Control Factors in the Evolution of Holocene Coastal Barriers in Southern Brazil. *Geomorphology* **2020**, *360*, 107180. [CrossRef]
60. Bitencourt, V.J.B.; Dillenburg, S.R.; Barboza, E.G.; Rosa, M.L.C.D.C.; Manzolli, R.P. Padrões De Empilhamento Estratigráfico E Seus Reflexos Na Morfologia Da Barreira Costeira Holocênica No Litoral Médio Do Rio Grande Do Sul, Brasil. *Rev. Bras. De Geomorfol.* **2020**, *21*, 529–548. [CrossRef]
61. Bitencourt, V.J.B.; Dillenburg, S.R. Application of Multivariate Statistical Techniques in Alongshore Differentiation of Coastal Barriers. *Mar. Geol.* **2020**, *419*, 106077. [CrossRef]
62. Toldo, E.E., Jr.; Almeida, L.E.S.B.; Nicolodi, J.L.; Absalonsen, L.; Gruber, N.L.S. O Controle Da Deriva Litorânea No Desenvolvimento Do Campo de Dunas e Da Antepraia No Litoral Médio Do Rio Grande Do Sul. *Pesqui. Em Geociências* **2006**, *33*, 35–42. [CrossRef]

63. Dillenburg, S.R.; Barboza, E.G.; Tomazelli, L.J.; Hesp, P.A.; Clerot, L.C.P.; Ayup-Zouain, R.N. The Holocene Coastal Barriers of Rio Grande Do Sul. In *Geology and Geomorphology of Holocene Coastal Barriers of Brazil*; Dillenburg, S.R., Hesp, P.A., Eds.; Springer: Berlin/Heidelberg, Germany, 2009; pp. 53–91.
64. Fryberger, S.G.; Dean, G. Dune Forms and Wind Regime. In *A Study of Global Sand Seas*; McKee, E.D., Ed.; US Geological Survey Professional Paper: Washington, DC, USA, 1979; pp. 137–169.
65. Bagnold, R.A. *The Physics of Blown Sand and Desert Dunes*; Methuen: London, UK, 1941.
66. Alcántara-Carrió, J.; Alonso, I. Measurement and Prediction of Aeolian Sediment Transport at Jandía Isthmus (Fuerteventura, Canary Islands). *J. Coast. Res.* **2002**, *18*, 300–315.
67. Revollo Sarmiento, G.; Cipolletti, M.; Perillo, M.; Delrieux, C.; Perillo, G. Methodology for Classification of Geographical Features with Remote Sensing Images: Application to Tidal Flats. *Geomorphology* **2015**, *257*, 10–22. [CrossRef]
68. Portz, L.; Rockett, G.C.; Franchini, R.A.L.; Manzolli, R.P.; Gruber, N.L.S. Gestão de Dunas Costeiras: O Uso de Sistema de Informações Geográficas (SIG) Na Implantação de Planos de Gestão No Litoral Do Rio Grande Do Sul, Brasil. *Rev. Gestão Costeira Integr.* **2014**, *14*, 517–534. [CrossRef]
69. Ludwig, A.; Meyer, H.; Nauss, T. Automatic Classification of Google Earth Images for a Larger Scale Monitoring of Bush Encroachment in South Africa. *Int. J. Appl. Earth Obs. Geoinf.* **2016**, *50*, 89–94. [CrossRef]
70. Lorenz, R.D.; Gasmi, N.; Radebaugh, J.; Barnes, J.W.; Ori, G.G. Dunes on Planet Tatooine: Observation of Barchan Migration at the Star Wars Film Set in Tunisia. *Geomorphology* **2013**, *201*, 264–271. [CrossRef]
71. Lipp-Nissinen, K.H.; Piñeiro, B.D.S.; Miranda, L.S.; Alves, A.d.P. Temporal Dynamics of Land Use and Cover in Paurá Lagoon Region, Middle Coast of Rio Grande Do Sul (RS), Brazil. *Rev. Gestão Costeira Integr.* **2018**, *18*, 25–39. [CrossRef]
72. Liang, J.; Gong, J.; Li, W. Applications and Impacts of Google Earth: A Decadal Review (2006–2016). *ISPRS J. Photogramm. Remote Sens.* **2018**, *146*, 91–107. [CrossRef]
73. Hu, Q.; Wu, W.; Xia, T.; Yu, Q.; Yang, P.; Li, Z.; Song, Q. Exploring the Use of Google Earth Imagery and Object-Based Methods in Land Use/Cover Mapping. *Remote Sens.* **2013**, *5*, 6026–6042. [CrossRef]
74. Vaz, D.A.; Sarmento, P.T.K.; Barata, M.T.; Fenton, L.K.; Michaels, T.I. Object-Based Dune Analysis: Automated Dune Mapping and Pattern Characterization for Ganges Chasma and Gale Crater, Mars. *Geomorphology* **2015**, *250*, 128–139. [CrossRef]
75. Xia, J.; Dong, P. A GIS Add-in for Automated Measurement of Sand Dune Migration Using LiDAR-Derived Multitemporal and High-Resolution Digital Elevation Models. *Geosphere* **2016**, *12*, 1316–1322. [CrossRef]
76. Rubin, D.M. A Unifying Model for Planform Straightness of Ripples and Dunes in Air and Water. *Earth-Sci. Rev.* **2012**, *113*, 176–185. [CrossRef]
77. Zimbelman, J.R.; Scheidt, S.P. Precision Topography of a Reversing Sand Dune at Bruneau Dunes, Idaho, as an Analog for Transverse Aeolian Ridges on Mars. *Icarus* **2014**, *230*, 29–37. [CrossRef]
78. Jackson, D.W.T.; Cooper, A.; Green, A.; Beyers, M.; Guisado-Pintado, E.; Wiles, E.; Benallack, K.; Balme, M. Reversing Transverse Dunes: Modelling of Airflow Switching Using 3D Computational Fluid Dynamics. *Earth Planet. Sci. Lett.* **2020**, *544*, 116363. [CrossRef]
79. Cooper, J.A.G. Mesoscale Geomorphic Change on Low Energy Barrier Islands in Chesapeake Bay, U.S.A. *Geomorphology* **2013**, *199*, 82–94. [CrossRef]
80. Selby, M.J.; Rains, R.B.; Palmer, R.W.P. Eolian Deposits of the Ice-Free Victoria Valley, Southern Victoria Land, Antarctica. *N. Z. J. Geol. Geophys.* **1974**, *17*, 543–562. [CrossRef]
81. Burkinshaw, J.R.; Rust, I.C. Aeolian Dynamics on the Windward Slope of a Reversing Transverse Dune, Alexandria Coastal Dunefield, South Africa. In *Aeolian Sediments*; 1993; pp. 13–21; ISBN 9781444303971. Available online: https://onlinelibrary.wiley.com/doi/abs/10.1002/9781444303971.ch2 (accessed on 5 November 2019).
82. Rubin, D.M. Lateral Migration of Linear Dunes in the Strzelecki Desert, Australia. *Earth Surf. Process. Landf.* **1990**, *15*, 1–14. [CrossRef]
83. Marín, L.; Forman, S.L.; Valdez, A.; Bunch, F. Twentieth Century Dune Migration at the Great Sand Dunes National Park and Preserve, Colorado, Relation to Drought Variability. *Geomorphology* **2005**, *70*, 163–183. [CrossRef]
84. Rockett, G.C.; Hesp, P.; Portz, L.; Barboza, E.G. Aeolian Geodiversity of the Itapeva Dunefield (Brazil) and Geoconservation in the Management of Protected Areas. *Geoheritage* **2022**, *14*, 111. [CrossRef]
85. Parise, C.K.; Calliari, L.J.; Krusche, N. Extreme storm surges in the south of brazil: Atmospheric conditions and shore erosion. *Braz. J. Oceanogr.* **2009**, *57*, 175–188. [CrossRef]
86. Machado, A.A.; Calliari, L.J. Synoptic Systems Generators of Extreme Wind in Southern Brazil: Atmospheric Conditions and Consequences in the Coastal Zone. *J. Coast. Res.* **2016**, *75*, 1182–1186. [CrossRef]
87. Martínez, M.L.; Landgrave, R.; Silva, R.; Hesp, P. Shoreline Dynamics and Coastal Dune Stabilization in Response to Changes in Infrastructure and Climate. *J. Coast. Res.* **2019**, *92*, 6. [CrossRef]
88. Levin, N.; Kidron, G.J.; Ben-Dor, E. The Spatial and Temporal Variability of Sand Erosion across a Stabilizing Coastal Dune Field. *Sedimentology* **2006**, *53*, 697–715. [CrossRef]
89. Costas, S.; Gallego-Fernández, J.B.; Bon de Sousa, L.; Kombiadou, K. Ecogeomorphic Response of a Coastal Dune in Southern Portugal Regulated by Extrinsic Factors. *Catena* **2023**, *221*, 106796. [CrossRef]
90. Lopez, A.C.B.; Hesp, P.A. Evolution of a Coastal Transgressive Dunefield to a Parabolic Dunefield, Canunda Dunes, South Australia. *Geomorphology* **2023**, *430*, 108653. [CrossRef]

91. Hesp, P.A. Conceptual Models of the Evolution of Transgressive Dune Field Systems. *Geomorphology* **2013**, *199*, 138–149. [CrossRef]
92. Tsoar, H. Sand Dunes Mobility and Stability in Relation to Climate. *Phys. A Stat. Mech. Its Appl.* **2005**, *357*, 50–56. [CrossRef]
93. Belly, P.Y. *Sand Movement by Wind*; Washington, DC, USA, 1964. Available online: https://repository.tudelft.nl/islandora/object/uuid:f5051163-7a83-440d-9d02-92683a442549/datastream/OBJ/download (accessed on 5 November 2019).
94. Sherman, D.J. Evaluation of Aeolian Sand Transport Equations Using Intertidal-Zone Measurements, Saunton Sands, England. *Sedimentology* **1990**, *37*, 385–392. [CrossRef]
95. Logie, M. Influence of Roughness Elements and Soil Moisture on the Resistance of Sand to Wind Erosion. Acidic soils and geomorphic processes. In Proceedings of the International Conference of the International Society of Soil Science, Jerusalem, Israel, 19 March–4 April 1981.
96. Sarre, R.D. Evaluation of Aeolian Sand Transport Equations Using Intertidal Zone Measurements, Saunton Sands, England. *Sedimentology* **1988**, *35*, 671–679. [CrossRef]
97. Guimarães, L.S. Morfodinâmica e Migração Das Dunas Eólicas Na Reserva Ecológica Do Taim, Litoral Sul Do RS. Master's Dissertation, Universidade Federal do Rio Grande do Sul, Porto Alegre, Brasil, 2005.
98. Puhl, P.R.; Dillenburg, S.R. Controls in the Development of Transgressive Dune Fields of the Holocene Coastal Barrier, Northern Coast of Rio Grande Do Sul. *Pesqui. Em Geocienc.* **2018**, *45*, 1–22. [CrossRef]
99. Rockett, G.C.; Telles, P.; Barboza, E.G.; Gruber, N.L.S.; Simão, C.E. Análise Espaço-Temporal Dos Ventos No Extremo Norte Da Planície Costeira Do Rio Grande Do Sul, Brasil. *Pesqui. Em Geocienc.* **2017**, *44*, 203–220. [CrossRef]

remote sensing

MDPI

Article

An Integrated Approach between Multispectral Satellite Images and Geophysical and Morpho-Topographic Surveys for the Detection of Water Stress Associated with Coastal Dune Erosion

Giovanni Scardino [1,2,*], Saverio Mancino [1], Gerardo Romano [1], Domenico Patella [1] and Giovanni Scicchitano [1,2]

1 Department of Earth and Geo-Environmental Sciences, University of Bari Aldo Moro, 70125 Bari, Italy; s.mancino2@studenti.uniba.it (S.M.); gerardo.romano@uniba.it (G.R.); domenico.patella@uniba.it (D.P.); giovanni.scicchitano@uniba.it (G.S.)

2 Interdepartmental Research Center for Coastal Dynamics, University of Bari Aldo Moro, 70125 Bari, Italy

* Correspondence: giovanni.scardino@uniba.it

Abstract: Coastal erosion occurs due to different processes involving physical and ecological systems. One of these factors is the degree of water stress experienced by dune vegetation. While healthy dune vegetation can help to stabilize the dune systems, water-stressed vegetation can instead enhance dune erosion. In this study, remote sensing techniques were used to monitor the water stress affecting the dune vegetation in dune systems along the alluvial plain of the Chiatona coast (Apulia, Southern Italy) located on the Ionian Arc. Multispectral satellite data from Landsat 8/9 and Sentinel-2 were used to assess the water stress at different spatial scales over a 4-year monitoring period from 2019 to 2023. The normalized difference vegetation index (NDVI) and the normalized difference moisture index (NDMI) were used to identify dune surfaces that were experiencing water stress. Furthermore, a terrestrial laser scanner and LiDAR data were acquired at different temporal ranges in areas affected by water stress to highlight coastal changes in areas associated with unhealthy dune vegetation. A large drop in NDVI values was observed in May 2020 due to the occurrence of coastal fires in some parts of the Chiatona coast. Geoelectrical surveys were conducted to investigate if coastal fires were capable of saline groundwater contamination, potentially enhancing dune erosion in these areas. The joint analysis of remote sensing, topographical, and geoelectric data showed that water stress reduced the amount of healthy dune vegetation, triggering dune deflation processes that resulted in increased coastal erosion rates, while also leading to the saline contamination of groundwater.

Keywords: water stress; vegetation; foredune; multispectral images; coastal erosion; groundwater saline contamination

Citation: Scardino, G.; Mancino, S.; Romano, G.; Patella, D.; Scicchitano, G. An Integrated Approach between Multispectral Satellite Images and Geophysical and Morpho-Topographic Surveys for the Detection of Water Stress Associated with Coastal Dune Erosion. *Remote Sens.* **2023**, *15*, 4415. https://doi.org/10.3390/rs15184415

Academic Editor: José Juan de Sanjosé Blasco

Received: 15 August 2023
Revised: 2 September 2023
Accepted: 6 September 2023
Published: 7 September 2023

1. Introduction

Coastal environments are important and complex ecosystems that require modern monitoring techniques for the development of reliable intervention strategies. Remote sensing technologies allow for the collection of large amounts of data at high temporal and spatial frequencies, which are useful for the detection and monitoring of the vegetation cover of coastal dunes [1,2]. The spectral signature of healthy, green vegetation exhibits a "peak-and-valley" curve that is exclusively related to the chlorophyll absorption band centered at wavelengths of approximately 450 and 670 nm [3,4]. The reflectance of healthy vegetation increases in the near-infrared portion of the electromagnetic spectrum at a spectral range of approximately 680–750 nm, depending on the species and environmental conditions [5]. In particular, chlorophyll production may decrease or even cease if the vegetation is stressed. Optical satellite data are frequently used for the analysis of vegetational cover and long-term changes in vegetation indices, such as the normalized difference vegetation index (NDVI) [6,7]. The NDVI is the most common method for evaluating the health of vegetation in coastal habitats [8] and provides the necessary phenological profiles used for the identification of the different types of vegetation [1]. NDVI time series data has been frequently used to monitor the productivity of seasonal vegetation, providing information that allows researchers to estimate the seasonal and inter-annual responses recorded by vegetation on coastal dune ecosystems [9]. Most studies reporting on the linear relation between rainfall events and NDVI have been carried out in arid or semi-arid areas [10,11]. Furthermore, an NDVI-based study carried out in Poland [12] confirms the suitability of NDVI in areas with sufficient water supplies. Relationships between NDVI and hydrological values have also been used to evaluate the response of forest ecosystems to changes in groundwater [13] as well as to evaluate groundwater runoff in wetland ecosystems to model environmental degradation [14]. NDVI has also been highlighted as a promising method for identifying groundwater-dependent vegetation [15], and there is an established relationship between NDVI and climate, surface water, and groundwater levels [7].

This study utilized a multidisciplinary approach comprising remote sensing and morpho-topographic and geophysical surveys to characterize water stress along the coastal dunes of the Chiatona coast (Apulia, Southern Italy). The aim of this study is to highlight the physical processes that could enhance coastal erosion. Remote sensing technologies, such as multispectral satellite images, integrated with morpho-topographic and geophysical data could represent a useful tool to automatically map coastal areas subjected to erosion processes and water stress. Evidence was collected directly from the study area, where, in recent years, both the shrubs and herbaceous vegetation of the foredunes and the arboreal vegetation of the secondary dunes have been observed to be suffering from a lack of water. To map the areas affected by water stress, we used multispectral satellite images across a 4-year time window (2019–2023) to analyze the study area in both the visible and infrared spectral ranges using optical satellite images. The analysis of multispectral optical satellite images allowed for the estimation of normalized vegetation indices, including the NDVI and the normalized difference moisture index (NDMI), which are particularly useful for characterizing the status of vegetative health with respect to water stress. Coastal changes were highlighted by conducting morpho-topographic surveys using terrestrial laser scanner (TLS) systems [16–20] as well as the analysis of LiDAR data [21–24] across different time periods. Geoelectrical surveys were also used to show the influence of groundwater flow on the root systems of the dune vegetation and were eventually used to highlight seawater intrusions in the study area. A drop in the NDVI observed in the study area was due to a coastal fire that occurred in the same period, resulting in a water deficit that affected the health of the dune vegetation. The NDMI was used to map the areas affected by the water deficit. At present, the dune vegetation is experiencing a slow natural recovery; however, dune erosion processes can still be observed in the present day.

2. Study Area: Chiatona Coastline, Apulia, Southern Italy

This study area is located in the southwestern sector of the Apulia region (southern Italy), about 1.2 km east of the Chiatona settlement, a seaside district of the municipality of Palagiano (TA), located on the coastal plain of the Gulf of Taranto (Figure 1).

The sediment cover on the coastal plain is the result of a long sedimentary accumulation and progradation process that began about 7000 years BP [25,26]. The sediments are primarily derived from siliciclastic deposits carried by the main Lucanian rivers (Bradano, Basento, Agri, Cavone, and Sinni). These siliciclastic sediments are redistributed by littoral drift trending SSW to NNE for the entire coastal stretch of the plain [27–29]. The small watercourses (Lato, Lenne, Patemisco, and Tara) that cross the sandy coast close to Palagianello and the surrounding towns do not have sufficient extension and flow rates to supply beaches [26]. There are different orders of dune ridges present along the mobile coastal system, with some extending more than 1 km inland and reaching elevations between 8 m and 17 m above sea level; these dunes are primarily stabilized by the presence of *macchia mediterranea* vegetation [28,30]. The beaches continued to accrete until the mid-20th century; later, however, they were subjected to erosion driven by a negative sediment balance due to the construction of dams and water collection facilities as well as the construction of tourist resorts, which led to the leveling of the foredunes [29,31]. It is clear that the erosive influence of coastal dynamics was amplified by anthropogenic influences, especially in the area of Chiatona and Castellaneta Marina. The current morphology of the coastal area is thus the consequence of intense anthropogenic modification [32] as well as wave-driven erosional phenomena, resulting in the retreat of the dune body due to the weakening of the foot of the dune, which can lead to blowouts; consequently, more and more secondary and tertiary dunes are affected by erosional processes [33–35].

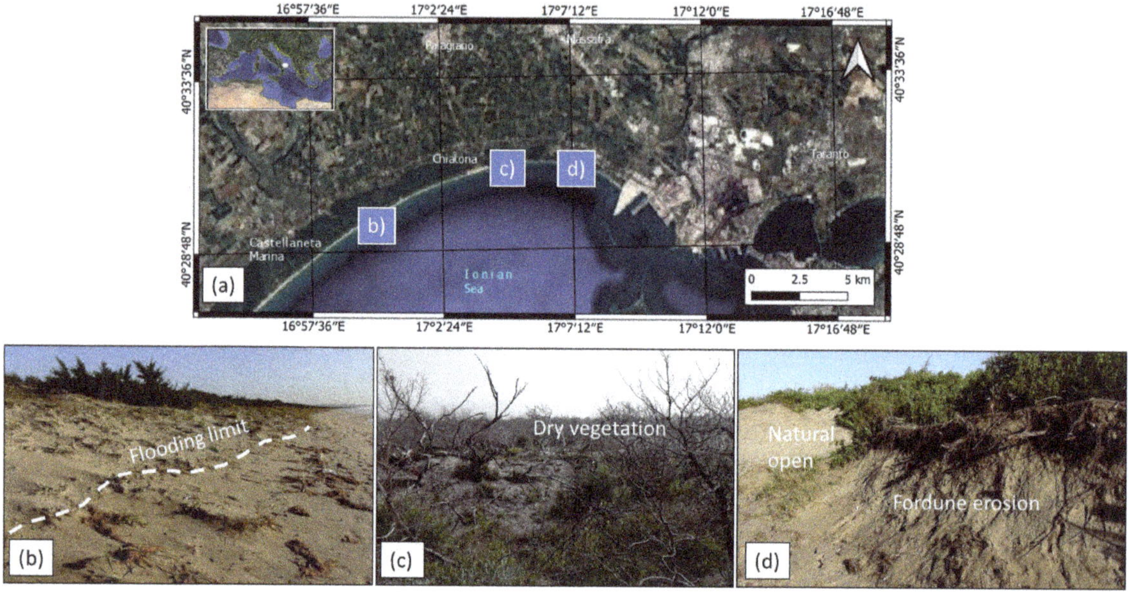

Figure 1. The study area is located on the coastal plain of the Gulf of Taranto. (**a**) Orthophoto of the coastal plain as acquired from Google Satellite images taken in 2023. (**b**) Evidence of dune erosion on the sandy coast of Pino di Lenne. (**c**) Dry vegetation in the Chiatona coastal area. (**d**) Foredune erosion and naturally open areas on Lido Azzurro.

Additional vegetation analysis allowed for the identification of the different types of vegetation present in the area. The significant environmental diversity observed in the study area is mainly a result of the geomorphological and microclimatic characteristics of the gravine formed by the erosional furrows. The current vegetation is of biogeographic interest because the Mediterranean-Eastern (Balkan) floristic component is associated with a moderate Western component [36]. The vegetation of the embryonic dunes is primarily represented by an agro-prairie ecosystem, composed of vegetation dominated by beach crabgrass (*Agropyron junceum*), a perennial plant that uses its high stoloniferous capacity to spread and avoid burial. *A. junceum* exhibits root lengths that range between 10.5 cm and 16.5 cm and heights ranging from 100.7 cm to 102.4 cm [37]. The pioneer plants are followed by massive vegetation colonization; here, sand begins to accumulate due to the cohesion provided by the stems of plants, with small deposits of embryonic dunes reaching elevations that are a few centimeters high. The development of the foredune occurs simultaneously with the appearance of European beachgrass (*Ammophila littoralis*), a perennial psammophilous grass with erect culms up to 1.5 m tall. These plants exhibit dense leaves that form thick and tall tufts; this is unlike *A. junceum*, which is characterized by isolated culms and well-spaced leaves. *A. littoralis* grows on dunes far above the water table, with most possessing roots that extend to about 1 m in depth; in some cases, however, the roots of this plant can be found at depths of 2 m and even up to 5 m [38].

The secondary and tertiary dunes are mainly covered by *Pinus halepensis*. Root length of Pinus halepensis ranges between 8 m and 5 m [39]. In the Chiatona area the saline groundwater contamination can influence the concentration of nitrogen, useful for root and shoot growth [39].

3. Materials and Methods

Analysis on the dune vegetation was followed after considering the areas affected by water stress detected through satellite data and an in-situ survey (Figure 2). The areas greatly affected by water stress have been surveyed through morpho-topographic techniques, using Terrestrial Laser Scanner and LiDAR data. The morpho-topographic and geophysical data allowed us to assess the coastal changes that occurred from 2009 to 2023 and to determine the entity of the sediment loss in the foredune area.

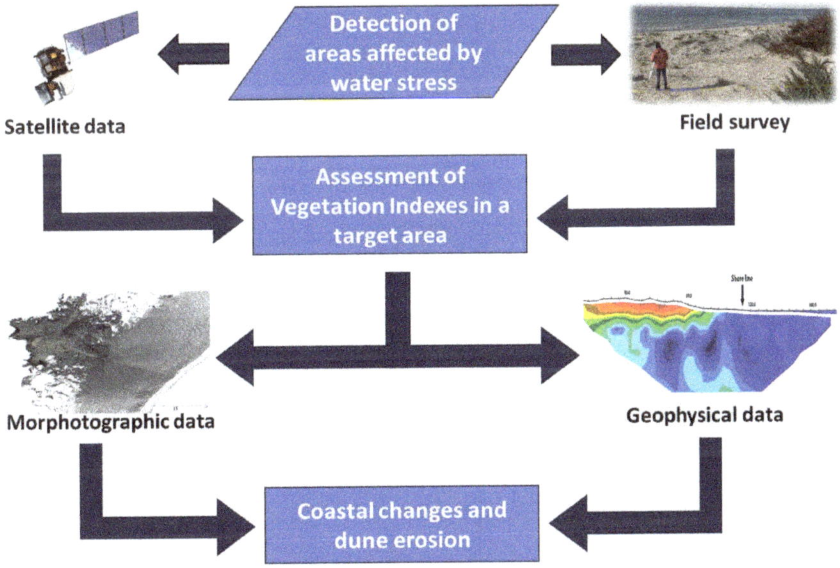

Figure 2. Flowchart of the work process followed to assess the coastal changes and dune erosion in the areas affected by water stress.

3.1. Satellite Data

The spectral data collected by satellite sensors provide a significant amount of information about the physical features of the landscape. The two main datasets used in this work were acquired from Landsat 8/9 and Sentinel-2.

These satellites are equipped with both optical and thermal sensors. The Operational Land Imager optical sensor (OLI; Ball Aerospace & Technologies Corporation, Broomfield, CO, USA) produces images across a total of nine spectral bands that encompass the visible, near-infrared, and microwave ranges, with a spectral resolution ranging from 443 nm to 2200.5 nm. In contrast, the Thermal Infrared Sensor (TIRS; NASA Goddard Space Flight Center) consists of two thermal bands that record ground surface temperatures, with a spectral resolution ranging from 10,895 nm to 12,005 nm. The Landsat 8/9 Level 1 missions provide panchromatic images at a 15-m spatial resolution and multispectral images at a 30-m spatial resolution along a 185 km swath, while the thermal bands provide 100 m resolution images at an acquisition frequency of 16 days.

The Sentinel-2 mission is a constellation of two polar-orbiting satellites, Sentinel-2A and Sentinel-2B, placed along the same sun-synchronous orbit with a phase offset of 180°. The optical sensors on these satellites were built by Astrium SAS (Paris, France) and have 13 spectral bands: four bands at a 10 m spatial resolution (with a spectral resolution ranging from 490 nm to 842 nm), six bands at a 20 m spatial resolution (with a spectral resolution ranging from 705 nm to 2190 nm), and three bands at a 60 m spatial resolution (with a spectral resolution ranging from 443 nm to 1375 nm). The orbital swath is 290 km with an acquisition frequency of 5 days.

The temporal ranges of the different satellite images are reported in Table S1. Cloud-free scenes were selected to minimize uncertainties due to heterogeneous atmospheric conditions; radiometric calibration and atmospheric correction were also applied [40].

The spectral contents of the datasets were analyzed to highlight the typical spectral signature of chlorophyll absorption. The spectral signatures of dune vegetation were extracted using semi-automatic classification algorithms in QGIS, a geospatial analysis software [41]. The spectral separability of signatures was assessed in the primary, secondary, and tertiary dunes in 500 m^2 squares.

Application of the Normalized Difference Vegetation Index (NDVI) in ecological studies has enabled the quantification and mapping of green vegetation and estimating the health vegetation status. NDVI is based on differences in reflectance in the red bands of visible spectra (due to vegetation pigment absorption) and maximum reflectance in the near infrared (NIR) bands (caused by cellular structure) [13,42]. In order to assess the soil moisture connected to the health vegetation status, the Normalized Difference Moisture Index was used. The NDMI is based on NIR and SWIR bands to display moisture. The SWIR band reflects changes in both the vegetation water content and the spongy mesophyll structure in vegetation canopies, while the NIR reflectance is affected by leaf internal structure and leaf dry matter content but not by water content [43–45].

The NDVI was calculated from satellite images without cloud cover to assess the health of the dune vegetation (Appendix A). The NDVI analysis allowed us to map the areas of the dune ridges that were influenced by water stress as well as extract time series data on the health of the vegetation. The NDMI was calculated from Sentinel 2 images to assess the water content of the vegetation (Appendix B).

3.2. Morpho-Topographic Data

Ground-based TLS and Airborne Laser Scanner (ALS) systems were used to obtain morpho-topographic data on the areas affected by water stress. ALS data were acquired for the inland areas from the former Italian Environmental Ministry (Ministero dell'Ambiente) between 2008 and 2009; these data had a vertical accuracy of 0.15 m and spatial resolution of 4 points/m^2 over inland areas and were georeferenced in the WGS84/UTM zone 33N coordinate reference system. The TLS surveys were performed in 2022 and 2023 using a Faro Focus X130 TLS and covered the littoral area of the emerged beach, ranging from the foreshore to the tertiary dunes. An outside acquisition setting was configured with a resolution of 28.9×10^6 points, an accuracy of 2 mm, an acquisition speed of 976,000 points/s, and a point distance of 3.068 mm/10 m.

The TLS point cloud was georeferenced using the GPS-Real Time Kinematic mode of the ITALPOS GNSS stations [46,47]. TLS data were filtered to remove vegetation and were interpolated using a natural neighbor algorithm to obtain digital terrain models (DTMs) that were representative of the coastal changes over time, with a cell width of 1 m. The main geomorphological features (e.g., shorelines, dune scarps, and dune ridges) were mapped and exported to a GIS environment. Furthermore, a difference of the DTMs was performed through a raster calculator in order to highlight the areas that experienced variations in the landforms and sediment loss.

3.3. Geoelectrical Surveys

The electrical resistivity tomography (ERT) methodology allows for the reconstruction of the distribution of electrical resistivity in the subsoil. Resistivity values depend on several factors, such as the porosity, the degree of saturation of the rocks, the nature of the fluids, and the mineralogy of the media being assessed. Consequently, ERT surveys are capable of acquiring key information in scenarios where significant resistivity contrasts are expected. They have been particularly successful in the detection of karst-related cavities and in the characterization of hydrogeological settings in coastal environments [48–51].

To investigate the subsoil in the study area, a single land–marine survey was conducted using a Syscal Pro 48 ch. (IRIS instruments, Orléans, France) connected to a 24-channel multielectrode land cable and a 13-channel multielectrode marine cable with an electrode spacing of 5 m. The total length of the surveys was 185 m (land: 115 m; marine: 70 m). The electrical connection between cables and ground was established through the use of integrated stainless steel and graphite electrodes on the land and marine cables, respectively. Where necessary, the contact resistance of the land cables was lowered through the use of salty water. Different electrode configurations were adopted to highlight the resistivity distribution patterns within the subsoil [52].

Specifically, data were acquired using the following configurations:

1. The Wenner–Schlumberger (WS) configuration due to its high signal-to-noise ratio as well as its moderate to high sensitivity to variations in vertical resistivity;
2. The dipole–dipole (DD) configuration (both in direct and reverse mode) due to its high sensitivity to lateral resistivity variations as well as to avoid experimental errors during the inversion procedure;
3. Multi gradient (GR) configuration due to its high spatial coverage in the shallower portions of the subsoil.

Topographical information was also included in the datasets; these were inverted using the RES2DINV program (Geotomo Software ver. 3.71.118; Loke and Barker, [53]). Resistivity models were produced by following the methods described by [53]; they were obtained using an L2-norm inversion while directly inverting the apparent resistivity values that better converged. Water column characteristics were also included in the inversion procedure by adding a stratum of 0.3 $\Omega \cdot$m of resistivity (mean resistivity of the seawater in general conditions) to the marine section of the ERT survey with a vertical extension calculated from the topographic information.

4. Results

4.1. Spectral Signature and Water Stress

The analysis of the spectral signature of dune vegetation revealed the presence of the characteristic "peak-and-valley" curve before the occurrence of coastal fires in the Chiatona area on 26 and 27 May 2020 (Figure 3a,b). Following the occurrence of these coastal fires, an area of water stress was observed, extending from the foredune to the tertiary dunes; this resulted in a drastic change in the typical spectral signature of dune vegetation, with a significant reflectance loss in the red and NIR bands (Figure 3c). The health of the vegetation appeared to be recovering slowly over the past year, evidenced by the restoration of the spectral signature associated with dune vegetation (Figure 3d).

Coastal fires occurred due to the aridity of the dune vegetation in some areas of the sandy Chiatona coast. An analysis of the NDVI values of these areas revealed seasonal variations from 2019 to May 2020 (Figure 4a,b), with values ranging between 0.4 and 0.8, which are consistent with chlorophyll absorption associated with healthy vegetation. Between May and June 2020, there was a significant drop in NDVI to values lower than 0.2 (Figure 4c). This drop was detected in the Landsat 8/9 images between 22 May 2020 and 7 June 2020 and in the Sentinel-2 images between 24 May 2020 and 13 June 2020. The coastal fires affected a surface area of approximately 300,750 m^2; these fires caused the vegetation in the region, primarily composed of *Pinus* spp., *A. junceum*, and *A. littoralis*, to be subjected to significant amounts of water stress (Figure 4). NDMI values lower than −0.25 were recorded during this time; these values usually represent areas with low canopy cover with high water stress or very low canopy cover with low water stress (Appendix B). It should be noted that the NDVI time series data revealed a positive trend that could be associated with an improvement in the health of the dune vegetation.

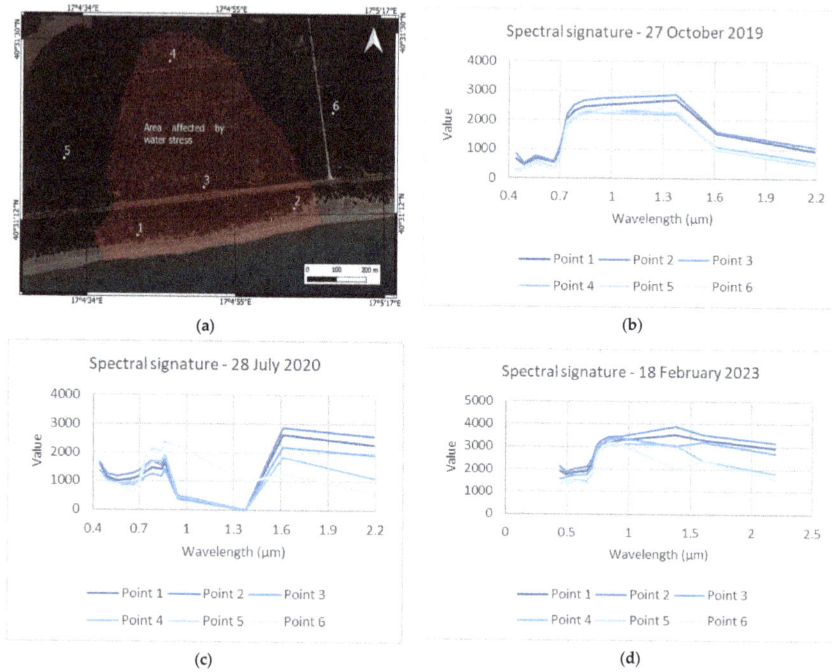

Figure 3. The spectral signatures of the dune vegetation in the study area: (**a**) sampling points used to assess the spectral signature of the dune vegetation; (**b**) spectral signature of the dune vegetation before the occurrence of coastal fires; (**c**) spectral signature of the dune vegetation after the occurrence of coastal fires; (**d**) restoration of vegetative health as highlighted by the changes in the spectral signature of the dune vegetation.

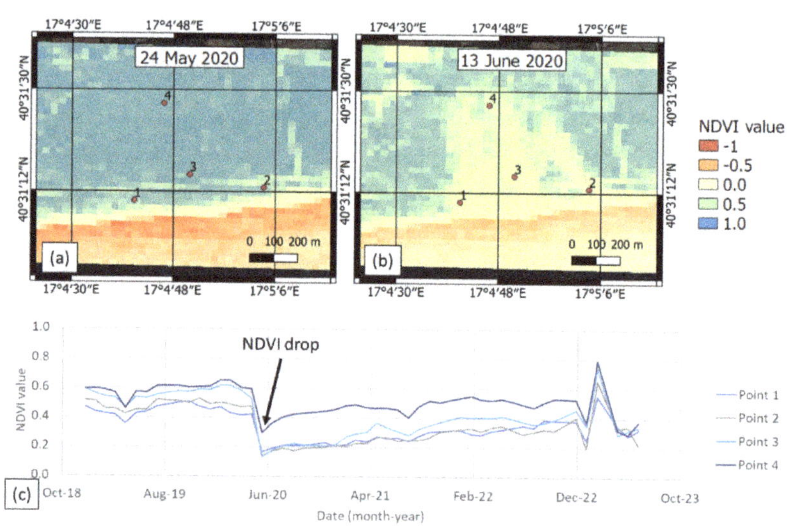

Figure 4. Map of NDVI values as calculated from Sentinel-2 images along the Chiatona coast: (**a**) NDVI map extracted before the coastal fire; (**b**) NDVI map after the coastal fires; (**c**) NDVI values over time. There is a clear drop in NDVI values associated with the coastal fire event.

4.2. Coastal Changes and Dune Erosion

The analysis of the morpho-topographic data revealed that dune erosion generally increased between 2020 and 2023. LiDAR data from 2009 revealed lower foredune elevations (Figure 5) compared to data from 2020, highlighting the general accretion experienced by the coastal system. However, TLS data acquired from 2022 to 2023 highlighted the extent of erosional processes on the foredunes at a present rate of -0.38 ± 0.1 m/year. DTMs were comparatively analyzed to calculate the sediment loss as a function of dune erosion. The loss in sediment resulted in the migration of the foredune scarp at a rate of 9 m^3/year (Figure 6).

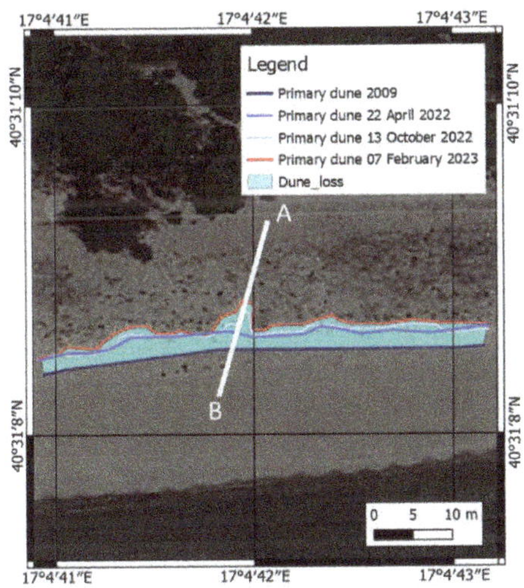

(a)

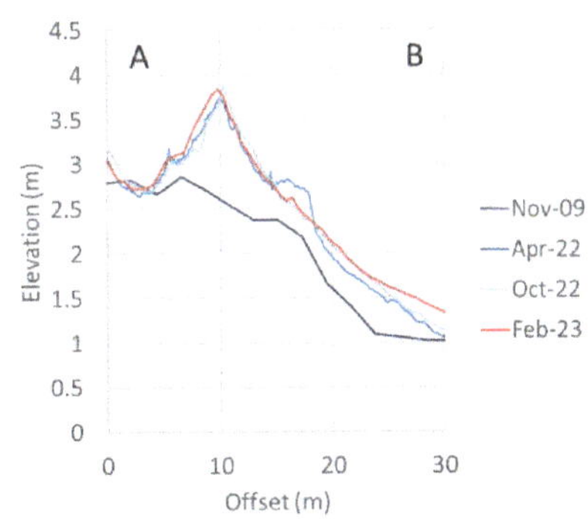

(b)

Figure 5. Dune erosion as identified using TLS and LiDAR data: (**a**) dune scarps mapped along the Chiatona coast; (**b**) change in topographic profiles of the foredunes over time.

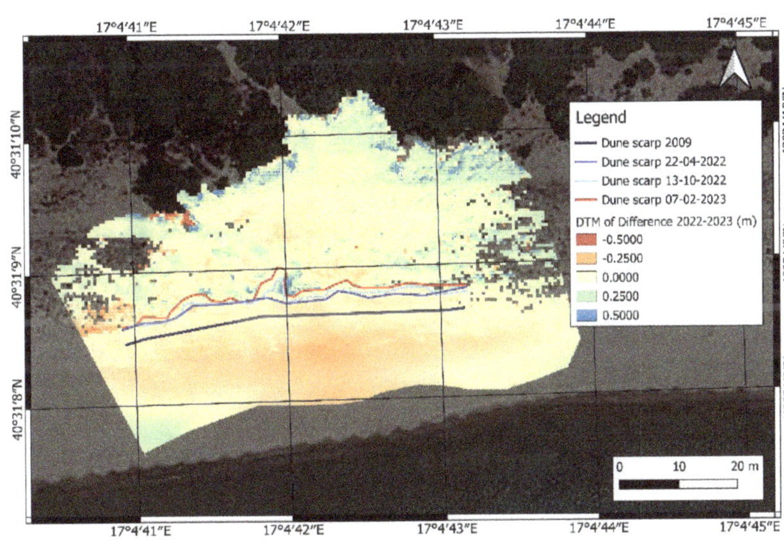

Figure 6. Coastal and dune erosion as highlighted by the difference in DTMs from 2022 and 2023. The red pixels in the raster image highlight the areas most affected by the loss of sediment.

Data from the ERT surveys revealed resistivity values that were consistent with the lithology of the study area as well as saltwater intrusion in the dune systems. The overall quality of the collected data was good, and no particular filtering procedures were adopted except for the removal of a few data points with negative resistivity values. The inversion of the geoelectrical data, which were acquired through the use of different electrode configurations (WS, DD, and GR), produced similar models. This was consistent with the rather simple geological setting of the study area, where resistivity variations are primarily due to the different degrees of saturation in the sediments. Figure 7a,b present the resistivity model obtained from the WS configuration and one possible interpretation of the model involving saltwater intrusion, respectively. This interpretation assumes that dry sands are characterized by high resistivities due to their high air content, while the resistivity of the saturated sands is strongly dependent on the nature of the pore-filling fluids. Consequently, more conductive areas are associated with seawater-saturated sands, while other parts of the model, which exhibit intermediate resistivity values, could represent areas saturated by fresh or brackish waters.

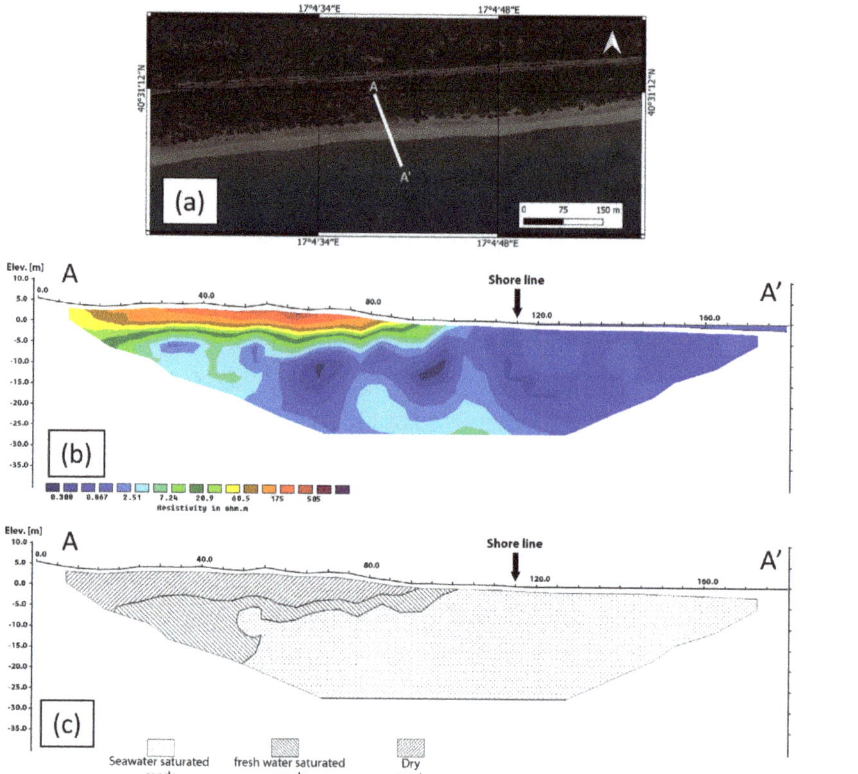

Figure 7. ERT profile: (**a**) the trace of the ERT survey; (**b**) the ERT model obtained from the inversion of the WS dataset (final r.m.s. = 7.8%); (**c**) a possible simplified interpretation of the ERT model in terms of saltwater intrusion.

The upper limit of saline wedge intrusion does not affect the roots of the dune vegetation (Figure 7).

5. Discussion

Dune vegetation water stress is a major factor that drives the destabilization of dune bodies, resulting in enhanced erosional processes [54,55]. Morpho-topographic data and in situ observations of the Chiatona coast revealed that portions of the foredunes were affected by deflation processes and sometimes even exhibited dune levelling (Figure 8) [26]. The deflation and erosional processes appeared to be enhanced in areas where the health of the dune vegetation was poor. Furthermore, the water stress experienced by the dune vegetation decreases the mechanical strength of the non-cohesive sediments [56]. Geotechnical tests have shown that vegetated dune systems are better at counteracting erosional processes driven by wave and surge impacts compared to dunes without vegetation [56].

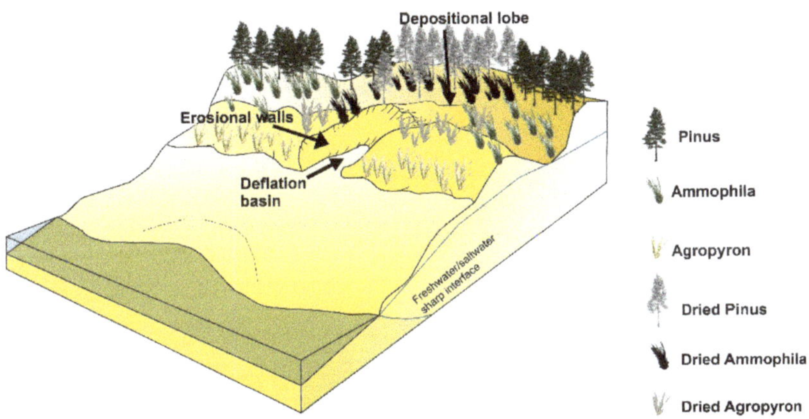

Figure 8. A diagram describing the erosional features on the sandy coast of the Ionian Arc. Erosional processes are primarily expressed in the form of deflation basins and erosion on the dunes affected by water stress and dry vegetation.

The relationship between the resilience of dunes and vegetation can be expressed in two main interactions: above-ground and below-ground interactions. Above-ground interactions involve the relationship between wave energy and surge impacts and plant height, flexibility, and the extent of the vegetated area [57,58]. Below-ground interactions involve the contribution of root density, depth, and size as well as mycorrhizal colonization to the mechanical strength of the dunes [56,59,60]. In this study area, *Pinus halepensis* and *A. littoralis* are the main species that contribute to the stabilization of the dune due to the below-ground interactions associated with their root system. In contrast, *A. littoralis* and *A. junceum* contribute to above-ground interactions by increasing the hydrodynamic drag and decreasing the energy imparted by wave impacts (Figure 9). Consequently, the poor health of dune vegetation decreases the ability of these vegetated dune systems to counteract the impact of storm events.

NDVI and NDMI analyses revealed that dune vegetation showed a slow recovery after the coastal fire occurred on 26 May 2020. However, LiDAR and TLS data revealed the presence of significant amounts of ongoing coastal erosion, with a shoreline erosion rate of -0.36 ± 0.18 m/year [26], consistent with the erosion rates observed on the primary dune scarp. The apparent contradiction between the rapid erosion rate and the slow restoration of dune vegetation is due to the different elastic responses of mobile coastal systems, which describes the resilience response of the dune system [61].

Multispectral satellite images were found to be better at describing the extent of vegetation compared to an orthomosaic RGB image (Figure 10). The red and NIR spectral bands of the Sentinel-2 and Landsat 8/9 data were much better at highlighting the distribution of vegetation characterized by low chlorophyll absorption (Figure 10b).

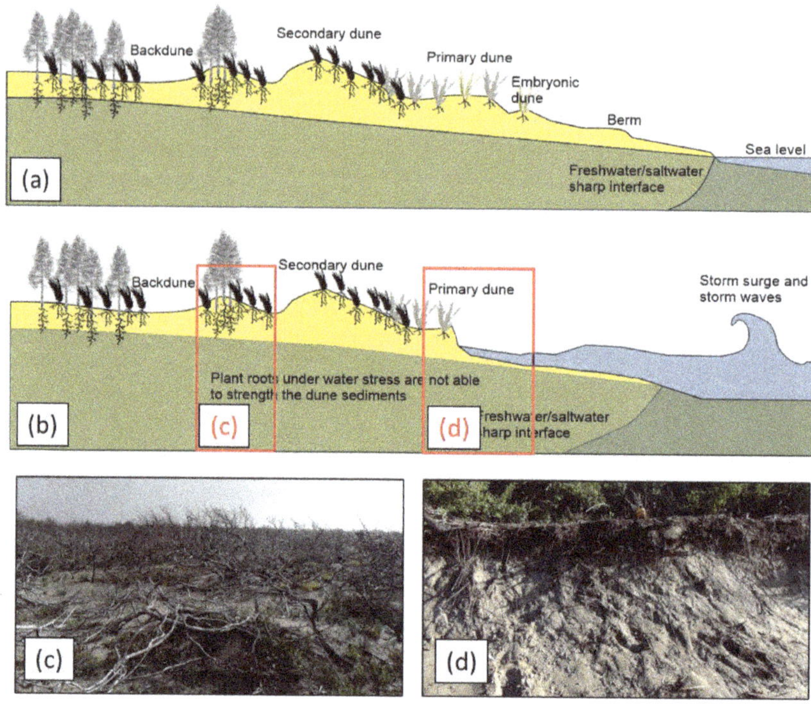

Figure 9. Coastal erosion and water stress in a vegetated dune system: (**a**) a steady-state mobile coastal system with low hydrodynamism; (**b**) dune erosion due to storm events and the limited contribution of dry vegetation to the shear strength of the dunes (red rectangles are referred to the areas affected by water stress on the vegetation); (**c**) dry vegetation on the tertiary and secondary dunes in Chiatona; (**d**) erosion of the primary dune scarp in Palagiano. Clear evidence of dry vegetation can also be observed.

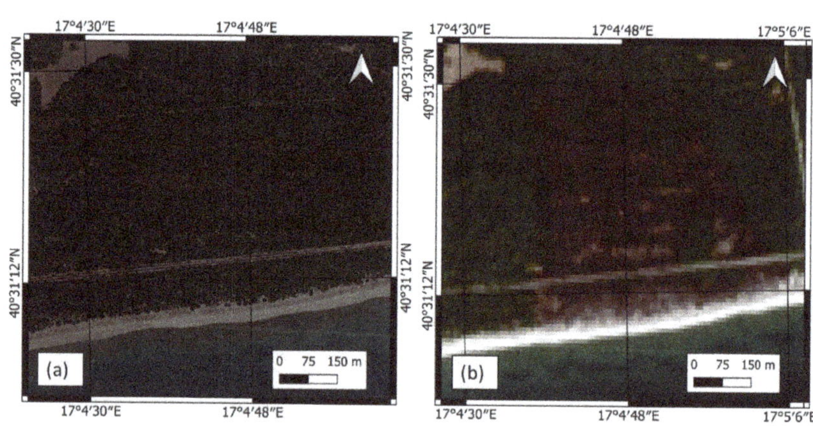

Figure 10. A comparison between an RGB orthomosaic and Landsat 8: (**a**) an orthomosaic image taken on June 2022; (**b**) Landsat 8 image taken on June 2022 with RGB channels highlighting the difference in spectral bands for the areas affected by coastal fires.

Multispectral images are widely used in vegetation-monitoring applications, especially in coastal environments [6,62]. Satellite images provide reliable records of the NDVI values of coastal vegetation due to their characteristic spectral signature [7]. Many studies have attempted to identify a correlation between the NDVI values of the dune vegetation and the height of the groundwater table [14,55,63], while other studies have attempted to integrate geophysical methods to correlate the height of the groundwater table with the health of the dune vegetation [64–67]. Many studies have also used satellite data to identify the different types of vegetation present in coastal dune systems through an analysis of NDVI and multispectral satellite imagery [22,68,69]. The height of the groundwater table is an important aspect that strongly influences vegetative health [9,15,70–72]. Many coastal management authorities are currently considering the use of dune vegetation as an eco-sustainable means of protecting coastlines against erosional processes triggered by climate change [55,73–77].

6. Conclusions

Mobile coastal systems are greatly affected by changes in climate. Among the many consequences of climate change, deficits in sediment balances are reflected in the erosion of coastal dunes. Dune vegetation can stabilize these dune systems, offsetting some of the effects of negative sedimentary balances. This study analyzed the extent of water stress in the dune systems of Chiatona (Apulia, Southern Italy). The main results obtained are the following:

- NDVI and NDMI analyses of multispectral satellite images revealed that the total surface area affected by water stress was approximately 300,750 m^2. This water stress was primarily due to a coastal fire event that occurred on 26 May 2020.
- Morpho-topographic and geoelectrical surveys were used to provide insights into the coastal dynamics of this stretch of coastline, as well as examine the response of the coastline to water stress, showing a rate of foredune erosion equal to -0.38 ± 0.1 m/year.

Furthermore, the following phenomena were observed:

- Coastal dune accretion occurred along the Chiatona coast from 2009 to 2020.
- On 26 May 2020—14:18 h UTC, coastal fires occurred on the Chiatona and Palagiano coasts.
- From June 2020 to February 2023, the Chiatona coast was subjected to significant erosional processes, primarily expressed by the retreat of the foredune scarp.

The multidisciplinary approach described in this work allowed us to assess the specific responses of this mobile coastal system in terms of its coastal resilience. In particular, the NDVI and NDMI analyses revealed that the health of the dune vegetation has been slowly recovering following the coastal fire event. However, this restoration has not been able to sufficiently stabilize the dune system, which is still subject to significant erosion. The remote sensing techniques described in this study can also be applied to coastal dune management to highlight the areas affected by water stress due to coastal fires.

Supplementary Materials: The following supporting information can be downloaded at: https://www.mdpi.com/article/10.3390/rs15184415/s1, Table S1: Dataset of multispectral images used for the assessment of water stress.

Author Contributions: Conceptualization, G.S. (Giovanni Scardino); methodology, G.S. (Giovanni Scardino), S.M. and G.R.; software, G.S. (Giovanni Scardino), G.R. and D.P.; validation, G.S. (Giovanni Scicchitano) and G.S. (Giovanni Scardino); formal analysis, G.S. (Giovanni Scardino), G.R. and D.P.; investigation, G.S. (Giovanni Scardino), S.M., G.R. and D.P.; resources, G.S. (Giovanni Scicchitano); data curation, G.S. (Giovanni Scardino), S.M., G.R. and D.P.; writing—original draft preparation, G.S. (Giovanni Scardino) and S.M.; writing—review and editing, G.R., D.P. and G.S. (Giovanni Scicchitano); visualization, G.S. (Giovanni Scicchitano); supervision, G.S. (Giovanni Scardino); project administration, G.S. (Giovanni Scicchitano); funding acquisition, G.S. (Giovanni Scicchitano). All authors have read and agreed to the published version of the manuscript.

Funding: This work was developed as part of the activities of the Research Agreement stipulated between the University of Bari Aldo Moro and the Agenzia Regionale Strategica per lo Sviluppo Ecosostenibile del Territorio (ASSET, Italy) (Scientific Coordinator G. Scicchitano).

Data Availability Statement: Suggested Data Availability Statements are available on request.

Acknowledgments: This research was partially funded by the Apulia Region (Italy) under the European Regional Development Fund and the European Social Fund (POR Puglia FESR-FSE 2014-2020-Action 10.4 "Research for Innovation" (REFIN): F675E915. An initial version of this work was developed by S. Mancino in the framework of his master thesis at the University of Bari Aldo Moro.

Conflicts of Interest: The authors declare no conflict of interest.

Appendix A

Some studies have used the normalized difference vegetation index (NDVI) as an indicator for vegetative health [78–80] (Caturegli et al., 2015; 2016; Volterrani et al., 2017). The NDVI is the ratio between the difference and the sum of reflected near-infrared and visible red radiation and describes the vigor of the vegetation being analyzed. The NDVI index was applied to multispectral satellite images using the following equation:

$$NDVI = \frac{R_{NIR} - R_{RED}}{R_{NIR} + R_{RED}} \tag{A1}$$

where R_{NIR} represents the reflectance value in the near-infrared (*NIR*) band and R_{RED} represents the reflectance value in the visible red band. The *NDVI* was calculated using the *NIR* and visible red bands from Landsat 8 and Sentinel-2 images in QGIS. Different *NDVI* values represent different degrees of vegetative vigor (Table A1).

Table A1. The relationship between NDVI values and the vigor of the vegetation.

NDVI	Type of Vegetation Vigor
<0.1	Bare ground or clouds
0.1–0.2	Almost no plant cover
0.2–0.3	Very low plant cover
0.3–0.4	Low canopy cover with low vigor or very low canopy cover with high vigor
0.4–0.5	Medium-low canopy cover with low vigor or very low canopy cover with high vigor
0.5–0.6	Medium canopy cover with low vigor or medium-low canopy cover with high vigor
0.6–0.7	Medium-high canopy cover with low vigor or medium canopy cover with high vigor
0.7–0.8	High plant cover with high vigor
0.8–0.9	Very high canopy cover with very high vigor
0.9–1.0	Total vegetative cover with very high vigor

Appendix B

The Normalized Difference Moisture Index (NDMI) [81] was used to evaluate the water stress experienced by the vegetation. The NDMI is the ratio between the difference and the sum of the reflected radiations in the NIR and in the short-wave infrared (SWIR) spectra:

$$NDMI = \frac{R_{NIR} - R_{SWIR}}{R_{NIR} + R_{SWIR}} \tag{A2}$$

where R_{NIR} represents the reflectance value in the *NIR* band and R_{SWIR} represents the reflectance value in the *SWIR* band. The *NDMI* was calculated using the *NIR* and *SWIR* bands from Landsat 8 and Sentinel-2 images in QGIS. *NDMI* values are representative of the degree of water stress and the type of vegetation (Table A2).

Table A2. The relationship between NDMI values and the degree of water stress and the type of vegetation.

NDMI	Type of Water Stress
−1–−0.8	Bare ground
−0.8–−0.6	Almost no plant cover
−0.6–−0.4	Very low plant cover
−0.4–−0.2	Low canopy cover with high water stress or very low canopy cover with low water stress
−0.2–0	Medium-low canopy cover with high water stress or low canopy cover with low water stress
0–0.2	Medium canopy cover with high water stress or medium-low canopy cover with low water stress
0.2–0.4	Medium-high canopy cover with high water stress or medium canopy cover with low water stress
0.4–0.6	High plant cover and no water stress
0.6–0.8	Very high plant cover and no water stress
0.8–1.0	Total plant cover and no water stress or stagnant water or clouds

References

1. Silvestri, S.; Marani, M.; Marani, A. Hyperspectral Remote Sensing of Salt Marsh Vegetation, Morphology and Soil Topography. *Phys. Chem. Earth Parts A/B/C* **2003**, *28*, 15–25. [CrossRef]
2. Yousefi Lalimi, F.; Silvestri, S.; Moore, L.J.; Marani, M. Coupled Topographic and Vegetation Patterns in Coastal Dunes: Remote Sensing Observations and Ecomorphodynamic Implications. *J. Geophys. Res. Biogeosci.* **2017**, *122*, 119–130. [CrossRef]
3. Zarco-Tejada, P.J.; Miller, J.R.; Morales, A.; Berjón, A.; Agüera, J. Hyperspectral Indices and Model Simulation for Chlorophyll Estimation in Open-Canopy Tree Crops. *Remote Sens. Environ.* **2004**, *90*, 463–476. [CrossRef]
4. Yadav, T. *Habitable Exoplanets: A Literature Review of Potential Signatures of Life*; ResearchGate: Berlin, Germany, 2018.
5. O'Malley-James, J.T.; Kaltenegger, L. The Vegetation Red Edge Biosignature Through Time on Earth and Exoplanets. *Astrobiology* **2018**, *18*, 1123–1136. [CrossRef] [PubMed]
6. Marzialetti, F.; Giulio, S.; Malavasi, M.; Sperandii, M.G.; Acosta, A.T.R.; Carranza, M.L. Capturing Coastal Dune Natural Vegetation Types Using a Phenology-Based Mapping Approach: The Potential of Sentinel-2. *Remote Sens.* **2019**, *11*, 1506. [CrossRef]
7. Fu, B.; Burgher, I. Riparian Vegetation NDVI Dynamics and Its Relationship with Climate, Surface Water and Groundwater. *J. Arid Environ.* **2015**, *113*, 59–68. [CrossRef]
8. Shalaby, A.; Tateishi, R. Remote Sensing and GIS for Mapping and Monitoring Land Cover and Land-Use Changes in the Northwestern Coastal Zone of Egypt. *Appl. Geogr.* **2007**, *27*, 28–41. [CrossRef]
9. Jackson, D.; Costas, S.; González-Villanueva, R.; Cooper, A. A Global 'Greening' of Coastal Dunes: An Integrated Consequence of Climate Change? *Glob. Planet. Change* **2019**, *182*, 103026. [CrossRef]
10. Peng, J.; Dong, W.; Yuan, W.; Zhang, Y. Responses of Grassland and Forest to Temperature and Precipitation Changes in Northeast China. *Adv. Atmos. Sci.* **2012**, *29*, 1063–1077. [CrossRef]
11. Groeneveld, D.P. Remotely-Sensed Groundwater Evapotranspiration from Alkali Scrub Affected by Declining Water Table. *J. Hydrol.* **2008**, *358*, 294–303. [CrossRef]
12. Dabrowska-Zielinska, K.; Kogan, F.; Ciolkosz, A.; Gruszczynska, M.; Kowalik, W. Modelling of Crop Growth Conditions and Crop Yield in Poland Using AVHRR-Based Indices. *Int. J. Remote Sens.* **2002**, *23*, 1109–1123. [CrossRef]
13. Aguilar, C.; Zinnert, J.C.; Polo, M.J.; Young, D.R. NDVI as an Indicator for Changes in Water Availability to Woody Vegetation. *Ecol. Indic.* **2012**, *23*, 290–300. [CrossRef]
14. Šimanauskienė, R.; Linkevičienė, R.; Bartold, M.; Dąbrowska-Zielińska, K.; Slavinskienė, G.; Veteikis, D.; Taminskas, J. Peatland Degradation: The Relationship between Raised Bog Hydrology and Normalized Difference Vegetation Index. *Ecohydrology* **2019**, *12*, e2159. [CrossRef]
15. Páscoa, P.; Gouveia, C.M.; Kurz-Besson, C. A Simple Method to Identify Potential Groundwater-Dependent Vegetation Using NDVI MODIS. *Forests* **2020**, *11*, 147. [CrossRef]
16. Caldara, M.; Capolongo, D.; Damato, B.; Pennetta, L. Can the Ground Laser Scanning Technology Be Useful for Coastal Defenses Monitoring? *Ital. J. Eng. Geol. Environ.* **2006**, *1*, 35–49.

17. Boeder, V.; Kersten, T.; Hesse, C.; Thies, T.; Sauer, A. Initial Experience with the Integration of a Terrestrial Laser Scanner into the Mobile Hydrographic Multi Sensor System on a Ship. In Proceedings of the ISPRS Istanbul Workshop 2010 on Modeling of Optical Airborne and Spaceborne Sensors, Istanbul, Turkey, 11–13 October 2010; Volume 38.
18. Lapietra, I.; Lisco, S.; Mastronuzzi, G.; Milli, S.; Pierri, C.; Sabatier, F.; Scardino, G.; Moretti, M. Morpho-Sedimentary Dynamics of Torre Guaceto Beach (Southern Adriatic Sea, Italy). *J. Earth Syst. Sci.* **2022**, *131*, 64. [CrossRef]
19. Lapietra, I.; Lisco, S.; Capozzoli, L.; De Giosa, F.; Mastronuzzi, G.; Mele, D.; Milli, S.; Romano, G.; Sabatier, F.; Scardino, G.; et al. A Potential Beach Monitoring Based on Integrated Methods. *J. Mar. Sci. Eng.* **2022**, *10*, 1949. [CrossRef]
20. O'Dea, A.; Brodie, K.L.; Hartzell, P. Continuous Coastal Monitoring with an Automated Terrestrial Lidar Scanner. *J. Mar. Sci. Eng.* **2019**, *7*, 37. [CrossRef]
21. Fabris, M. Monitoring the Coastal Changes of the Po River Delta (Northern Italy) since 1911 Using Archival Cartography, Multi-Temporal Aerial Photogrammetry and LiDAR Data: Implications for Coastline Changes in 2100 A.D. *Remote Sens.* **2021**, *13*, 529. [CrossRef]
22. Frati, G.; Launeau, P.; Robin, M.; Giraud, M.; Juigner, M.; Debaine, F.; Michon, C. Coastal Sand Dunes Monitoring by Low Vegetation Cover Classification and Digital Elevation Model Improvement Using Synchronized Hyperspectral and Full-Waveform LiDAR Remote Sensing. *Remote Sens.* **2021**, *13*, 29. [CrossRef]
23. Arshad, B.; Barthelemy, J.; Perez, P. Autonomous Lidar-Based Monitoring of Coastal Lagoon Entrances. *Remote Sens.* **2021**, *13*, 1320. [CrossRef]
24. Schmidt, A.; Rottensteiner, F.; Soergel, U. Monitoring Concepts for Coastal Areas Using Lidar Data. *ISPRS Int. Arch. Photogramm. Remote Sens. Spat. Inf. Sci.* **2013**, *XL-1/W1*, 311–316. [CrossRef]
25. Tropeano, M.; Cilumbriello, A.; Sabato, L.; Gallicchio, S.; Grippa, A.; Longhitano, S.G.; Bianca, M.; Gallipoli, M.R.; Mucciarelli, M.; Spilotro, G. Surface and Subsurface of the Metaponto Coastal Plain (Gulf of Taranto—Southern Italy): Present-Day- vs LGM-Landscape. *Geomorphology* **2013**, *203*, 115–131. [CrossRef]
26. Scardino, G.; Sabatier, F.; Scicchitano, G.; Piscitelli, A.; Milella, M.; Vecchio, A.; Anzidei, M.; Mastronuzzi, G. Sea-Level Rise and Shoreline Changes Along an Open Sandy Coast: Case Study of Gulf of Taranto, Italy. *Water* **2020**, *12*, 1414. [CrossRef]
27. Cilumbriello, A.; Sabato, L.; Tropeano, M.; Gallicchio, S.; Grippa, A.; Maiorano, P.; Mateu-Vicens, G.; Rossi, C.A.; Spilotro, G.; Calcagnile, L.; et al. Sedimentology, Stratigraphic Architecture and Preliminary Hydrostratigraphy of the Metaponto Coastal-Plain Subsurface (Southern Italy). *Mem. Descr. Carta Geol. d'It* **2010**, *XC*, 67–84.
28. Sabato, L.; Longhitano, S.G.; Gioia, D.; Cilumbriello, A.; Spalluto, L. Sedimentological and Morpho-Evolution Maps of the 'Bosco Pantano Di Policoro' Coastal System (Gulf of Taranto, Southern Italy). *J. Maps* **2012**, *8*, 304–311. [CrossRef]
29. Bonora, N.; Immordino, F.; Schiavi, C.; Simeoni, U.; Valpreda, E. Interaction between Catchment Basin Management and Coastal Evolution (Southern Italy). *J. Coast. Res.* **2002**, *36*, 81–88. [CrossRef]
30. Longhitano, S.G. Short-Term Assessment of Retreating vs. Advancing Microtidal Beaches Based on the Backshore/Foreshore Length Ratio: Examples from the Basilicata Coasts (Southern Italy). *Open J. Mar. Sci.* **2015**, *5*, 123–145. [CrossRef]
31. Caldara, M.; Centenaro, E.; Mastronuzzi, G.; Sansò, P.; Sergio, A. Features and Present Evolution of Apulian Coast (Southern Italy). *J. Coast. Res.* **1998**, *SI*, 55–64.
32. Rizzo, A.; De Giosa, F.; Di Leo, A.; Lisco, S.; Moretti, M.; Scardino, G.; Scicchitano, G.; Mastronuzzi, G. Geo-Environmental Characterisation of High Contaminated Coastal Sites: The Analysis of Past Experiences in Taranto (Southern Italy) as a Key for Defining Operational Guidelines. *Land* **2022**, *11*, 878. [CrossRef]
33. Hesp, P.A. Ecological Processes and Plant Adaptations on Coastal Dunes. *J. Arid Environ.* **1991**, *21*, 165–191. [CrossRef]
34. Hesp, P. Foredunes and Blowouts: Initiation, Geomorphology and Dynamics. *Geomorphology* **2002**, *48*, 245–268. [CrossRef]
35. Hesp, P.; Martinez, M.; da Silva, G.M.; Rodríguez-Revelo, N.; Gutierrez, E.; Humanes, A.; Laínez, D.; Montaño, I.; Palacios, V.; Quesada, A.; et al. Transgressive Dunefield Landforms and Vegetation Associations, Doña Juana, Veracruz, Mexico. *Earth Surf. Process. Landf.* **2011**, *36*, 285–295. [CrossRef]
36. Biondi, E.; Guerra, V. Vegetazione e Paesaggio Vegetale Delle Gravine Dell'arco Jonico. *Fitosociologia* **2008**, *45*, 57–125.
37. Romano, G.; Ricci, G.F.; Leronni, V.; Venerito, P.; Gentile, F. Soil Bioengineering Techniques for Mediterranean Coastal Dune Restoration Using Autochthonous Vegetation Species. *J. Coast. Conserv.* **2022**, *26*, 71. [CrossRef]
38. Huiskes, A.H.L. *Ammophila arenaria* (L.) Link (*Psamma arenaria* (L.) Roem. et Schult.; *Calamgrostis arenaria* (L.) Roth). *J. Ecol.* **1979**, *67*, 363–382. [CrossRef]
39. Cuesta, B.; Vega, J.; Villar-Salvador, P.; Rey-Benayas, J.M. Root Growth Dynamics of Aleppo Pine (*Pinus halepensis* Mill.) Seedlings in Relation to Shoot Elongation, Plant Size and Tissue Nitrogen Concentration. *Trees* **2010**, *24*, 899–908. [CrossRef]
40. Yuan, J.; Niu, Z. Evaluation of Atmospheric Correction Using FLAASH. In Proceedings of the 2008 International Workshop on Earth Observation and Remote Sensing Applications, Beijing, China, 30 June–2 July 2008; pp. 1–6.
41. Congedo, L. Semi-Automatic Classification Plugin: A Python Tool for the Download and Processing of Remote Sensing Images in QGIS. *J. Open Source Softw.* **2021**, *6*, 3172. [CrossRef]
42. Gamon, J.A.; Field, C.B.; Goulden, M.L.; Griffin, K.L.; Hartley, A.E.; Joel, G.; Peñuelas, J.; Valentini, R. Relationships Between NDVI, Canopy Structure, and Photosynthesis in Three Californian Vegetation Types. *Ecol. Appl.* **1995**, *5*, 28–41. [CrossRef]
43. Nejad, M.F.; Zoratipour, A. Assessment of LST and NDMI Indices Using MODIS and Landsat Images in Karun Riparian Forest. *J. For. Sci.* **2019**, *65*, 27–32. [CrossRef]

44. Mihai, B.; Horoias, R. NDMI use in recognition of water stress issues, related to winter wheat yields in Southern Romania. *Sci. Pap. Ser. Manag. Econ. Eng. Agric. Rural. Dev.* **2022**, *22*, 105–112.
45. Strashok, O.; Ziemiańska, M.; Strashok, V. Evaluation and Correlation of Normalized Vegetation Index and Moisture Index in Kyiv (2017–2021). *J. Ecol. Eng.* **2022**, *23*, 212–218. [CrossRef] [PubMed]
46. Serpelloni, E.; Casula, G.; Galvani, A.; Anzidei, M.; Baldi, P. Data Analysis of Permanent GPS Networks in Italy and Surrounding Region: Application of a Distributed Processing Approach. *Ann. Geophys.* **2006**, *49*, 897–928. [CrossRef]
47. Serpelloni, E.; Faccenna, C.; Spada, G.; Dong, D.; Williams, S.D.P. Vertical GPS Ground Motion Rates in the Euro-Mediterranean Region: New Evidence of Velocity Gradients at Different Spatial Scales along the Nubia-Eurasia Plate Boundary. *J. Geophys. Res. Solid Earth* **2013**, *118*, 6003–6024. [CrossRef]
48. Romano, G.; Capozzoli, L.; Abate, N.; De Girolamo, M.; Liso, I.S.; Patella, D.; Parise, M. An Integrated Geophysical and Unmanned Aerial Systems Surveys for Multi-Sensory, Multi-Scale and Multi-Resolution Cave Detection: The Gravaglione Site (Canale Di Pirro Polje, Apulia). *Remote Sens.* **2023**, *15*, 3820. [CrossRef]
49. Muzzillo, R.; Zuffianò, L.E.; Rizzo, E.; Canora, F.; Capozzoli, L.; Giampaolo, V.; De Giorgio, G.; Sdao, F.; Polemio, M. Seawater Intrusion Proneness and Geophysical Investigations in the Metaponto Coastal Plain (Basilicata, Italy). *Water* **2021**, *13*, 53. [CrossRef]
50. Niculescu, B.M.; Andrei, G. Application of Electrical Resistivity Tomography for Imaging Seawater Intrusion in a Coastal Aquifer. *Acta Geophys.* **2021**, *69*, 613–630. [CrossRef]
51. Kazakis, N.; Pavlou, A.; Vargemezis, G.; Voudouris, K.S.; Soulios, G.; Pliakas, F.; Tsokas, G. Seawater Intrusion Mapping Using Electrical Resistivity Tomography and Hydrochemical Data. An Application in the Coastal Area of Eastern Thermaikos Gulf, Greece. *Sci. Total Environ.* **2016**, *543*, 373–387. [CrossRef]
52. Dahlin, T.; Zhou, B. A Numerical Comparison of 2D Resistivity Imaging with 10 Electrode Arrays. *Geophys. Prospect.* **2004**, *52*, 379–398. [CrossRef]
53. Loke, M.H.; Barker, R.D. Rapid Least-Squares Inversion of Apparent Resistivity Pseudosections by a Quasi-Newton Method. *Geophys. Prospect.* **1996**, *44*, 131–152. [CrossRef]
54. Nordstrom, K.F.; Hartman, J.; Freestone, A.L.; Wong, M.; Jackson, N.L. Changes in Topography and Vegetation near Gaps in a Protective Foredune. *Ocean Coast. Manag.* **2007**, *50*, 945–959. [CrossRef]
55. Taminskas, J.; Šimanauskienė, R.; Linkevičienė, R.; Volungevičius, J.; Slavinskienė, G.; Povilanskas, R.; Satkūnas, J. Impact of Hydro-Climatic Changes on Coastal Dunes Landscape According to Normalized Difference Vegetation Index (the Case Study of Curonian Spit). *Water* **2020**, *12*, 3234. [CrossRef]
56. Sigren, J.; Figlus, J.; Armitage, A. Coastal Sand Dunes and Dune Vegetation: Restoration, Erosion, and Storm Protection. *Shore Beach* **2014**, *82*, 5–12.
57. Bouma, T.J.; De Vries, M.B.; Low, E.; Peralta, G.; Tánczos, I.C.; van de Koppel, J.; Herman, P.M.J. Trade-Offs Related to Ecosystem Engineering: A Case Study on Stiffness of Emerging Macrophytes. *Ecology* **2005**, *86*, 2187–2199. [CrossRef]
58. Augustin, L.N.; Irish, J.L.; Lynett, P. Laboratory and Numerical Studies of Wave Damping by Emergent and Near-Emergent Wetland Vegetation. *Coast. Eng.* **2009**, *56*, 332–340. [CrossRef]
59. Miller, R.M.; Jastrow, J.D. Hierarchy of Root and Mycorrhizal Fungal Interactions with Soil Aggregation. *Soil Biol. Biochem.* **1990**, *22*, 579–584. [CrossRef]
60. De Baets, S.; Poesen, J.; Reubens, B.; Wemans, K.; De Baerdemaeker, J.; Muys, B. Root Tensile Strength and Root Distribution of Typical Mediterranean Plant Species and Their Contribution to Soil Shear Strength. *Plant Soil* **2008**, *305*, 207–226. [CrossRef]
61. Westman, W.E. Resilience: Concepts and Measures. In *Resilience in Mediterranean-Type Ecosystems*; Dell, B., Hopkins, A.J.M., Lamont, B.B., Eds.; Tasks for Vegetation Science; Springer: Dordrecht, The Netherlands, 1986; pp. 5–19. ISBN 978-94-009-4822-8.
62. Suo, C.; McGovern, E.; Gilmer, A. Coastal Dune Vegetation Mapping Using a Multispectral Sensor Mounted on an UAS. *Remote Sens.* **2019**, *11*, 1814. [CrossRef]
63. Alessio, G.A.; De Lillis, M.; Brugnoli, E.; Lauteri, M. Water Sources and Water-Use Efficiency in Mediterranean Coastal Dune Vegetation. *Plant Biol.* **2004**, *6*, 350–357. [CrossRef]
64. Abd El-Dayem, M.; Abd El-Gawad, A.; Bedair, S.; Farag, K.S.I. Groundwater Resource Evaluation Using Geoelectrical Resistivity Survey in the Ghard El-Hunishat Area of New Delta Project Province, North Western Desert, Egypt. *Groundw. Sustain. Dev.* **2023**, *21*, 100918. [CrossRef]
65. Alexopoulos, J.; Dilalos, S.; Poulos, S.; Ghionis, G.; Mavroulis, S. *Application of Geoelectrical Techniques in the Investigation of a Coastal Sand Dune Field*; European Association of Geoscientists & Engineers: Utrecht, The Netherlands, 2014.
66. Frohlich, R.K.; Urish, D.W.; Fuller, J.; O'Reilly, M. Use of Geoelectrical Methods in Groundwater Pollution Surveys in a Coastal Environment. *J. Appl. Geophys.* **1994**, *32*, 139–154. [CrossRef]
67. Urish, D.W.; Frohlich, R.K. Surface Electrical Resistivity in Coastal Groundwater Exploration. *Geoexploration* **1990**, *26*, 267–289. [CrossRef]
68. Medina Machín, A.; Marcello, J.; Hernández-Cordero, A.I.; Martín Abasolo, J.; Eugenio, F. Vegetation Species Mapping in a Coastal-Dune Ecosystem Using High Resolution Satellite Imagery. *GIScience Remote Sens.* **2019**, *56*, 210–232. [CrossRef]
69. Murphy, J.T.; Owensby, C.E.; Ham, J.; Coyne, P. Estimation of Vegetative Characteristics by Remote Sensing. *Acad. Res. J. Agric. Sci. Res.* **2014**, *2*, 34–46. [CrossRef]
70. Silva, F.G.; Wijnberg, K.M.; de Groot, A.V.; Hulscher, S.J.M.H. The Influence of Groundwater Depth on Coastal Dune Development at Sand Flats Close to Inlets. *Ocean Dyn.* **2018**, *68*, 885–897. [CrossRef]

71. Lammerts, E.J.; Maas, C.; Grootjans, A.P. Groundwater Variables and Vegetation in Dune Slacks. *Ecol. Eng.* **2001**, *17*, 33–47. [CrossRef]
72. Van Bodegom, P.M.; Oosthoek, A.; Broekman, R.; Bakker, C.; Aerts, R. Raising Groundwater Differentially Affects Mineralization and Plant Species Abundance in Dune Slacks. *Ecol. Appl.* **2006**, *16*, 1785–1795. [CrossRef] [PubMed]
73. Zarnetske, P.L.; Ruggiero, P.; Seabloom, E.W.; Hacker, S.D. Coastal Foredune Evolution: The Relative Influence of Vegetation and Sand Supply in the US Pacific Northwest. *J. R. Soc. Interface* **2015**, *12*, 20150017. [CrossRef]
74. Lawlor, P.; Jackson, D.W.T. A Nature-Based Solution for Coastal Foredune Restoration: The Case Study of Maghery, County Donegal, Ireland. In *Human-Nature Interactions: Exploring Nature's Values Across Landscapes*; Misiune, I., Depellegrin, D., Egarter Vigl, L., Eds.; Springer International Publishing: Cham, Switzerland, 2022; pp. 417–429. ISBN 978-3-031-01980-7.
75. Scicchitano, G.; Scardino, G.; Monaco, C.; Piscitelli, A.; Milella, M.; De Giosa, F.; Mastronuzzi, G. Comparing Impact Effects of Common Storms and Medicanes along the Coast of South-Eastern Sicily. *Mar. Geol.* **2021**, *439*, 106556. [CrossRef]
76. Scardino, G.; Scicchitano, G.; Chirivì, M.; Costa, P.J.M.; Luparelli, A.; Mastronuzzi, G. Convolutional Neural Network and Optical Flow for the Assessment of Wave and Tide Parameters from Video Analysis (LEUCOTEA): An Innovative Tool for Coastal Monitoring. *Remote Sens.* **2022**, *14*, 2994. [CrossRef]
77. Scardino, G.; Anzidei, M.; Petio, P.; Serpelloni, E.; De Santis, V.; Rizzo, A.; Liso, S.I.; Zingaro, M.; Capolongo, D.; Vecchio, A.; et al. The Impact of Future Sea-Level Rise on Low-Lying Subsiding Coasts: A Case Study of Tavoliere Delle Puglie (Southern Italy). *Remote Sens.* **2022**, *14*, 4936. [CrossRef]
78. Caturegli, L.; Casucci, M.; Lulli, F.; Grossi, N.; Gaetani, M.; Magni, S.; Bonari, E.; Volterrani, M. GeoEye-1 Satellite versus Ground-Based Multispectral Data for Estimating Nitrogen Status of Turfgrasses. *Int. J. Remote Sens.* **2015**, *36*, 2238–2251. [CrossRef]
79. Caturegli, L.; Corniglia, M.; Gaetani, M.; Grossi, N.; Magni, S.; Migliazzi, M.; Angelini, L.; Mazzoncini, M.; Silvestri, N.; Fontanelli, M.; et al. Unmanned Aerial Vehicle to Estimate Nitrogen Status of Turfgrasses. *PLoS ONE* **2016**, *11*, e0158268. [CrossRef]
80. Volterrani, M.; Minelli, A.; Gaetani, M.; Grossi, N.; Magni, S.; Caturegli, L. Reflectance, Absorbance and Transmittance Spectra of Bermudagrass and Manilagrass Turfgrass Canopies. *PLoS ONE* **2017**, *12*, e0188080. [CrossRef]
81. Jin, S.; Sader, S.A. Comparison of Time Series Tasseled Cap Wetness and the Normalized Difference Moisture Index in Detecting Forest Disturbances. *Remote Sens. Environ.* **2005**, *94*, 364–372. [CrossRef]

remote sensing MDPI

Article

High-Resolution Real-Time Coastline Detection Using GNSS RTK, Optical, and Thermal SfM Photogrammetric Data in the Po River Delta, Italy

Massimo Fabris [1,*], Mirco Balin [2] and Michele Monego [1]

1 Department of Civil, Environmental and Architectural Engineering, University of Padua, 35131 Padua, Italy; michele.monego@unipd.it
2 Brenta Lavori srl, 35010 Vigonza, Italy; mircobalin@gmail.com
* Correspondence: massimo.fabris@unipd.it

Abstract: High-resolution coastline detection and monitoring are challenging on a global scale, especially in flat areas where natural events, sea level rise, and anthropic activities constantly modify the coastal environment. While the coastline related to the 0-level contour line can be extracted from accurate Digital Terrain Models (DTMs), the detection of the real-time, instantaneous coastline, especially at low tide, is a challenge that warrants further study and evaluation. In order to investigate an efficient combination of methods that allows to contribute to the knowledge in this field, this work uses topographic total station measurements, Global Navigation Satellite System Real-Time Kinematic (GNSS RTK) technique, and the Structure from Motion (SfM) approach (using a low-cost drone equipped with optical and thermal cameras). All the data were acquired at the beginning of 2022 and refer to the areas of Boccasette and Barricata, in the Po River Delta (Northeastern of Italy). The real-time coastline obtained from the GNSS data was validated using the topographic total station measurements; the correspondent polylines obtained from the photogrammetric data (using both automatic extraction and manual restitutions by visual inspection of orhophotos) were compared with the GNSS data to evaluate the performances of the different techniques. The results provided good agreement between the real-time coastlines obtained from different approaches. However, using the optical images, the accuracy was strictly connected with the radiometric changes in the photos and using thermal images, both manual and automatic polylines provided differences in the order of 1–2 m. Multi-temporal comparison of the 0-level coastline with those obtained from a LiDAR survey performed in 2018 provided the detection of the erosion and accretion areas in the period 2018–2022. The investigation on the two case studies showed a better accuracy of the GNSS RTK method in the real-time coastline detection. It can be considered as reliable ground-truth reference for the evaluation of the photogrammetric coastlines. While GNSS RTK proved to be more productive and efficient, optical and thermal SfM provided better results in terms of morphological completeness of the data.

Keywords: GNSS RTK; optical and thermal images; SfM photogrammetry; real-time coastline; Po River Delta; automatic extraction; multi-temporal comparison

Citation: Fabris, M.; Balin, M.; Monego, M. High-Resolution Real-Time Coastline Detection Using GNSS RTK, Optical, and Thermal SfM Photogrammetric Data in the Po River Delta, Italy. *Remote Sens.* **2023**, *15*, 5354. https://doi.org/10.3390/rs15225354

Academic Editors: Ramón Blanco Chao, Germán Flor-Blanco and José Juan de Sanjosé Blasco

Received: 25 September 2023
Revised: 7 November 2023
Accepted: 8 November 2023
Published: 14 November 2023

1. Introduction

Coastal areas are a highly dynamic and complex environment, hosting approximately 10% and 44% of the Earth's population at elevations of 0–10 m and up to 150 km from the coast, respectively. The population density in these areas is expected to increase by 25% by 2050 [1]. These areas, characterized by significant socio-economic activities, are continuously exposed to natural risks (storms, storm surges, cyclones, hurricanes, extreme precipitation, and flooding) and sea level rise generated by climate change [2–4]. In many areas, mainly in river deltas, these effects are exacerbated by land subsidence [5–7].

These challenging environments are characterized by large topographic changes, differences in bed cover (rough surfaces alternating with textureless and reflective surfaces), the presence of water as channels, thin patches and deeper puddles, variations in the coastline due to the combined effects of human activities on land and sea, regional environmental modification, and climate change. All these effects can represent important challenging aspects for the monitoring of these areas [8].

Many studies were conducted by different researchers in the field of high-resolution deformations monitoring in coastal areas due to erosion and accretion phenomena. Structure from Motion (SfM) photogrammetry, LiDAR (Light Detection And Ranging), and Global Navigation Satellite System (GNSS) techniques were used to extract high-resolution and high-precision digital terrain models (DTMs). The comparison between these data acquired at different times and co-registered in the same reference system provided the areas and volumes involved in the deformation processes. In particular, for coastal areas, erosion and/or accretion can be measured by comparing the 0-level contour lines extracted from the DTMs [9–11].

In this context, the high-resolution coastline identification becomes a crucial challenge when analyzing flat areas. Reconstructing past coastlines can be achieved by using archival data: cartographies, aerial photogrammetry, and satellite imagery [12,13]. In many cases, more so for historical data, georeferenced digital models cannot be obtained due to the absence of Ground Control Points (GCPs) coeval with the past surveys and/or the lack of information about the tidal level when images were acquired. In these cases, the restitution can be performed directly on 2D orthophotomaps where the unknown tide level, with the water that can cover large portions when flat areas are involved, can have a very negative impact on the result of the real coastline [14].

The extraction of coastline from optical images can be performed automatically by means of various methods developed in recent decades. Supervised [15] and unsupervised [16,17] image classification methods, specific tools (Automatic Coastal Extraction Tool [18]) and suitable indices [19] can be used. In the latter approach, many indices are available to detect the coastline in an easy way using satellite images [20–25].

Furthermore, drones equipped with optical and thermal cameras are increasingly used in environmental surveys. The applications of these sensors on drones are available for the detection of peat fires [26], wildlife detections and monitoring [27,28], building audits, forest monitoring [29], agriculture parameters estimations [30], etc. Due to the general low resolution of thermal images, thermal and optical images can be combined to generate higher-resolution and high-quality thermal mapping models [31,32]. These data, easily acquired by low-cost drones, can also find applications in the definition and monitoring of the coastline.

Several researchers conducted studies for the coastline detection and monitoring using different methods. Zanutta et al. [33] performed 3-D surveys of the coast in the Emilia Romagna region (Northern Italy) using the photogrammetric SfM by drone equipped with optical camera and GNSS (PPK, post-processed kinematics). They demonstrated the ability of these methodologies for coastline monitoring by using 3D digital models extracted from the surveys. However, the authors refer to the 0-level coastline extracted from digital models, but do not investigate the applicability of these methods in the real-time coastline detection. Michałowska et al. [34] used aerial photographs and orthophotomaps over a 65-year time interval for the coastline monitoring of the southern coast of the Baltic Sea (Slowinski National Park) in order to study the erosion and accumulation phenomena. They took advantage of the absence of anthropogenic impact and non-tidal sea with periodic sea level fluctuations. In this case, the authors compared the foredune toe line, the natural line formed by vegetation on the beach obtained on the basis of the visual interpretation of the orthophotomaps. However, with this approach they did not detect the real ground–water separation. Romagnoli et al. [35] analyzed the evolution of the long-term patterns of coastal change of Lipari (Aeolian Islands, Italy) in the last 60 to 70 years using a multidisciplinary approach, which includes aerial photogrammetric images, drone surveys, and satellite data, high-resolution multi-beam bathymetry and field observations. The authors digitized the coastline from orthomosaics for the multi-temporal comparison without analyzing the aspects linked to the tide level, since the study was focused on the long-term scale.

In the above-mentioned works, the authors focused their analyses on the restitution and comparison of the 0-level coastline, but the real-time ground–water separation was not investigated.

Typically, the restitution of the 0-level contour line can be performed by acquiring high-resolution data (from aerial optical images, LiDAR, GNSS, classical topography, etc.) of the area under investigation in low tide. In this way, after the generation of the 3D model (point cloud) and the Digital Terrain Model (DTM), the automatic 0-level contour line is assumed to represent the estimation of the coastline. On the contrary, the detection of the real-time coastline is a challenge due to several reasons: (i) high tide and low tide real-time contour levels can be extracted from DTMs covering areas below the mean sea level, which may not be available in many cases; (ii) the tidal stations, which provide the tidal elevation, can be far from the area under investigation, making the information related to the tidal data less accurate in the study area or with delays or advances compared to the forecasts. In this context, while the above-mentioned works focused on estimating and monitoring the coastline related to the 0-level contour line, the real-time coastline detection has received limited attention.

This work focuses on the identification of the real-time (instantaneous) coastline, in low tide elevation using high-resolution geomatics techniques. For this aim, the main objective is to analyze the performances of the different techniques in terms of coverage, resolution, and accuracy. In addition to the more commonly used techniques, the extraction of the real-time coastline from thermal imagery by means of a supervised classification method is performed. Data related to the instantaneous coastline are used for the definition of the involved transition surfaces from land to sea, taking advantage of several aims: (i) accurate definition of the free beach area; (ii) boundaries detection of the areas assigned to bathing facilities and/or for fishing-related activities; (iii) strip of surface interested by the periodic water fluctuation (periodically flooded, from low tide to high tide and vice versa) that can be used in the multi-temporal comparisons, etc.

In detail, the real-time detection of the coastline was investigated by means of GNSS RTK (Real-Time Kinematic) measurements, topographic total station, and SfM photogrammetric surveys using a low-cost drone acquiring both optical and thermal images. The test sites were located in the flat area of the Po River Delta (PRD) (an area subject to the phenomenon of land subsidence in northern Italy) [6]. The Boccasette and Barricata beaches were investigated at the beginning of 2022 (Figure 1).

In both cases, the coastline in low tide conditions was surveyed simultaneously with the different techniques, reducing the field work to one hour to avoid different tide elevations in the acquired data. Thermal images were acquired only for a portion of Barricata Beach.

The comparison between the different techniques applied in these sample sites was used to evaluate the accuracy and reliability of the methods. This approach could represent a valid tool for analyzing the multi-temporal evolution of the coastline, which is fundamental for long-term monitoring and risk-mitigation activities in the area.

The procedure followed the subsequent phases (Figure 2):

1. Validation of the GNSS RTK real-time coastlines using the polylines measured with the total station;
2. Extraction of the DTMs and orthophotos from optical and thermal photogrammetric data;
3. Georeferencing and validation of the photogrammetric data;
4. Restitution of the real-time coastlines on the optical and thermal orthophotos;
5. Extraction of the automatic real-time coastline from the thermal orthophoto;

6. Comparison between the reference GNSS RTK polylines with those obtained from the photogrammetric orthophotos, both in terms of distances and surfaces generated by polyline intersections;
7. Evaluation of accuracies and performances of the different techniques;
8. Extraction of the 0-level contour lines from the DTMs;
9. Extraction of the 0-level contour lines from the DTM generated using an ALS (Airborne Laser Scanning) LiDAR survey conducted in 2018;
10. Comparison between the obtained 0-level contour lines to evaluate modifications of the coastlines in terms of erosion and/or accretion in the 2018–2022 period.

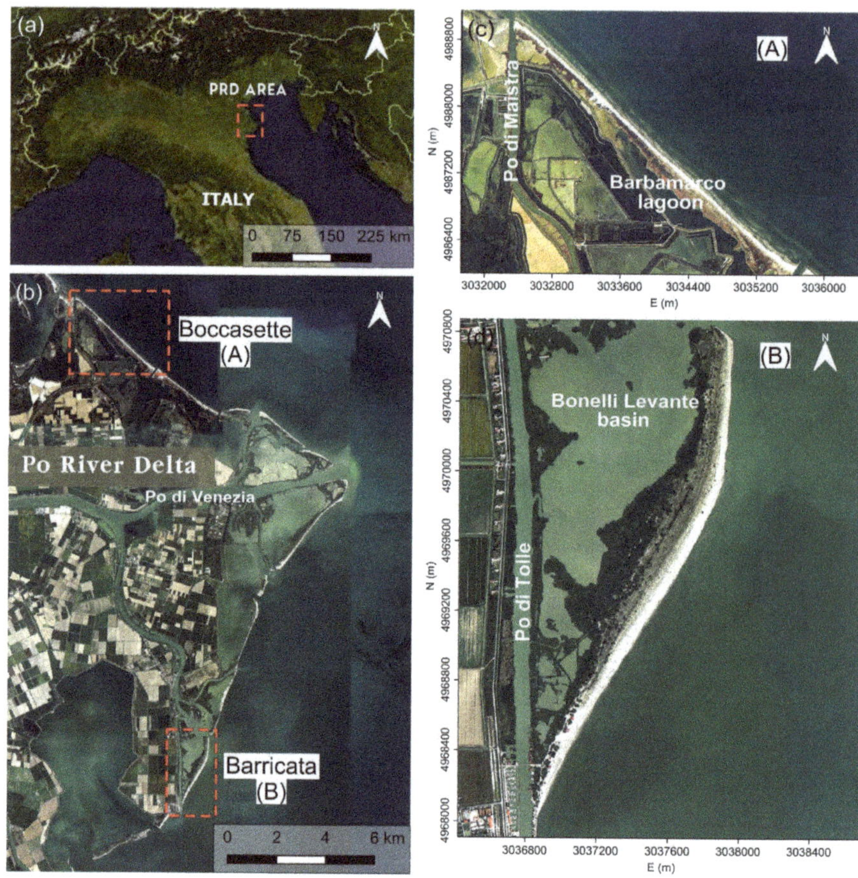

Figure 1. (**a**) Location of the Po River Delta (PRD) in northern Italy; (**b**) location of the studied areas; Boccasette beach (**c**) in the Barbamarco lagoon and Barricata beach (**d**) in the Bonelli Levante basin.

This work is organized as follows. Section 2 introduces the two study areas in the PRD. Section 3 describes the surveys performed using the different techniques, the available LiDAR data collected in 2018, the procedures adopted in the processing, and the approaches used in the coastline comparisons. Section 4 provides the experimental results focusing on real-time and 0-level coastlines multi-temporal comparisons. Section 5 discusses the obtained results, and Section 6 summarizes the work and provides some conclusions.

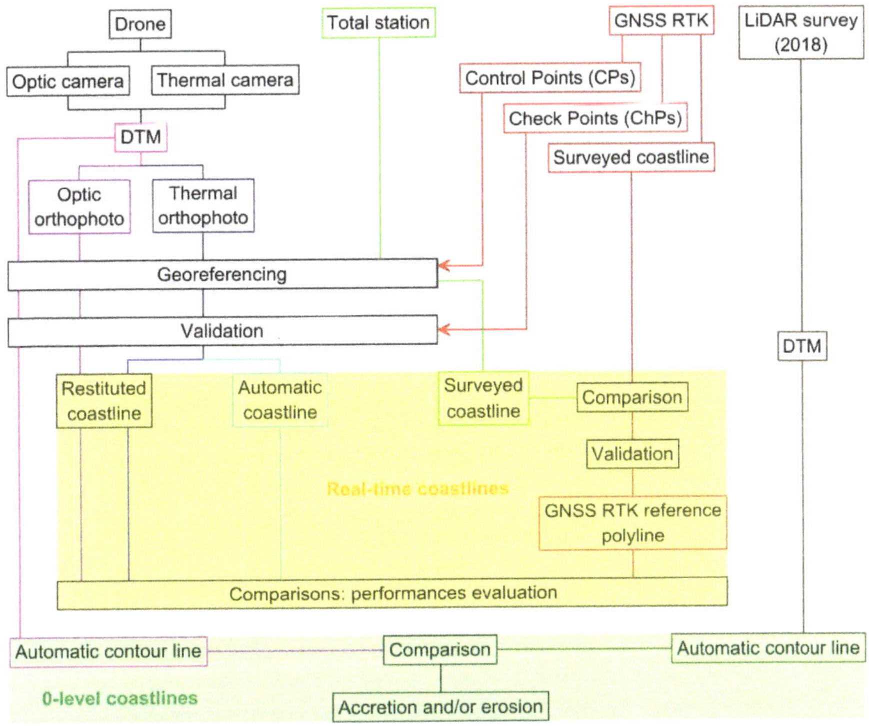

Figure 2. Flowchart of the methodology proposed in this work.

2. The Study Areas

The coastal areas under investigation, the Boccasette and Barricata beaches, are located in the PRD. It is the last portion of the Po River basin, subdivided into seven branches that flow into the Adriatic Sea. In the past, PRD was affected by high values of land subsidence rates ([14,36,37] and references therein). Currently, the phenomenon, even strongly reduced, is still ongoing [38–40], and most of the area lies below the mean sea level (in the order of 2–3 m), protected by earthen levees for hydraulic safety [41–43].

The Boccasette beach is a flat coastal portion in the northern part of the main branch of the Po River (Po di Venezia). It is characterized by fine sand and extends for about 4.4 km in the southeast-northwest direction, from the northern mouth of the Barbamarco lagoon to the Po di Maistra mouth (Figure 1). Similarly, Barricata beach is a flat coastal portion in the southern part of the Po di Venezia branch: it is characterized by fine sand and extends for about 3.2 km in the north-south direction, from the mouth in the center of the Bonelli Levante basin to the Po di Tolle mouth (Figure 1). Both areas are located outside the embankments, on islets connected to the earthen levees with two bridges. That of Barricata is only open in the summer. Additionally, bathing establishments are active in both areas during the summer, which occupy about 200 m of the beaches around the access bridges.

3. Materials and Methods

3.1. The Surveys

3.1.1. GNSS RTK and Classical Topographic Measurements

The surveys of the coastline were carried out under low tide conditions to reconstruct the morphology of the studied areas below the mean sea level. Information about tide elevation was provided by the Forecasting and Tide Reporting Centre (Centro Previsioni e Segnalazioni Maree) service of the Venice Municipality [44]. Based on these data and considering the delay reported in the PRD, the survey was conducted on 27 January and 25

February 2022 for the Boccasette and Barricata beaches, respectively. Measurements were carried out in winter to avoid interferences with (i) bathing establishments and (ii) other anthropic activities (fishing, and similar) that are mainly active in summer. In addition, due to fact that the motion of waves on flat areas greatly complicate the estimation of the ground–water separation by the operator, the measurements were performed during minimum wind speed, which minimizes the sea waves motion on the beaches, simplifying the operator's choices. The surveys start far from bathing establishments, from the southeastern corner and from the northern corner in the Boccasette and Barricata areas, respectively (Figure 1).

In both cases, taking into account the minimum tide level (−0.5 m at 13:55 for Boccasette and −0.3 m at 13:35 for Barricata), the measurements started half an hour earlier and ended half an hour later (from 13:25 to 14:25 for Boccasette and from 13:05 to 14:05 for Barricata), in order to carry out the survey with the three different techniques in the same tide conditions.

In the GNSS RTK survey a Leica Viva GS 15 GNSS receiver was used. The GNSS antenna was fixed to the operator's backpack by measuring the height from the phase center to the ground (Figure 3a). The sampling rate was set to one second and the operator walked along the ground–sea transition estimating the real-time coastline in the planned hour. A total of 3600 and 3035 points were acquired and distances of 4.78 and 3.99 km were traveled for the Boccasette and Barricata beaches, respectively, with average distance between points of 1.3 m (in the Barricata the survey finished early due to the minor length of the beach). Data were registered in the UTM reference system (EPSG:6876 RDN2008/Zone 12, N-E). At the same time, a total station Leica TCR1201 was used. The station was stationed on a GCP (used for the low-cost drone survey) and oriented to a second GCP for the co-registration of the topographic data in the same reference system of the photogrammetric and GNSS RTK surveys. The measurements were made with a second operator equipped with a prism that walked along the ground–sea border, estimating the real-time coastline and acquiring a point every two steps (Figure 3b). After about 1 h, 278 points were acquired covering 0.644 km in Boccasette and 230 points covering 0.568 km in Barricata (average distance between points of 2.4 m). Subsequently, the ellipsoidal elevations obtained with the GNSS measurements were converted to orthometric elevations using parameters provided by the IGMI (Istituto Geografico Militare Italiano).

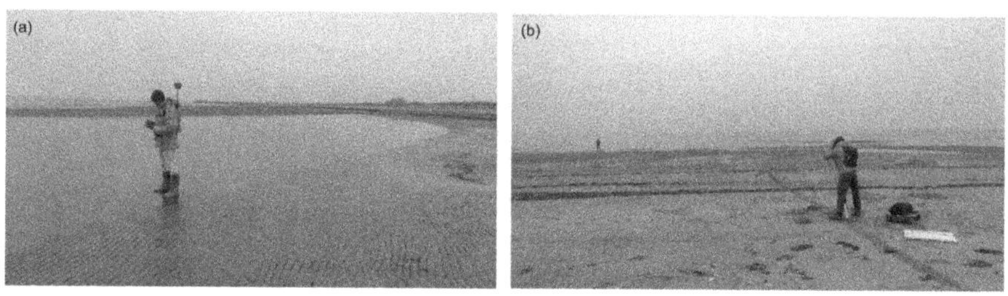

Figure 3. Acquisition of (**a**) real-time kinematic and (**b**) topographic points in the Boccasette area. The photographs highlight the cloudy day, which makes colors uniform (resulting in poor radiometric changes in the optical images) and presents in the ground–water estimation.

3.1.2. The 3D Photogrammetric Survey Using a Low-Cost Drone

During the survey period, a low-cost drone, Parrot Anafi, equipped with an optical camera (Sony IMX230) with a CMOS sensor of 1/2,4″ and diagonal of 7.83 mm, resolution of 21 MP, focal length of 4 mm (equivalent focal of 23 mm, 4608 × 3456 pixels in the image plane), and low-resolution thermal imaging (FLIR camera Lepton 3.5) with 3224 × 2448 pixels in the image plane was used. Before the measurement time, GCPs were uniformly distributed along the investigated areas (Figure 4a); their location was measured using the Leica Viva GS 15 GNSS receiver by applying the RTK approach, co-registering the photogrammetric data in the same reference system of the data acquired with the other techniques. In detail, 32 and 29 GCPs were measured in Boccasette and Barricata, respectively. In the hour of the measurements, for each investigated area, 4 optical photogrammetric surveys longitudinal to the coast were carried out with an overlap greater than 15% (Figure 4c); each flight was set with an elevation of 60 m, acquiring 6 strips, with each covering 420 m of the longitudinal coast in about 15 min (the runtime of a single battery) and with an overlap between the subsequent images greater than 60% (Figure 4).

Figure 4. (**a**) SfM (Structure from Motion) photogrammetric targets used during the surveys; (**b**) images acquisition by the low-cost drone; (**c**) covered surface by the four photogrammetric flights in the Barricata area; the image (**b**) highlights the better weather conditions compared to the Boccasette one (Figure 3), featuring a sunny day that improves the colors brightness (resulting in good radiometric changes in the optical images).

Finally, 960 optical images were acquired, covering a surface of about 1.68 × 0.21 km, longitudinal and transversal to the coastline of the Boccasette area. The experience developed on Boccasette beach allowed us to optimize the time in the acquisition of optical images in the Barricata area: for this reason, in addition to 1065 optical images (covering 1.58 × 0.21 km) of the ground–water transition area, a new flight was performed only for Barricata beach during the same hour of survey using the thermal camera. A total of 55 images were acquired in a strip, setting a flight altitude of 110 m and covering an area of about 0.32 × 0.07 km of ground–water separation. The thermal camera on the low-cost drone provided RGB images where the radiometry of each pixel is linked to a scale of temperature (Figure 5). In this way, the performances of the coastline extraction from thermal images can be evaluated.

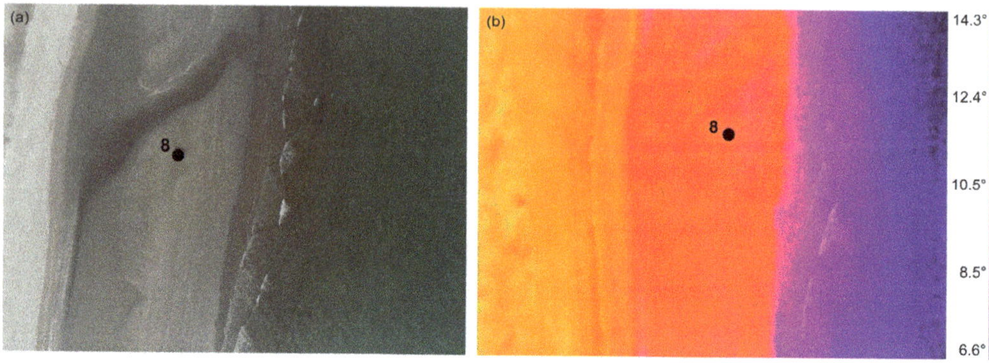

Figure 5. Optical (**a**) and thermal (**b**) images acquired with the low-cost drone, representing the same portion of the Barricata coastal area (around the GCP 8); the acquired coastal portion can be subdivided into three main classes. From left to right: the dry sand, the wet sand, and the sea water. In terms of radiometric changes of the RGB images, the ground–water border is more evident in the thermal image.

3.2. *The 2018 LiDAR Data*

An ALS LiDAR survey was carried out in the PRD coastal area, outside the levees, on 14 April 2018, using the Optech ALTM Galaxy sensor. These data were available at the Veneto Region (Direzione Pianificazione Territoriale and Unità Organizzativa Genio Civile di Rovigo) and the Local Authority of "Parco Regionale Veneto del Delta del Po". The survey was included in the monitoring activities of the sand islets (where Boccasette and Barricata beaches are located), considerable storm surge barriers that protect the levees from the erosive action generated by the motion of the sea waves. The 3D points were acquired together with ortho-images characterized by GSD (Ground Sample Distance) of 20 cm. Measurements were performed using an integrated GNSS/INS (Inertial Navigation System) system for georeferencing the data during the low tide elevation, so as to cover as much as possible of the ground–sea transition area (using a flight altitude of about 1500 m). Finally, orthometric elevations of the acquired 3D points were obtained from the ellipsoidal one by using the geoid model grids provided by the IGMI. In this way, these data were co-registered in the same reference system of the surveys previously described (Section 3.1.1). In a previous work [14], the 0-level coastline was extracted for the PRD area, including the Boccasette and Barricata case studies (for more details, see Ref. [14]).

3.3. *Processing and Comparisons*

3.3.1. SfM Photogrammetric Images Processing

The SfM photogrammetric technique was used to generate the 3D point cloud for both Boccasette and Barricata beaches. Agisoft Metashape software version 1.8.4 [45,46] was applied together with the coordinates of the available GCPs to georeference the data. In detail, the GCPs were subdivided in Control Points (CPs, 24 and 22 for Boccasette and Barricata, respectively) used in the processing, and Check Points (ChPs, 8 and 7 for Boccasette and Barricata, respectively) used to evaluate the accuracy of the extracted 3D point clouds [47] (Figure 6).

Subsequently, a DTM and the corresponding orthophoto were generated for both study areas to extract the 0-level contour line, which is assumed to be the real coastline that can be used in the multi-temporal comparisons. The survey in low tide elevation provided clear advantages in the reconstruction of the ground morphology below the mean sea level. Furthermore, the orthophotos were used to generate the temporary coastline (real-time) during the surveys, corresponding to a tide elevation of −0.5 m and −0.3 m in Boccasette and Barricata, respectively. In this case, the restitution of the polylines was performed by the operator using the visual estimation of the ground–sea boundary, due to the high noise of the automatic contour levels extracted from the DTMs in the ground–water transition.

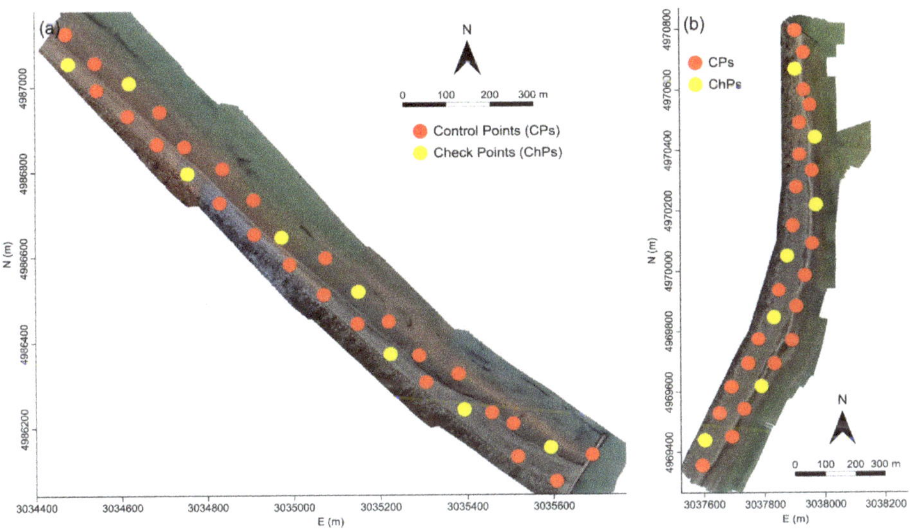

Figure 6. Location of Control Points (CPs) (used in the processing) and Check Points (ChPs) (used to evaluate the accuracy of the extracted 3D models) in the Boccasette (**a**) and Barricata (**b**) beaches.

3.3.2. Automatic Real-Time Coastline Extraction from Thermal Images

In the first phase, because the low-cost drone used in the survey provides the thermal images only in terms of RGB bands, they were used in Agisoft Metashape software version 1.8.4 to extract the orthophoto of the acquired area, which was georeferenced using the available GCPs detectable on the thermal images.

The obtained data were used in the QGIS software version 3.18 to extract the automatic polyline representing the real-time coastline. In this way, a pixel-based classification technique, which includes supervised and unsupervised approaches, was used. In general, supervised classification provides more accurate results, but is time-consuming and requires greater use of resources due to the identification of training samples [48]. On the other hand, supervised classification is commonly adopted in coastline extraction, especially when high-resolution images are used [49]. For this reason, in this work supervised classification was applied to estimate land–water separation. This approach subdivides the spectral range of an image into regions that are linked to a type of land cover. The method requires a priori knowledge of the land cover type to be classified to correctly choose the training samples.

Using the extracted thermal orthophoto, 10 training samples, equally subdivided in terms of the number of pixels, were defined to separate water and ground regions (this procedure was also applied to the optical orthophotos). This way, the orthophotos were transformed into black and white raster images, and the real-time coastlines were extracted based on the separation polyline between the pixels belonging to the two regions.

Subsequently, the polylines were optimized by simplifying the form and eliminating pixel irregularities by introducing new points along the geometry with a step of 2 m (points-to-path algorithm). This value was chosen to align with the order of the coastline estimation, resulting in more regular polylines.

From the thermal orthophoto, the restitution of the real-time coastline was performed by the operator by visual inspection of the land–water boundary.

3.3.3. Coastline Comparisons

The obtained real-time coastlines were compared to each other, both in terms of distances and surfaces generated by polyline intersections. This allowed us to evaluate the following: (i) performances, including accuracies, of the different real-time coastlines generated using the GNSS RTK, total station, manual and automatic restitution on optical and thermal orthophotos with the same tide elevation, in both study areas; (ii) multi-temporal changes using the 0-level contour line derived from the DTMs extracted using the SfM approach (drone DTM 2022) and the ALS LiDAR survey performed in 2018.

In the first series of comparisons, the GNSS RTK coastline was assumed as the ground truth due to the accuracy of this technique and the spatial coverage, which is greater than that of the polylines obtained with the other methods. However, to evaluate the accuracies linked to the land–water estimation by the operator (depending on the sensitivity of the operator in the identification of the coastline), which can be very complex on flat areas influenced by the sea waves motion, the GNSS RTK coastline was compared with the topographic coastline surveyed by other operators in the overlapped areas, in order to validate the GNSS RTK data.

Subsequently, the ground truth GNSS RTK coastline was compared with the following: (i) the restitution on the visible orthophotos; (ii) the restitution on the thermal ortophoto; (iii) the automatic polyline extracted from the thermal orthophoto.

The comparisons were performed by calculating the 2D distances between the obtained polylines in the overlapped areas, providing averages and standard deviations. However, due to the complexity of several portions, which are difficult to evaluate with distances perpendicular to the polylines, the comparisons between the different coastlines were performed by calculating the positive/negative areas of the polygons generated by the polyline intersections. In this study, taking into account the reference GNSS RTK polyline, positive values were obtained in the sea direction, while negative values were attributed in the land direction, both for distances and areas. In the comparison of GNSS RTK and total station polylines using QGIS software version 3.18, perpendicular distances were calculated from the measured topographic points to the GNSS RTK polyline. Areas were calculated considering the polygons generated by the intersection between the GNSS RTK and topographic polylines. The same approach was used when manual and automatic restitutions were involved.

In this case, the uncertainties derived from the comparison between the reference polylines with those obtained by the total station and manual and automatic restitutions were evaluated using the Ratio Index (RI) and the Distributed Ratio Index (DRI) [50]. The first is obtained by dividing the total areas of the generated polygons with the length of the reference GNSS RTK polyline. The latter is calculated by taking into account each obtained polygon, dividing the i-area with the length of the related polyline. This way, statistical parameters such as average and standard deviation can be calculated in order to provide the degree of accuracy.

3.3.4. Accretion/Erosion in the 2018–2022 Period

The multi-temporal comparison involved the contour line of 0-level obtained from the DTMs extracted using the ALS LiDAR measurements performed in 2018 and the photogrammetric surveys carried out with the low-cost drone in 2022. In the first phase, the LiDAR data (in the EPSG: 32632 WGS 84/UTM Zone 32N reference system) were converted into the reference system used in this work.

Similarly to the procedure previously described (Section 3.3.2), starting with the 2018 survey, which is assumed as reference, the comparison was carried out by calculating areas of the polygons generated by the intersection between the multi-temporal coastlines. Accretion (positive) and erosion (negative) were linked to the location of the 2022 coastline compared to the 2018 coastline.

4. Results

4.1. Photogrammetric 3D Models and Orthophotos

Three-dimensional models were generated by means of the SfM approach for Boccasette and Barricata beaches using optical images and the Agisoft Metashape software version 1.8.4. CPs were used in the processing and ChPs were used to validate the 3D models.

Table 1 summarizes the accuracies in terms of RMSE (Root Mean Square Error). The obtained values, in the order of a few centimeters, are in agreement with those reported by Vecchi et al. [9] and Zanutta et al. [33], who worked in similar environmental contexts and with similar settings of photogrammetric surveys using drones.

Table 1. Number of CPs used in processing, available ChPs (see Figure 6), and comparisons between the 3D coordinates and the extracted 3D photogrammetric SfM models in terms of RMSE.

3D Model	N. CPs	N. ChPs	RMSE (cm)	
			CPs	ChPs
Boccasette	24	8	4.1	4.9
Barricata	22	7	3.5	3.8

Subsequently, dense clouds were generated and orthophotos were extracted (Figures 7 and 8).

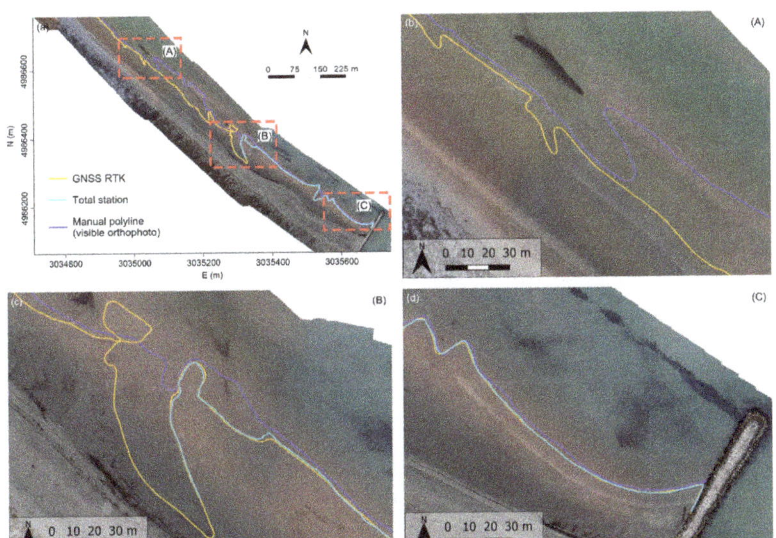

Figure 7. (**a**) Polylines representing the real-time coastlines obtained from the GNSS RTK and topographic surveys and the manual restitution by visual inspection of the optical orthophoto. The background is the same orthophoto of the Boccasette beach; (**b–d**) show the three details (A), (B), and (C) respectively.

For a portion of the Barricata beach, a different othophoto was generated using the thermal images. Six CPs were used in the processing based on the available visible targets. The 3D model was extracted with RMSE on CPs of 0.11 m. After the generation of the dense cloud, the corresponding orthophoto was obtained.

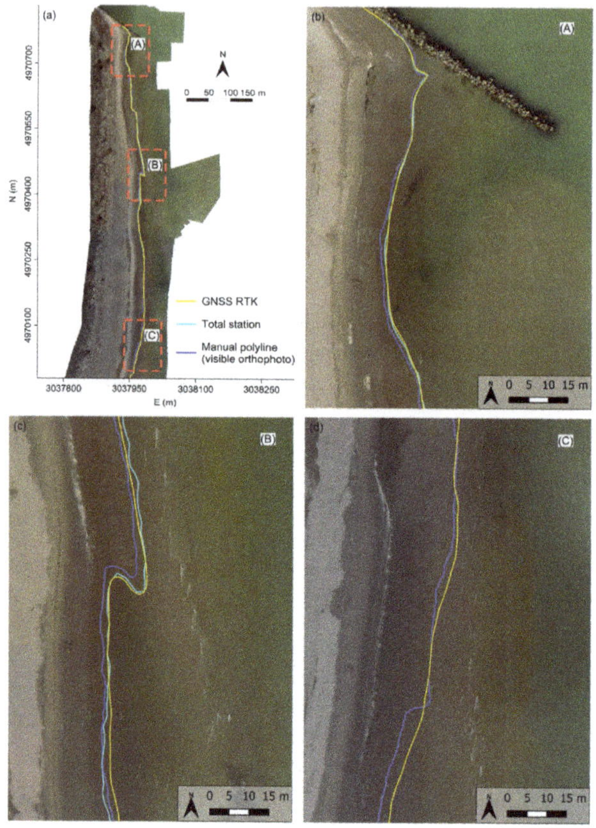

Figure 8. (**a**) Polylines representing the real-time coastlines obtained from the GNSS RTK and topographic surveys and the manual restitution by visual inspection of the optical orthophoto. The background is the same orthophoto of the Barricata beach; (**b–d**) show the three details (A), (B), and (C) respectively.

4.2. Restitution of Real-Time Coastlines by Visual Inspection

The restitution of the ground–water polyline separation was performed by the operator using visual inspection. This operation was carried out using both optical (for Boccasette and Barricata) and thermal (only for Barricata) orthophotos. Figures 7 and 8 show the obtained polylines overlapped with the optical orthophotos together with the real-time coastlines obtained with the GNSS RTK and topographic surveys for Boccasette and Barricata, respectively.

Details (A) and (B) of Figure 7b,c show the difficulties of the operator in the interpretation of the ground–water separation from the optical orthophoto with little variation in radiometry of the aerial images (see Figure 3). In these cases, large errors can be performed. On the contrary, details (A), (B), and (C) of Figure 8b–d show that manual restitution of the real-time coastline can provide better results when the bands of the aerial images cover a wide spectrum of the radiometric range.

4.3. Automatic Real-Time Coastline Extraction

Since the low-cost drone used here the thermal imagery from the drone was provided in terms of RGB bands and the land–water separation is radiometrically clearer compared to the optical images (Figure 5), we explored a procedure based on supervised classification to automatically extract the real-time coastline. The availability of only three bands did not allow us to apply commonly used indexes (for example NDWI, NDVI, etc.) developed for satellite images when other bands are available [20,21].

Figure 9 shows the optical and thermal orthophotos of the Barricata beach overlapped with the polylines generated by means of the following: (i) the GNSS RTK survey; (ii) the manual restitution using visual inspection on both optical and thermal orthophotos; (iii) the automatic polyline extracted from the supervised classification.

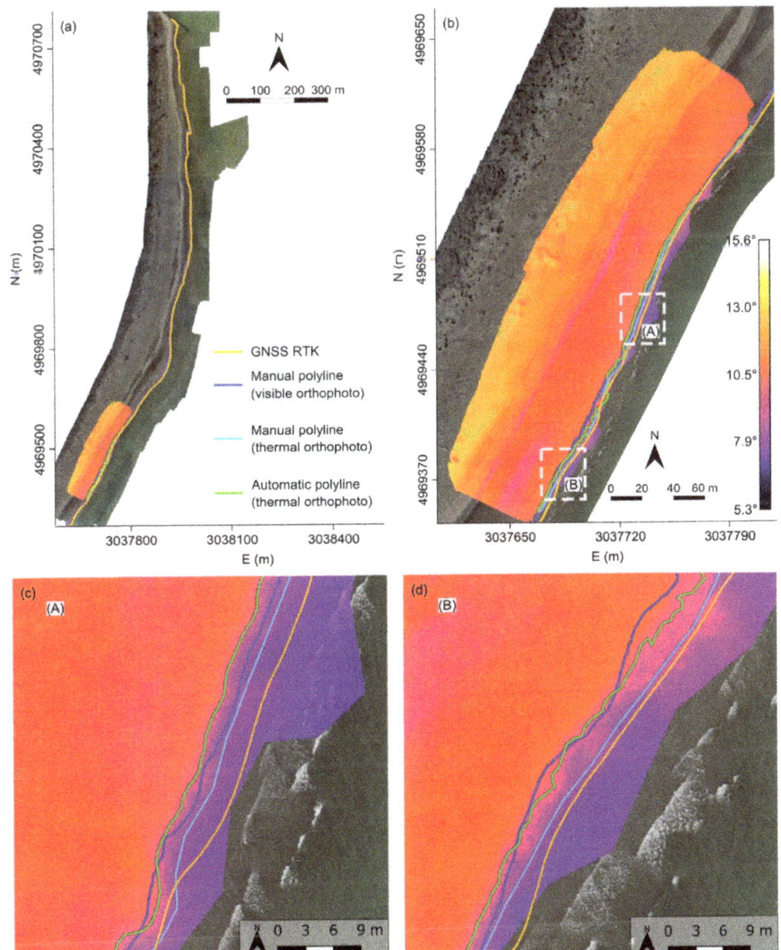

Figure 9. (**a**) Polylines representing the real-time coastlines obtained from the GNSS RTK (i), the manual restitution by visual inspection of the optical (ii) and thermal (iii) orthophotos, the automatic polyline obtained from the supervised classification of the thermal orthophoto (iv). The location of the thermal data on the optical orthophoto of the Barricata beach is also reported; (**b**) detail of the thermal orthophoto; (**c**,**d**) show the two details (A) and (B).

Details (A) and (B) of Figure 9c,d show that all the obtained polylines representing the real-time coastline are very closed, with maximum distances in the order of a few meters.

4.4. Real-Time Coastlines Comparisons

Real-time coastline comparisons were made in terms of distances and surfaces generated by polyline intersections. Figure 10 shows the distance comparisons between the polylines surveyed with the GNSS RTK technique and the total station topographic instrument, both for the Boccasette and Barricata case studies.

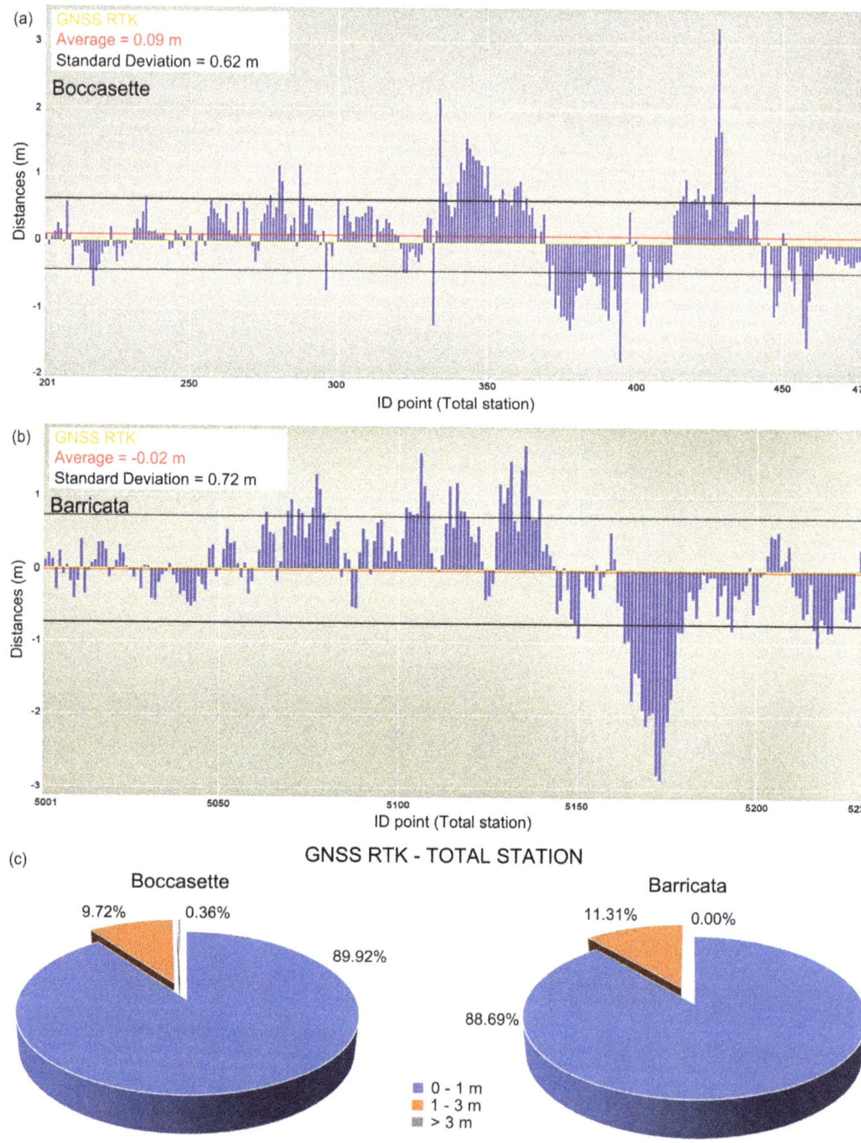

Figure 10. Comparison between GNSS RTK and the total station surveyed polylines representing the real-time coastline both for Boccasette and Barricata beaches: (**a**) distances assuming the GNSS polyline as reference in Boccasette; (**b**) distances assuming the GNSS polyline as reference in Barricata; (**c**) distribution of the calculated distances.

The calculated differences were classified into three groups: (i) less than 1 m; (ii) between 1 and 3 m; and (iii) greater than 3 m. These values were chosen based on the type of the analyzed surfaces (in this case flat areas) and the difficulties in the definition of the real-time coastline due to the motion of waves on the beach. While differences less than 1 m are not significant within the accuracy of the estimation, values between 1 and 3 m can be acceptable. On the contrary, differences greater than 3 m highlight errors in the detection of the real-time coastline.

The same approach was used in the comparison between the GNSS RTK technique and the polyline extracted by the visual inspection on the orthophotos. Figure 11 shows the obtained results together with the distribution of the distances in the three groups previously defined.

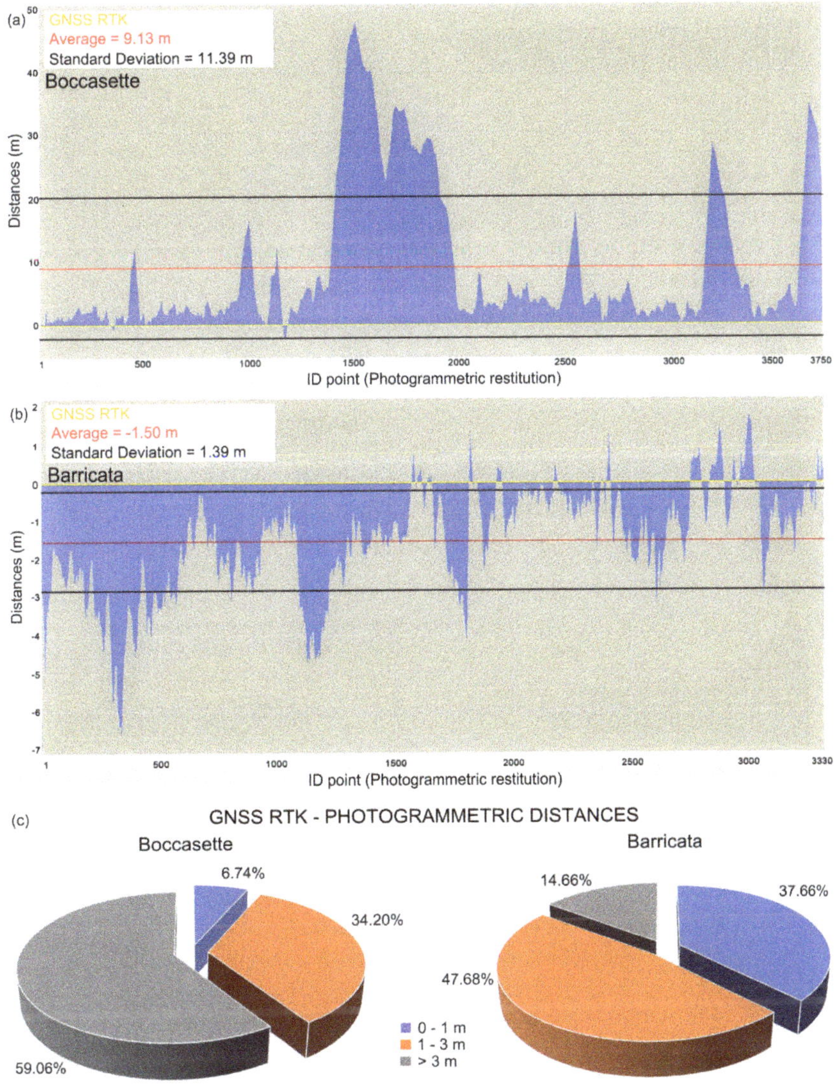

Figure 11. Comparison between GNSS RTK and photogrammetric polyline obtained from visual inspection of the optical orthophoto, representing the real-time coastline both for Boccasette and Barricata beaches: (**a**) distances assuming the GNSS polyline as reference in Boccasette; (**b**) distances assuming the GNSS polyline as reference in Barricata; (**c**) distribution of the calculated distances.

Figure 10a,b show a substantial equilibrium between generated positive and negative distances when the GNSS RTK and the polylines surveyed by the total station are compared, both on the beaches of Boccasette and Barricata. On the contrary, Figure 11a,b show significant positive distances obtained in Boccasette and negative distances in Barricata, highlighting that the photogrammetric restitution was very different compared to the GNSS RTK measurements.

In addition, in Boccasette 25.7% of the differences provided values greater than 10 m, while in Barricata only 0.01% of the distances provided values greater than 5 m, showing that the SfM technique, together with the restitution by visual inspection on the orthophoto, provided very different results in the two studied areas.

The accuracy of the comparison in terms of surfaces generated by the polyline intersections can be evaluated using the RI and DRI indexes. Tables 2 and 3 show the results obtained for the Boccasette and Barricata case studies.

Table 2. Length, RI, and DRI values derived from the comparison between the GNSS RTK reference polyline with: (i) the survey performed using the total station; (ii) the restitution by visual inspection of the optical orthophoto considering the overlap length with the total station polyline; (iii) the restitution by visual inspection of the optical orthophoto considering the whole dataset for the Boccasette case study.

Comparisons	Length (m)	RI	DRI			
			Min (m)	Max (m)	Average (m)	St. Dev. (m)
GNSS – Total station (i)	635.73	0.47	0.00	0.78	0.21	0.21
GNSS – Restitution (optical) (ii)	635.73	4.67	0.03	6.08	2.63	2.77
GNSS – Restitution (optical) (iii)	2628.52	8.81	0.03	10.60	4.09	3.75

Table 3. Length, RI, and DRI values derived from the comparison between the GNSS RTK reference polyline with: (i) the survey performed using the total station; (ii) the restitution by visual inspection of the optical orthophoto considering the overlap length with the total station polyline; (iii) the restitution by visual inspection of the optical orthophoto considering the whole dataset; (iv) the restitution by visual inspection of the optical orthophoto considering the overlap length with the thermal data; (v) the restitution by visual inspection of the thermal orthophoto; (vi) the automatic polyline extracted from the thermal orthophoto for the Barricata case study.

Comparisons	Length (m)	RI	DRI			
			Min (m)	Max (m)	Average (m)	St. Dev. (m)
GNSS – Total station (i)	563.79	0.52	0.01	0.99	0.22	0.22
GNSS – Restitution (optical) (ii)	563.80	0.93	0.02	1.43	0.46	0.38
GNSS – Restitution (optical) (iii)	1649.78	1.63	0.02	2.41	0.52	0.55
GNSS – Restitution (optical) (iv)	281.30	2.90	-	-	-	-
GNSS – Restitution (thermal) (v)	281.30	1.29	0.22	1.49	0.89	0.56
GNSS – Automatic (thermal) (vi)	281.30	2.76	0.08	3.53	1.18	1.36

4.5. Multi-Temporal Coastlines Comparisons

The comparisons between the 0-level coastline extracted from the ALS LiDAR survey performed in 2018 and the 0-level coastlines extracted using the DTMs obtained from surveys carried out with the low-cost drone in 2022, both for Boccasette and Barricata beaches, are shown in Figure 12.

Polygons generated by the intersection of polylines were classified as erosion (red, when the 2022 drone-based coastline falls in the land direction compared to the 2018 ALS-based coastline) and accretion (green, when the 2022 drone-based coastline falls in the sea direction compared to the 2018 ALS-based coastline).

On Boccasette beach, accretion of 13,751 m^2 was mainly located in the south of the analyzed area where the Barbamarco lagoon opens in the Adriatic Sea. On the contrary, erosion (14,551 m^2) was evident in the northern portion (Figure 12a).

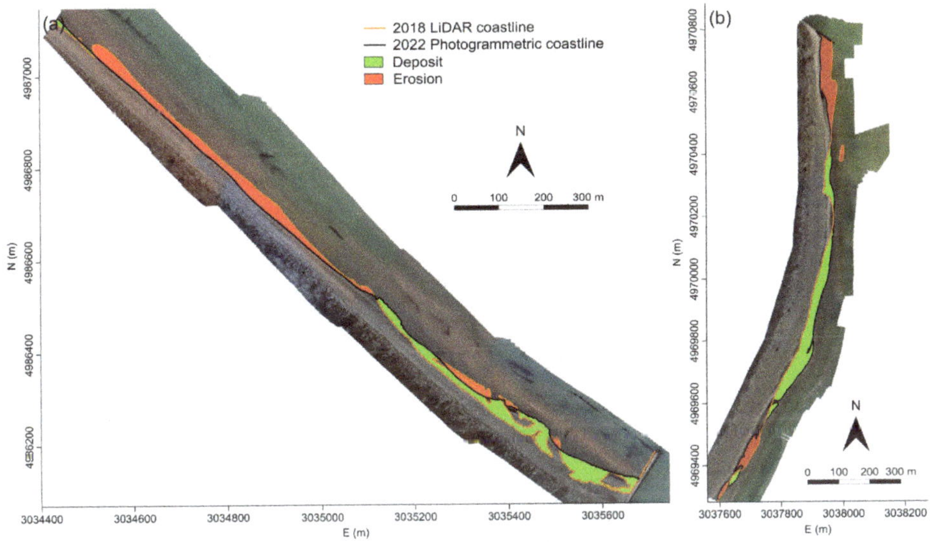

Figure 12. Multi-temporal comparison between the 0-level coastlines obtained from: (i) the survey performed in 2018 (ALS LiDAR); (ii) the DTMs generated using the surveys carried out in 2022 with the low-cost drone applying the SfM approach: (**a**) Boccasette and (**b**) Barricata study areas. The background are the optical orthophotos extracted using the data acquired in 2022.

For Barricata beach, accretion of 21,783 m^2 was detected almost completely in the central portion of the study area, while erosion (14,223 m^2) was located exclusively in the south and, mainly, in the north of the beach, where the Bonelli Levante basin opens into the sea (Figure 12b).

5. Discussion

5.1. Analysis of the Results

The detection of the real-time coastline performed by two operators and using the ground-based GNSS RTK and the total station was very similar, both for Boccasette and Barricata beaches (Figure 10). The differences provided average values of less than 10 cm and standard deviation in the order of 60–70 cm in very challenging areas (Figure 3). For 90% of the points, the differences were less than 1 m in both cases. However, while the GNSS RTK allowed the coverage of 4.78 km for Boccasette and 3.99 km for Barricata within the survey time, the total station had a more limited spatial coverage of 0.64 km and 0.57 km, respectively. For this reason, the GNSS RTK polyline, validated using the total station in the overlapped area, was assumed as the ground truth of the real-time coastline for both study areas.

Involving the restitution performed using visual inspection of the optical orthophotos, the results of the comparison with the GNSS RTK polyline for Boccasette and Barricata were very different. While for Barricata the differences provided an average of 1.5 m and a standard deviation of 1.4 m, these values in Boccasette drastically increased up to 9.1 m and 11.4 m, respectively. Moreover, the points with distances greater than 3 m increased from 15% in Barricata to about 60% in Boccasette (Figure 11). The difficulty in interpreting the land–water boundary by the operator is evident when the radiometry of the optical images presents very small variations (Figure 3). On the contrary, the perception of the land–sea separation is much clearer on sunny days (Figure 4b), where the radiometric changes of the optical images cover a wide range of the spectrum. However, the automatic real-time coastline extracted using the radiometric approach and the supervised classification of the optical orthophotos failed both for Boccasette and Barricata case studies. The radiometric changes along the ground–water separation were not enough for the automatic detection. This result is in agreement with previous studies conducted in the same areas using aerial photographs. The poor quality of the images increases the errors of manual and automatic restitution of the coastlines [11,14].

In addition, the ground-based GNSS RTK approach is preferable, compared to the drone-based SfM approach, also in terms of costs and spatial coverage, since in the hour of survey with the low-cost drone only 50% of the GNSS RTK polyline was covered.

Similar considerations can be developed by analyzing the RI and DRI indices, when surfaces generated by polyline intersections were generated (Tables 2 and 3). In particular, Table 3 (Barricata case study) shows that the comparison between the GNSS RTK polyline and the restitution by visual inspection on the optical orthophoto related to the same spatial coverage provided RI and DRI values of distance about two times greater than the GNSS RTK and total station comparison. Considering the problems related to the ground–water estimation in these areas, the deterioration of the results was still limited and acceptable for many applications, when the survey was performed in good weather conditions that provided significant radiometric changes of the optical images. On the other hand, the aerial images acquired in bad weather conditions were characterized by poor radiometric changes that provided unacceptable estimation of the coastline (Table 2).

The polyline obtained from the thermal orthophoto provided better results (Table 3). In the overlapped area the RI index was reduced from 2.9 m (GNSS RTK and restitution by visual inspection on the optical orthophoto) to 1.29 m (GNSS RT and restitution by visual inspection on the thermal orthophoto), due to the better radiometric separation of the ground–water boundary on thermal images (Figure 5). The automatic extraction of the real-time coastline slightly improves the results obtained with the optical orthophoto, but with clear advantages of the automatic approach when wide areas must be analyzed.

Considering the promising results of the supervised classification, in future developments the automatic real-time coastline extraction from thermal images will be analyzed more in detail by also using suitable indices developed for multispectral satellite images and adapted to RGB thermal images.

Multi-temporal comparisons of the 0-level contour line in the 2018–2022 period provided general erosion in Boccasette and accretion in Barricata. However, while the calculated erosion in Boccasette was very limited (the average retreat of the coastline was about 0.5 m), in Barricata the accretion was more evident (the average advancement of the coastline was about 4.7 m). The modifications in the analyzed period could be due to several factors such as the dynamics of the nearshore, the morphology of the lagoon behind the two studied beaches, the effects of the storm surges that occurred in the two areas, etc. However, these values are also strongly influenced by both the accuracies of the 3D models and seasonal changes (the ALS LiDAR survey was performed in April 2018 and the photogrammetric SfM surveys were performed in January and February 2022).

5.2. Comparison with Previous Research Works

The results of this work were compared with some others research studies conducted in a similar coastal environment with analogue aims.

Lee et al. [51] performed accuracy and efficiency tests using a total station and a GPS RTK for the measurements of spot points and continuous walking data by means of a backpack system at the Gosapo macro-tidal sand beach (Republic of Korea). They found accuracies of the data in the order of a few centimeters, in agreement with the results obtained in this work regarding the comparison between total station and the GNSS RTK data.

On the other hand, the accuracies of the coastlines obtained from optic and thermal orthophoto were also compared with those estimated by using multispectral satellite images.

Marchel and Specht [52] extracted the coastline at the public beach in Gdynia (Poland) using a DJI Matrice 300 RTK UAV. They generated the Digital Surface Model (DSM) and the orthophotomosaic of the studied area from which the ground–water separation was extracted. In addition, the authors marked out the coastline course using high-resolution Pléiades Neo satellite imagery (resolution 0.5 m) and Hexagon Europe satellite images (resolution 0.3 m). They calculated the accuracy of the obtained coastlines using a reference polyline measured with a GNSS RTK receiver. The results provided differences of less than 1 m between the UAV and GNSS RTK coastlines. However, this value, which is slightly better than the one obtained in this work, is strongly influenced by the radiometric characteristics of the optic images, depending mostly on the weather conditions and, consequently, the quality of visual inspection.

El Kafrawy et al. [53] compared and evaluated six coastline extraction methods applied to a Landsat8 2015 image related to the Ras El-Hekma (Egypt) coastal zone. They used the reference 2015 coastline extracted from high-resolution imagery of Pléiades B1 (resolution 0.50 m). The results of the comparisons provided by the authors showed 90% of shifting distances in 1 pixel (30 m), highlighting a close correlation between accuracy and pixel size of the satellite images.

Alcaras et al. [50] extracted the coastline of a coastal area close to the delta of the Nestos River, in the Northern Aegean Sea (Easter Macedonia and Thrace, Greece) from very high-resolution (VHR) Pléiades imagery. They extracted the coastline using the Normalized Difference Water Index (NDWI) obtained and processed from both initial images and pan-sharpened images. The authors compared the extracted polyline with the reference one, manually achieved from the panchromatic image. Results of the comparisons provided RI and DRI values comparable with those obtained in this study.

In these last works, the automatic extraction of the coastline was facilitated by the use of multiband satellite images and limited by the ground pixel size [23,50]. On the other hand, with low-cost thermal cameras on drones, the advantage was the acquisition of images with higher resolution, allowing the authors to obtain good results in the automatic coastline extraction—even with only three bands.

6. Conclusions

In this work the detection of the real-time coastline in the flat areas of Boccasette and Barricata beaches (PRD, northern Italy) was investigated. Performances of the GNSS RTK technique, total station topographic instrument, and the photogrammetric SfM approach, using both optical and thermal images acquired with a low-cost drone, were analyzed.

According to the comparisons, the best performances in the detection of the real-time coastline in terms of costs, accuracies, and spatial coverage were obtained using the GNSS RTK technique. Nevertheless, drone-based survey by means of optical and thermal cameras allows for the following: (i) the production of the 3D model of the study area using the optical images, with the extraction of the 0-level contour line, useful in the multi-temporal comparisons; (ii) the real-time coastline detection using automatic procedures based on supervised classification of the thermal images. However, the drone-based survey cannot cover the length of the GNSS RTK measurements in the same working time.

Finally, the ground truth real-time coastline obtained from images using visual inspection or the automatic approach can provide significant errors. This work demonstrated the advantages of the ground-based GNSS RTK survey of the real-time coastline as a reliable ground truth. It can be used as a high-accuracy reference to evaluate the performances of polylines extracted from optical and/or thermal images and acquired from different platforms.

Author Contributions: Conceptualization, M.F.; methodology, M.F., M.B. and M.M.; software, M.F., M.B. and M.M.; validation, M.F., M.B. and M.M.; formal analysis, M.F. and M.B.; investigation, M.F. and M.B.; resources, M.F.; data curation, M.F.; writing—original draft preparation, M.F. and M.B.; writing—review and editing, M.F. and M.M.; visualization, M.F., M.B. and M.M.; supervision, M.F.; project administration, M.F. All authors have read and agreed to the published version of the manuscript.

Funding: This research received no external funding.

Data Availability Statement: Data are contained within the article.

Acknowledgments: The authors would like to thank the following Institutions: The Laboratory of Geomatics and surveying staff of the University of Padua; the "Direzione Pianificazione Territoriale"and "U.O. Genio Civile di Rovigo" (Veneto Region).

Conflicts of Interest: The authors declare no conflict of interest.

References

1. Zhao, Q.; Pan, J.; Devlin, A.T.; Tang, M.; Yao, C.; Zamparelli, V.; Falabella, F.; Pepe, A. On the Exploitation of Remote Sensing Technologies for the Monitoring of Coastal and River Delta Regions. *Remote Sens.* **2022**, *14*, 2384. [CrossRef]
2. Ericson, J.P.; Vörösmarty, C.J.; Dingman, S.L.; Ward, L.G.; Meybeck, M. Effective sea-level rise and deltas: Causes of change and human dimension implications. *Glob. Planet. Chang.* **2006**, *50*, 63–82. [CrossRef]
3. Antonioli, F.; Anzidei, M.; Amorosi, A.; Lo Presti, V.; Mastronuzzi, G.; Deiana, G.; De Falco, G.; Fontana, A.; Fontolan, G.; Lisco, S.; et al. Sea-level rise and potential drowning of the Italian coastal plains: Flooding risk scenarios for 2100. *Quat. Sci. Rev.* **2017**, *158*, 29–43. [CrossRef]
4. Kulp, S.A.; Strauss, B.H. New elevation data triple estimates of global vulnerability to sea-level rise and coastal flooding. *Nat. Commun.* **2019**, *10*, 4844. [CrossRef] [PubMed]
5. Karami, E.; Alizadeh, N.; Farhadi, H.; Abdolazimi, H.; Maghsoudi, Y. Monitoring of land surface displacement based on SBAS-InSAR time-series and GIS techniques: A case study over the Shiraz Metropolis, Iran. *ISPRS Ann. Photogramm. Remote Sens. Spatial Inf. Sci* **2023**, *X-4/W1-202*, 371–378. [CrossRef]
6. Fiaschi, S.; Fabris, M.; Floris, M.; Achilli, V. Estimation of land subsidence in deltaic areas through differential SAR interferometry: The Po River Delta case study (Northeast Italy). *Int. J. Remote Sens.* **2018**, *39*, 8724–8745. [CrossRef]
7. Saleh, M.; Becker, M. New estimation of Nile Delta subsidence rates from InSAR and GPS analysis. *Environ. Earth Sci.* **2018**, *78*, 6. [CrossRef]
8. Tang, S.; Song, L.; Wan, S.; Wang, Y.; Jiang, Y.; Liao, J. Long-Time-Series Evolution and Ecological Effects of Coastline Length in Coastal Zone: A Case Study of the Circum-Bohai Coastal Zone, China. *Land* **2022**, *11*, 1291. [CrossRef]
9. Vecchi, E.; Tavasci, L.; De Nigris, N.; Gandolfi, S. GNSS and Photogrammetric UAV Derived Data for Coastal Monitoring: A Case of Study in Emilia-Romagna, Italy. *J. Mar. Sci. Eng.* **2021**, *9*, 1194. [CrossRef]
10. Eboigbe, M.A.; Kidner, D.B.; Thomas, M.; Thomas, N.; Aldwairy, H. Analysis of Low-Cost Uav Photogrammetry Solutions for Beach Modelling and Monitoring Using the Opensource Quantum Gis. In *The International Archives of the Photogrammetry, Remote Sensing and Spatial Information Sciences, Proceedings of the ASPRS 2022 Annual Conference, Denver, CO, USA, 6–8 February and 21–25 March 2022*; ISPRS: Hannover, Germany, 2022; Volume XLVI-M-2-2022.
11. Fabris, M. Coastline evolution of the Po River Delta (Italy) by archival multi-temporal digital photogrammetry. *Geomat. Nat. Hazards Risk* **2019**, *10*, 1007–1027. [CrossRef]
12. Alberico, I.; Cavuoto, G.; Di Fiore, V.; Punzo, M.; Tarallo, D.; Pelosi, N.; Ferraro, L.; Marsella, E. Historical maps and satellite images as tools for shoreline variations and territorial changes assessment: The case study of Volturno Coastal Plain (Southern Italy). *J. Coast. Conserv.* **2017**, *22*, 919–937. [CrossRef]
13. Laksono, F.A.T.; Borzì, L.; Distefano, S.; Di Stefano, A.; Kovács, J. Shoreline Prediction Modelling as a Base Tool for Coastal Management: The Catania Plain Case Study (Italy). *J. Mar. Sci. Eng.* **2022**, *10*, 1988. [CrossRef]
14. Fabris, M. Monitoring the Coastal Changes of the Po River Delta (Northern Italy) since 1911 Using Archival Cartography, Multi-Temporal Aerial Photogrammetry and LiDAR Data: Implications for Coastline Changes in 2100 A.D. *Remote Sens.* **2021**, *13*, 529. [CrossRef]
15. Goksel, C.; Senel, G.; Dogru, A.O. Determination of shoreline change along the Black Sea coast of Istanbul using remote sensing and GIS technology. *Desalination Water Treat.* **2019**, *177*, 242–247. [CrossRef]
16. Viana, R.D.; dos Reis, G.N.L.; Velame, V.M.G.; Körting, T.S. Shoreline extraction using unsupervised classification on Sentinel-2 imagery. In Proceedings of the XIX Brazilian Symposium on Remote Sensing, Santos, Brazil, 14–17 April 2019; Volume 19, p. 96606.
17. Alcaras, E.; Amoroso, P.P.; Baiocchi, V.; Falchi, U.; Parente, C. Unsupervised classification based approach for coastline extraction from Sentinel-2 imagery. In Proceedings of the 2021 International Workshop on Metrology for the Sea; Learning to Measure Sea Health Parameters (MetroSea), Virtual Conference, 4–6 October 2021; pp. 423–427.
18. Domazetović, F.; Šiljeg, A.; Marić, I.; Faričić, J.M.; Vassilakis, E.; Panđa, L. Automated Coastline Extraction Using the Very High-Resolution WorldView (WV) Satellite Imagery and Developed Coastline Extraction Tool (CET). *Appl. Sci.* **2021**, *11*, 9482. [CrossRef]
19. Karaman, M. Comparison of thresholding methods for shoreline extraction from Sentinel-2 and Landsat-8 imagery: Extreme Lake Salda, track of Mars on Earth. *J. Environ. Manag.* **2021**, *298*, 113481. [CrossRef] [PubMed]
20. McFeeters, S.K. The use of the Normalized Difference Water Index (NDWI) in the delineation of open water features. *Int. J. Remote Sens.* **1996**, *17*, 1425–1432. [CrossRef]
21. Ma, M.; Wang, X.; Veroustraete, F.; Dong, L. Change in area of Ebinur Lake during the 1998 2005 period. *Int. J. Remote Sens.* **2007**, *28*, 5523–5533. [CrossRef]

22. Wolf, A.F. Using WorldView-2 Vis–NIR Multispectral Imagery to Support Land Mapping and Feature Extraction Using Normalized Difference Index Ratios. In *SPIE 8390, Algorithms and Technologies for Multispectral, Hyperspectral, and Ultraspectral Imagery XVIII, International Society for Optics and Photonics*; SPIE: Bellingham, WA, USA, 2012; Volume 8390, p. 83900N. [CrossRef]
23. Braga, F.; Tosi, L.; Prati, C.; Alberotanza, L. Shoreline detection. Capability of COSMO—SkyMed and high resolution multispectral images. *Eur. J. Remote Sens.* **2013**, *46*, 837–853. [CrossRef]
24. Ma, H.; Guo, S.; Zhou, Y. Detection of Water Area Change Based on Remote Sensing Images. In *GeoInformatics in Resource Management and Sustainable Ecosystem, Communications in Computer and Information Science*; Bian, F., Xie, Y., Cui, X., Zeng, Y., Eds.; Springer: Berlin/Heidelberg, Germany, 2013; Volume 398.
25. Liu, Y.; Wang, X.; Ling, F.; Xu, S.; Wang, C. Analysis of coastline extraction from Landsat-8 OLI imagery. *Water* **2017**, *9*, 816. [CrossRef]
26. Burke, C.; Wich, S.; Kusin, K.; McAree, O.; Harrison, M.E.; Ripoll, B.; Ermiasi, Y.; Mulero-Pázmány, M.; Longmore, S. Thermal-Drones as a Safe and Reliable Method for Detecting Subterranean Peat Fires. *Drones* **2019**, *3*, 23. [CrossRef]
27. Burke, C.; Rashman, M.; Wich, S.; Symons, A.; Theron, C.; Longmore, S. Optimising observing strategies for monitoring animals using drone-mounted thermal infrared cameras. *Int. J. Remote Sens.* **2019**, *40*, 439–467. [CrossRef]
28. Povlsen, P.; Linder, A.C.; Larsen, H.L.; Durdevic, P.; Arroyo, D.O.; Bruhn, D.; Pertoldi, C.; Pagh, S. Using Drones with Thermal Imaging to Estimate Population Counts of European Hare (*Lepus europaeus*) in Denmark. *Drones* **2023**, *7*, 5. [CrossRef]
29. Maurya, N.K.; Tripathi, A.K.; Chauhan, A.; Pandey, P.C.; Lamine, S. Recent Advancement and Role of Drones in Forest Monitoring: Research and Practices. In *Advances in Remote Sensing for Forest Monitoring*; Arellano, P., Pandey, P.C., Eds.; Wiley: Hoboken, NJ, USA, 2022; pp. 223–254.
30. Rouze, G.; Neely, H.; Morgan, C.; Kustas, W.; Wiethorn, M. Evaluating unoccupied aerial systems (UAS) imagery as an alternative tool towards cotton-based management zones. *Precis. Agric.* **2021**, *22*, 1861–1889. [CrossRef]
31. Dahaghin, M.; Samadzadegan, F.; Javan, F.D. Precise 3D extraction of building roofs by fusion of UAV-based thermal and visible images. *Int. J. Remote Sens.* **2021**, *42*, 7002–7030. [CrossRef]
32. Hou, Y.; Chen, M.; Volk, R.; Soibelman, L. Investigation on performance of RGB point cloud and thermal information data fusion for 3D building thermal map modeling using aerial images under different experimental conditions. *J. Build. Eng.* **2022**, *45*, 103380. [CrossRef]
33. Zanutta, A.; Lambertini, A.; Vittuari, L. UAV Photogrammetry and Ground Surveys as a Mapping Tool for Quickly Monitoring Shoreline and Beach Changes. *J. Mar. Sci. Eng.* **2020**, *8*, 52. [CrossRef]
34. Michałowska, K.; Głowienka, E. Multi-Temporal Analysis of Changes of the Southern Part of the Baltic Sea Coast Using Aerial Remote Sensing Data. *Remote Sens.* **2022**, *14*, 1212. [CrossRef]
35. Romagnoli, C.; Bosman, A.; Casalbore, D.; Anzidei, M.; Doumaz, F.; Bonaventura, F.; Meli, M.; Verdirame, C. Coastal Erosion and Flooding Threaten Low-Lying Coastal Tracts at Lipari (Aeolian Islands, Italy). *Remote Sens.* **2022**, *14*, 2960. [CrossRef]
36. Teatini, P.; Tosi, L.; Strozzi, T. Quantitative evidence that compaction of Holocene sediments drives the present land subsidence of the Po Delta, Italy. *J. Geophys. Res.* **2011**, *116*, B08407. [CrossRef]
37. Corbau, C.; Simeoni, U.; Zoccarato, C.; Mantovani, G.; Teatini, P. Coupling land use evolution and subsidence in the Po Delta, Italy: Revising the past occurrence and prospecting the future management challenges. *Sci. Total Environ.* **2019**, *654*, 1196–1208. [CrossRef] [PubMed]
38. Farolfi, G.; Bianchini, S.; Casagli, N. Integration of GNSS and satellite InSAR data: Derivation of fine-scale vertical surface motion maps of Po Plain, Northern Apennines, and Southern Alps, Italy. *IEEE Trans. Geosci. Remote Sens.* **2019**, *57*, 319–328. [CrossRef]
39. Cenni, N.; Fiaschi, S.; Fabris, M. Monitoring of Land Subsidence in the Po River Delta (Northern Italy) Using Geodetic Networks. *Remote Sens.* **2021**, *13*, 1488. [CrossRef]
40. Fabris, M.; Battaglia, M.; Chen, X.; Menin, A.; Monego, M.; Floris, M. An Integrated InSAR and GNSS Approach to Monitor Land Subsidence in the Po River Delta (Italy). *Remote Sens.* **2022**, *14*, 5578. [CrossRef]
41. Carlo, C. Physical Processes and Human Activities in the Evolution of the Po Delta, Italy. *J. Coast. Res.* **1998**, *14*, 775–793.
42. Correggiari, A.; Cattaneo, A.; Trincardi, F. The modern Po Delta system: Lobe switching and asymmetric prodelta growth. *Mar. Geol.* **2005**, *222–223*, 49–74. [CrossRef]
43. Simeoni, U.; Corbau, C. A review of the Delta Po evolution (Italy) related to climatic changes and human impacts. *Geomorphology* **2009**, *107*, 64–71. [CrossRef]
44. Municipality of Venice, Centro Previsioni e Segnalazioni Maree. Available online: https://www.comune.venezia.it/it/content/centro-previsioni-e-segnalazioni-maree (accessed on 24 February 2022).
45. Agisoft LLC. *Agisoft Metashape User Manual*; Professional Edition, Version 1.8; Agisoft LLC: St. Petersburg, Russia, 2022.
46. Agisoft LLC. *Metashape Python Reference, Release 1.8.2*; Agisoft LLC: St. Petersburg, Russia, 2022.
47. Fabris, M.; Fontana Granotto, P.; Monego, M. Expeditious Low-Cost SfM Photogrammetry and a TLS Survey for the Structural Analysis of Illasi Castle (Italy). *Drones* **2023**, *7*, 101. [CrossRef]
48. Nagendra, H.; Gadgil, M. Biodiversity Assessment at Multiple Scales: Linking Remotely Sensed Data with Field Information. *Proc. Natl. Acad. Sci. USA* **1999**, *96*, 9154–9158. [CrossRef]
49. Maglione, P.; Parente, C.; Santamaria, R.; Vallario, A. Modelli Tematici 3D Della Copertura Del Suolo a Partire Da DTM Immagini Telerilevate Ad Alta Risoluzione WorldView-2. *Rend. Online Della Soc. Geol. Ital.* **2014**, *30*, 33–40. [CrossRef]

50. Alcaras, E.; Falchi, U.; Parente, C.; Vallario, A. Accuracy Evaluation for Coastline Extraction from Pléiades Imagery Based on NDWI and IHS Pan-Sharpening Application. *Appl. Geomat.* **2022**, *15*, 595–605. [CrossRef]
51. Lee, J.M.; Park, J.-Y.; Choi, J.-Y. Evaluation of sub-aerial topographic surveying techniques using total station and RTK-GPS for application in macrotidal sand beach environment. *J. Coast. Res.* **2013**, *65*, 535–540. [CrossRef]
52. Marchel, Ł.; Specht, M. Method for Determining Coastline Course Based on Low-Altitude Images Taken by a UAV. *Remote Sens.* **2023**, *15*, 4700. [CrossRef]
53. El Kafrawy, S.; Basiouny, M.; Ghanem, E.; Taha, A. Performance evaluation of shoreline extraction methods based on remote sensing data. *J. Geogr. Environ. Earth Sci. Int.* **2017**, *11*, 1–18. [CrossRef]

remote sensing

Article

Posidonia oceanica Cartography and Evolution of the Balearic Sea (Western Mediterranean)

Laura del Valle Villalonga [1,*], Guillem Xavier Pons [2] and Marcial Bardolet [3]

1 Earth Sciences Research Group, Universitat de les Illes Balears, Ctra. de Valldemossa, km 7.5, 07122 Palma, Spain
2 Department of Geography, Universitat de Les Illes Balears, Ctra. de Valldemossa, km 7.5, 07122 Palma, Spain; guillemx.pons@uib.es
3 Institut Balear de la Natura (IBANAT), 07009 Palma, Spain; marcialbardolet@gmail.com
* Correspondence: laura.del-valle@uib.es

Abstract: The Balearic coastline presents an environmental and biological heterogeneity, which confers great complexity on the marine environment and treasures important biodiversity, both at the level of species and marine communities. The endemic phanerogam of the Mediterranean Sea, *Posidonia oceanica*, holds a pivotal role in maintaining high biodiversity, warranting protection as stipulated in the Posidonia Decree 25/2018. The purpose of this study is to provide quantitative criteria that will allow the delimitation of areas with *Posidonia oceanica* for conservation and to aid planning and management of this species, contributing to the reduction of biodiversity loss caused by anthropogenic impacts and global change. Utilizing a comprehensive approach, the study employs photo interpretation of aerial photographs taken at depths between 0 and 5 m, data from Side-Scan Sonar (SSS) campaigns, reprocessing information from the LIFE Posidonia project at depths between 20 and 30 m, and targeted sampling using Remote Operated Vehicles (ROV) and ocular recognition at strategic points. The research not only seeks to assess the present state of the phanerogam but also analyzes its evolution, establishing a technological database for consultation and integrated analysis. This database facilitates effective management by tracking habitat changes, representing a significant contribution to the understanding of the impact of global change on ecosystems through Geographical Information Technologies (TIGs).

Keywords: *Posidonia oceanica*; Balearic Sea; TIG; SSS

Citation: del Valle Villalonga, L.; Pons, G.X.; Bardolet, M. *Posidonia oceanica* Cartography and Evolution of the Balearic Sea (Western Mediterranean). *Remote Sens.* **2023**, *15*, 5748. https://doi.org/10.3390/rs15245748

Academic Editors: Ramón Blanco Chao, Germán Flor-Blanco and José Juan de Sanjosé Blasco

Received: 2 October 2023
Revised: 10 December 2023
Accepted: 12 December 2023
Published: 15 December 2023

1. Introduction

Posidonia oceanica, an endemic phanerogam of the Mediterranean Sea [1], forms extensive meadows around the Balearic Islands, known as submarine forests. These meadows are essential for the ecosystem as they host a diverse range of species and provide crucial benefits. They serve as shelters for numerous organisms, produce oxygen daily while absorbing carbon dioxide, and generate a significant amount of biomass annually. Moreover, they can form long-lasting barrier reefs or seagrass. Additionally, these meadows play a vital role in defining and protecting beaches. During winter, the accumulation of leaves on the beach, known as banquettes, stabilizes the seafloor and mitigates the impact of waves on the coast. Furthermore, the shells of the epiphytic calcareous organisms on *P. oceanica* leaves contribute to the formation of whiter beaches [2–4].

P. oceanica is the dominant seagrass species in the Mediterranean, where this endemic species covers about 50,000 km^2 above 45 m depth (estimated by [5]). This species develops extensive meadows, which are structurally and topographically complex, as the growth of the vertical rhizomes of the plant, acting along millenary time scales [6], develops reefs. These meadows represent one of the most productive ecosystems in the Mediterranean Sea, with a net primary production of about 1000 g DW m′ year-1 [7]. The clear waters and shallow slopes characteristic of the Balearic Islands allow the development of extensive

P. oceanica beds, which probably represent the dominant marine ecosystem on the coast of the Balearic Islands. Yet, our knowledge on the extent, status, ecology, and functions of these important ecosystems in the Balearic Islands is still meager. Moreover, the body of knowledge from before the 21st century on the ecology of these meadows, including information on their presence in the Cabrera island [8] and the Bay of Palma [9], the impacts of aquaculture operations at Fornells Bay (Menorca [10]), carbonate production in Pollença Bay [11], and some aspects of their palaeodynamics [12], is scarce. Seagrass ecosystems are under pressure worldwide [13], including the Balearic Islands (e.g., [14]), so the conservation of these important ecosystems calls for increased efforts to understand their ecology and role in the ecosystem, as well as to formulate effective conservation strategies [15].

P. oceanica is highly sensitive to environmental changes, which makes it an excellent indicator of water quality. Recognizing its significance, it has been designated a habitat of priority interest for the EU since 1992 [16]. Various regulations, such as Decree 25/2018 [17], have been put in place to protect and conserve *P. oceanica*. However, despite these protective measures, this vital community faces threats due to an increase in activities in the Balearic waters, endangering its conservation status. Anchoring, uncontrolled mooting of boats, trawling, beach regeneration, desalination plants, and other direct impacts pose significant risks.

In some parts of the Balearic Islands, up to 40% of them have disappeared, mainly due to the damage caused by the anchors of recreational boats. According to the results of a study carried out by researchers, the extent of Posidonia has been reduced by between 13% and 38% since 1960, and since the 1990s, the density of Posidonia bundles has decreased by 50 %. That is why it is very important to have an updated, detailed cartography [18].

To address this situation, the Balearic government approved Decree 25/2018 [17], aiming to ensure the compatibility of human activities with the protection of *P. oceanica* and its habitat. In order to achieve this, it is necessary to conduct a comprehensive, continuous, and high-quality cartography of *P. oceanica* surrounding the Balearic Islands, utilizing the latest and most advanced technologies available. This includes photo interpretation of aerial photographs in shallow areas (0–5 m), SSS (Side-Scan Sonar) campaigns, sampling using ROVs, and others. The initial contribution, presented in "Mapping of the Marine Habitats of the Balearic Islands: Compilation of Layers and Benthic Communities [19]," underscores the imperative to enhance and expand current cartography. This pertains not only to the broader spectrum of bionomic considerations but also to the comprehensive evaluation of habitat conservation statuses. The present study emphasizes the lack of cartography information on the Serra de Tramuntana coastline and the necessity to enhance and update knowledge about habitat conservation in vulnerable areas such as the Bays of Pollença, Alcúdia, and Fornells, among others.

One of the first results obtained with this project is the continuous mapping of the seabed of the entire Balearic Islands and the approximation of the state of conservation of this seabed based on bibliographic information. Secondly, this work has allowed us to detect information gaps and discriminate information that is outdated, both at a cartographic and biological level.

The final map of the Balearic Islands includes a total of 55 habitats mapped over an area of 5067.67 km^2 of seabed, ranging from 0 to 50 m of practically all the seabed that surrounds the islands, up to 400 m in the channel of Menorca and up to 100 m in the Mallorca canal. This surface represents 22% of the surface of the Balearic Sea (up to 700 m deep) and 40% of the continental shelf of the Balearic Islands (up to 200 m). Partial results were also published for the island of Mallorca [20].

In this first global approximation, the predominant marine habitat is *P. oceanica*, with an area of 459.77 km^2. However, large areas continued without data mapping (for example, Serra de Tramuntana region).

Therefore, the primary objective of this work is to accurately determine the precise location of *P. oceanica* in the Balearic Sea, specifically between 0 and 35 m, including the Serra de Tramuntana region. This research also aims to evaluate the evolution and conservation status of *P. oceanica* meadows, ultimately leading to the development of an effective management plan.

Study Area

The Balearic Sea is a subdivision of the western basin of the Mediterranean Sea, located between the Balearic Islands (Mallorca, Menorca, Eivissa, and Formentera) and the Spanish coast (Figure 1). It is limited in the southwest by the cape of Sant Antoni (38°50′N–0°12′E) and extends to the cape of Barbaria in the extreme southwest of Formentera. In the southeast of Formentera coast, it extends from Punta Roja, the eastern end, to the southern Cabrera (39°7′N–2°54′E) and the islet of l'Aire at the extreme south of Menorca, and from the northwest of the east coast of Menorca to cape Favàritx (40°0′N–4°14′E) until the San Sebastián cape (14°54′N–3°10′E).

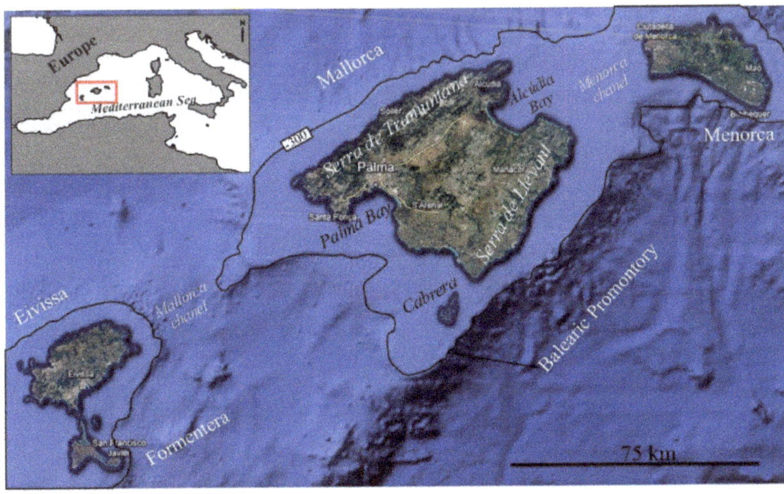

Figure 1. Location of the Balearic Islands. Red frame, location of the Balearic Islands in the western mediterranean.

The study area of the Balearic Sea includes the marine platform that surrounds the Balearic Islands from 0 to 35 m deep, including the islets. When *P. oceanica* is observed, this marine phanerogam is included in the List of Wild Species under the Special Protection Regime of the Balearic Islands [21] and the Spanish Catalog of Endangered Species [22]. It is also listed in Annex I of the Berne Convention [23], Annex II of the Barcelona Convention [24], and Annex I of Directive 92/43/EEC on the conservation of natural habitats and wild fauna and flora (as a habitat) [16].

P. oceanica is an endemic marine phanerogam of the Mediterranean, similar to terrestrial plants with leaves, flowers, and fruits, but it lives permanently submerged between the surface and 30 m deep, where there is still enough light to allow it to perform photosynthesis. In places where the transparency of the waters is greater, such as the Balearic Archipelago or the Eastern Mediterranean, the distribution of this species can reach depths of up to 40–45 m [25,26].

2. Methods

The *P. oceanica* cartography of all the Balearic Islands presented in this study is the result of the use of different techniques (Figure 2).

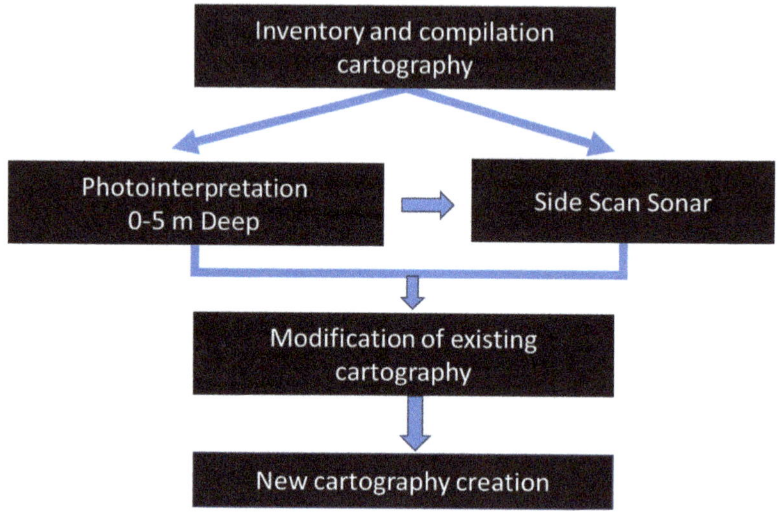

Figure 2. Methods flowchart.

Photo interpretation of aerial photographs between 0 to 5 m deep; digitalization and photo interpretation based on the orthophotos of the National Aerial Orthophotography Plan 2018/19, supported by all previous coverages available.

Side-Scan Sonar campaign (SSS) between 0 to 35 m: through 2 sonar models with a frequency range between 100–500 kHz and 500–900 kHz, and the use of SonarWiz v7 software for data acquisition.

Reprocessing of available information from the LIFE *Posidonia* project between 20 and 30 m deep: bionomic cartography carried out in the LIFE *Posidonia* project of the Balearic Islands (LIFE00/NAT/E/7303).

In the case of the islands of Eivissa and Formentera, a new cartography was also prepared based on the following works: collection and analysis of the information available in the study area of the Socioenvironmental Observatory of Menorca and Natural History Society of the Balearic Islands [19], Consell de Eivissa (Calas de Ibiza, 2018), OCEANSNELL (reference cartography in Formentera, 2016), LIFE Posidonia Project (2010) and Ecocartography (Ministerio de Agricultura, Pesca y Alimentación, 2008).

Finally, the government of the Balearic Islands carried out a punctual sampling campaign to check the correct identification of marine habitats in the priority areas for inconsistencies, etc., with a ROV (underwater Drone) and a bathyscope.

Therefore, an inventory and compilation of the existing cartography were carried out; photo interpretation between 0 and 5 m deep, a program of specific sampling, a Side-Scan Sonar campaign on the island of Majorca between 0 and 30 m deep (detailed information on the typology of the funds and the limits determined with a resolution of 1 m, carried by different companies contracted by the government of the Balearic Islands to carry out field work [27]); the reprocessing of the information available from the LIFE *Posidonia* Project in Mallorca, between 20 and 30 m deep and processing and integration of the cartography.

For prospecting work with Side-Scan Sonar, a DGPS was used. It is an optimal differential positioning receiver for bathymetric and geophysical surveys, thanks to its high resolution and 20 kHz position update. The SBAS, BEACON and OMNISTAR differential corrections are applied to this receiver. At the same time, it supports RTK corrections up to 50 cm. Navigation for plotting the campaign line plan was carried out using the Hypack 2018 v1.18.1.0 Software. This software provides position and route details on high-resolution color screens. The operator enters track indication details, seabed features and risk areas, which can also be recorded and viewed on screen to aid in vessel positioning. Navigation and positioning data are reviewed and evaluated before positions are calculated and quality indicators are generated. Said navigation was executed on a plan of lines of work generated in the office, which were distributed throughout the field of study to cover the greatest possible extension. It is of great importance to emphasize that this extension was carried out being aware of the limitations that work at sea entails, such as the maneuverability of the vessel, unforeseen underwater events (shoals or submerged rocks), lack of space in coves and bays, etc.

For the acquisition of indirect data, two 500–900 kHz dual-frequency Side-Scan Sonar (SSS) units were used to appreciate the different types of bottom and the limits between them with a high degree of detail and thus achieve a mosaic with resolution of pixels of the order of 0.1 × 0.1 m (Figure 3).

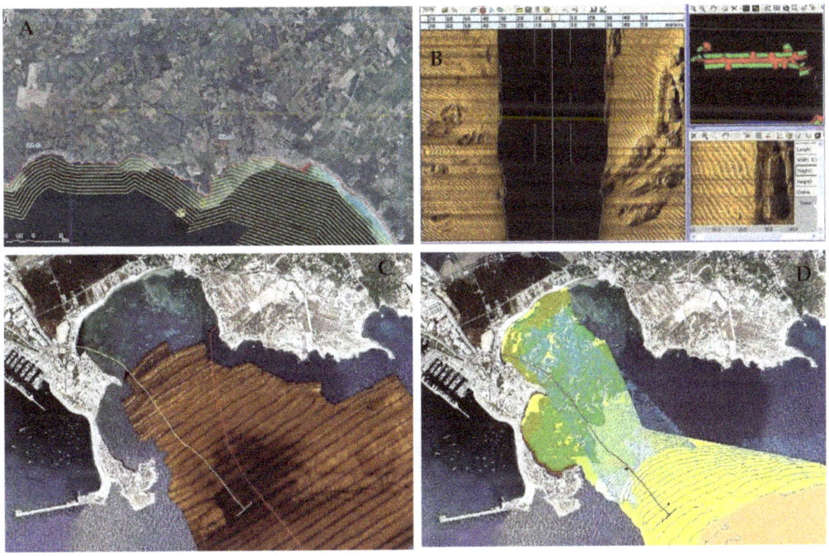

Figure 3. (**A**) Side-Scan Sonar Campaign Plan—Detail Example. (**B**) Image acquisition with SonarWiz 7 software. (**C**,**D**) Examples of the interpretation of geomorphological maps from SSS.

During field work, the SSS was coupled to a "fish" towed by a boat. In order for the geopositioning of the transects to correspond as closely as possible to reality, a correction was applied to suppress the distance between the "fish" and the boat, which is where the GPS system is located.

The sonar is composed of a torpedo-shaped underwater vehicle with two transducers, a Transceiver and Processing Unit (TPU) from which the acoustic pulses are generated and a Kevlar cable that tows the equipment and transmits the data. The records are sent to a PC for viewing and storage using the appropriate software. The reflection of the signal coming from the bottom is captured by the same transducers, amplified and transmitted through the towline to the recorder, where the corresponding signal is digitized and sent to the PC where the appropriate software processes and stores it.

The data acquisition was carried out with the specific software, SonarWiz 7, which allows the data to be acquired in real time and the data files to be recorded every minute for subsequent processing. While observing the computer screen with real-time acquisition, both lengths and heights of objects can be measured with great precision, while complex areas can be magnified for further study. For the correct use and geographical orientation of the sedimentary forms and other types of bottom, as well as other objects found, a post processing of the image was carried out. The depth of the SSS with respect to the bottom was regulated so that the maximum sweep range was 75 m per band, and it was ensured that there was only a 10–15% overlap between adjacent transects.

2.1. Side-Scan Sonar Data Processing

Once the raw Side-Scan Sonar files were obtained by the field teams in XTF (eXtended Triton Format) format, they were processed with the SonarWiz7 software. This software allows the visualization, edition and treatment of these files, therefore it was used to eliminate the sheet of water from all the data (Bottom track), correct and smooth navigation, apply image filters, and export of the data in the form of a georeferenced mosaic. The post processing was carried out with the SonarWiz7 software from Chesapeake, with the correction of the range, correction of the navigation, application of filters and finally, the composition of the mosaics (reflectivity map in photographic form of the background).

The integration of all the previously mentioned raw information was carried out with the specific software ArcGis 10.4. This software allows the superimposition of different sources of georeferenced information, which is necessary for the digitization of the different morphologies of the seabed. The digitization of the limits of the seagrass meadows was carried out with the highest possible precision and at a minimum spatial scale of 1:800. All bald spots, voids and channels of relevant size within the identified meadows were represented. When digitizing marine phanerogams observed as discrete spots, a minimum area of 100 m^2 was considered. However, in the majority of instances, the entire entity was meticulously represented, distinctly delineated from its surroundings, and characterized according to one of the established categories.

2.2. Reprocessing Methodology

In order to process and analyze the information correctly, the layers were projected to the ETRS89 UTM Zone 31 projection system, applying the transformation "ED50 to WGS84 NTv2 Baleares". Subsequently, the original topology was reviewed and corrected in all complete layers, covering the entire surface. In each layer, the original information was preserved, and the following fields were created in their corresponding attribute tables. All the geometry errors that the original layers had, such as overlaps and gaps between polygons, were debugged. Processing and integration of the cartography: All the cartography was integrated in a single layer by grids.

3. Results

3.1. Cartography of Posidonia oceanica

The final product is the generation of homogeneous, continuous, and high-quality cartography of marine habitats, especially *P. oceanica*, which surrounds the Islands of Mallorca, Menorca, Eivissa, and Formentera, ranging from the coastline to approximately 35 m deep (Figure 4A–D). In addition, the data obtained by SSS have a resolution of the order of 0.2 m, allowing differentiation of the different types of funds and the precise determination of their limits.

The results are shown in Table 1.

The total area of the habitats containing *P. oceanica* in the study area is 592.82 km^2, surpassing [27] the suggested area of 375 km^2 by 93.40%. The final cartography, a result of diverse methods, including SSS techniques, photo interpretation, and GIS corrections, reveals a mapped *P. oceanica* area of 553.68 km^2 between 0 and 35 m deep in the Balearic Sea around the Balearic Islands (Table 1). Especially, an area of 135.40 km^2 was mapped in the islands of Eivissa and Formentera (Figure 4A,B), 338.92 km^2 in Mallorca (Figure 4C), and 79.37 km^2 in Menorca (Figure 4D).

Figure 5 illustrates a substantial shift in the Alcanada area, previously dominated by *P. oceanica*, now characterized by a prevalence of photophilic algae. Numerous voids, notably elongated and some ellipsoid in shape, between 1 and 3 m by 2 and 3 m, are evident.

In the western Bay of Palma (Figure 6), a 2 km long by 1 km wide area of dead *P. oceanica* rhizome (03051203) is observed, surrounded by meadows on dead plants (rhizome) covering 3 by 3 km, indicative of degradation.

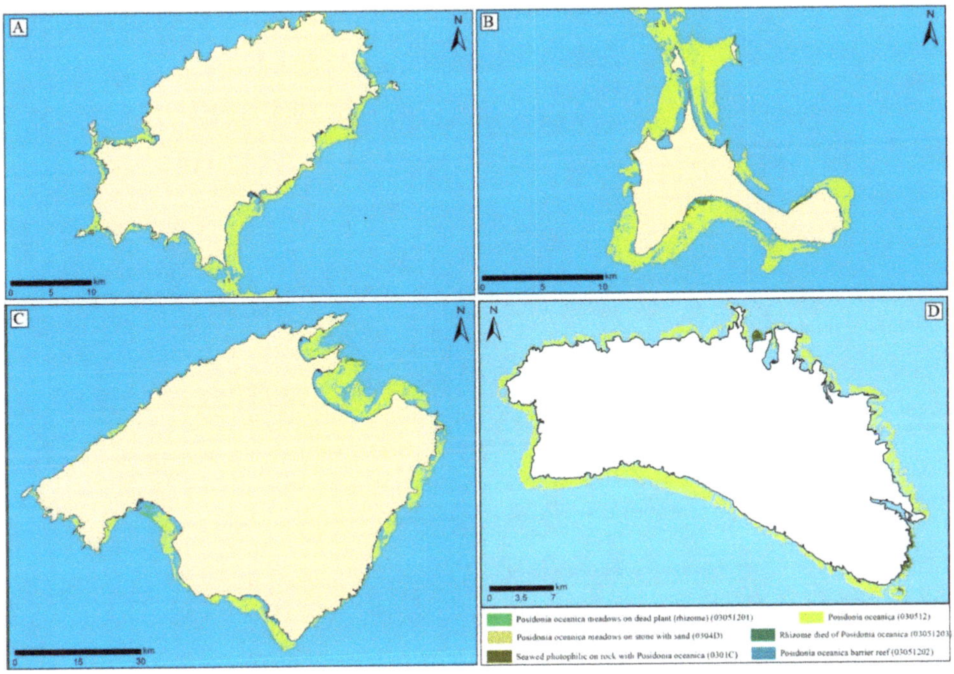

Figure 4. (**A**) Map of the habitats containing *P. oceanica* in Eivissa. (**B**) Map of the habitats containing *P. oceanica* in Formentera. (**C**) Map of the habitats containing *P. oceanica* in Mallorca. (**D**) Map of the habitats containing *P. oceanica* in Menorca.

Table 1. Summary table of the surface of the habitats containing *P. oceanica* in the Balearic Islands, from 0 to 35 m deep (in km^2).

	Area (km^2)			
Atlas Posidonia	**Ibiza/Formentera**	**Mallorca**	**Menorca**	**Total**
Photophilic algae on stone with *Posidonia oceanica* (0301C)	2.2465	6.1898	3.3025	11.7388
Posidonia oceanica barrier reef (03051202)	0.0897	0.3048	0.1089	0.5034
Dead plant of *Posidonia oceanica* (03051203)	0.0026	0.3535	0.1375	0.4936
Posidonia oceanica (030512)	135.4006	338.9158	79.3673	553.6837
Posidonia oceanica on stone with sand (0304D)	0.7505	20.0409	0.2852	21.0765
Posidonia oceanica meadows on dead plant (rhizome) (03051201)	0.0964	5.2298	0.0015	5.3278
Total	138.5863	371.0346	83.2030	592.8238

The cartography in Cala Pudent unveils a distinct 1.17 km long and 1 m wide line and voids parallel to the coast in the western sector of the Bay of Palma, about 2 km long and 2 m wide.

For the first time, a detailed mapping of the Serra de Tramuntana marine platform between 0 and 35 m depth was conducted (Figures 7 and 8). Figure 7 displays the *P. oceanica* map in Pollença Bay and part of the Serra de Tramuntana, showcasing metric-sized holes within the mapped area. Notably, there are 4 km-sized voids with very rectilinear shapes. The cartography also identifies areas where Posidonia returns to colonize on top of dead rhizomes (03051201). The cartography of the Serra de Tramuntana reveals a significant presence of photophilic seaweed on rocks with *P. oceanica* (0301C), and *P. oceanica* (030512) is observed in isolated spots or continuous meadows on stone with sand (0304D). Notably, *P. oceanica* (030512) presence is low in the Port of Sóller.

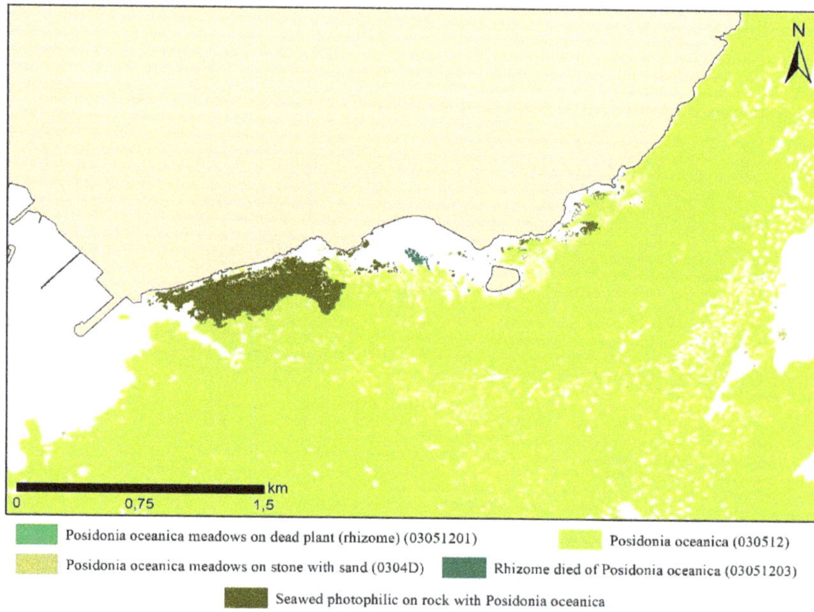

Figure 5. Map of the habitats containing *P. oceanica* in Alcanada, specifically the northwest area of Mallorca.

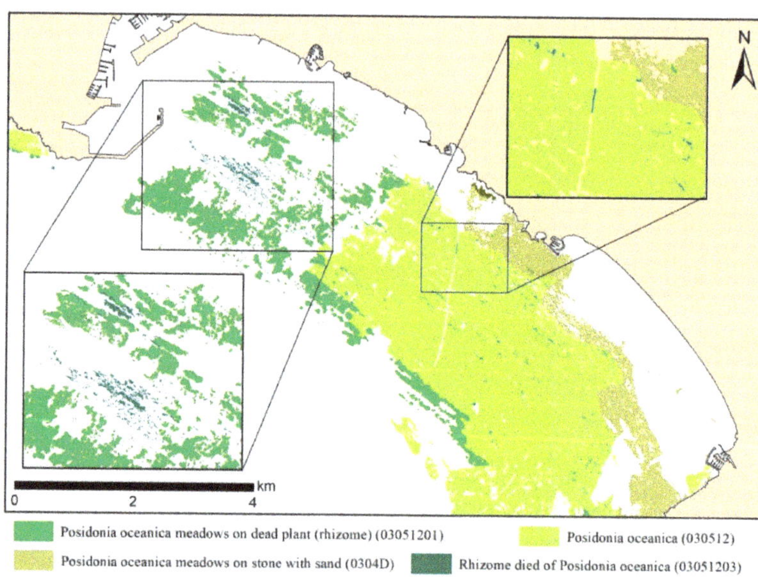

Figure 6. Map of the habitats containing *P. oceanica* in Palma Bay, especially the southwest area of Mallorca.

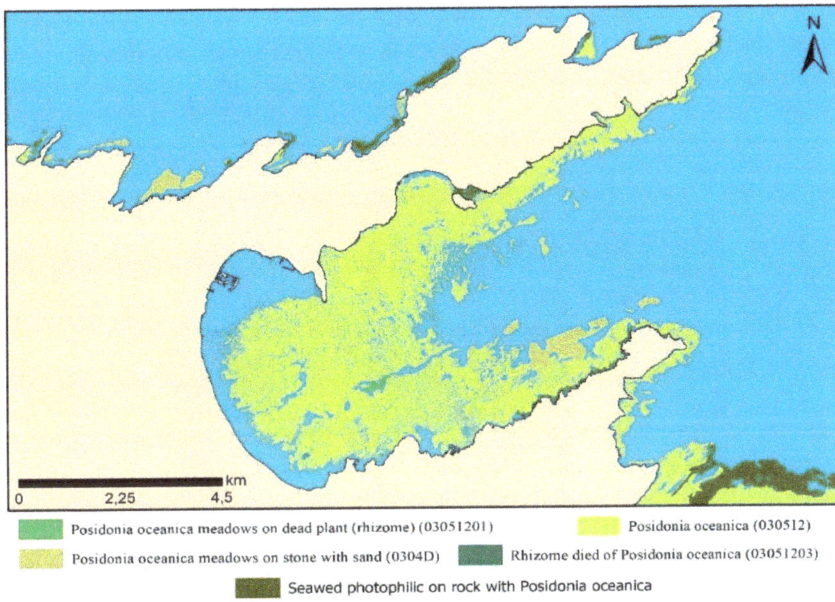

Figure 7. Map of the habitats containing *P. oceanica* in Pollença Bay, especially the north area of Mallorca.

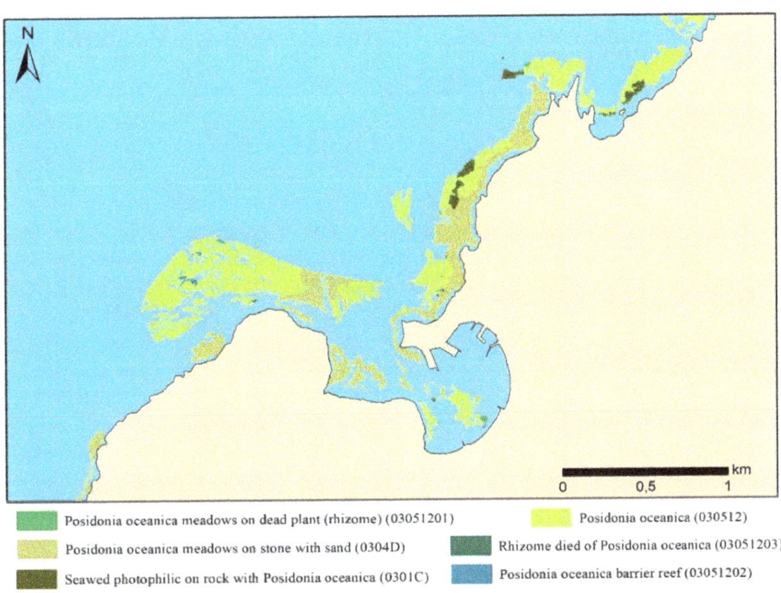

Figure 8. Map of the habitats containing *P. oceanica* in the Sóller Port, especially the north area of Mallorca.

3.2. Conservation of *Posidonia oceanica*

An evaluation of the conservation status of *P. oceanica* was conducted based on the sampling program results for dead bushes (Figure 9A–D). The presence of dead bushes or areas of dead rhizome without foliar apparatus indicates meadow regression, providing crucial information about conservation and evolution.

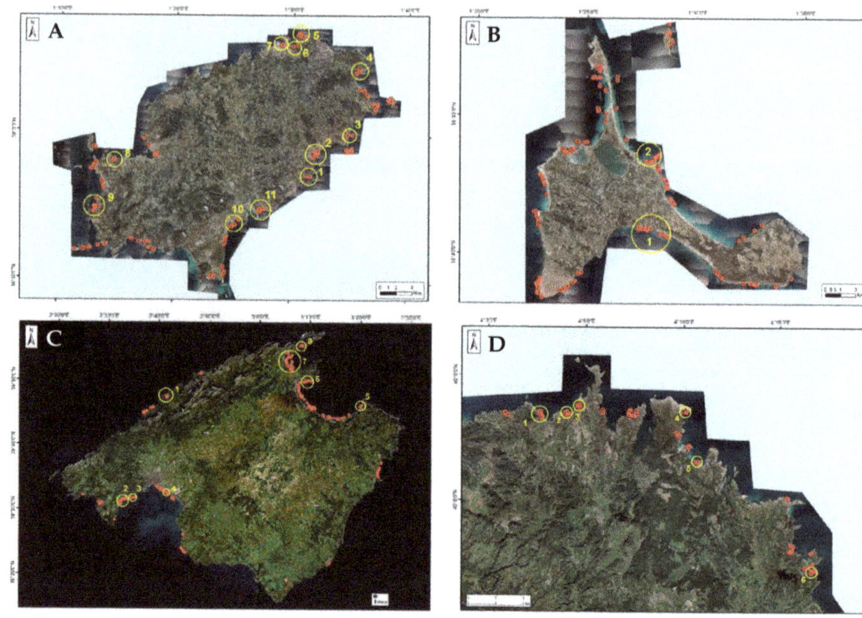

Figure 9. (**A**). Location of the sampling stations with the presence of dead plant in Eivissa Island. 1 Cala Llonga, 2 Santa Eulària Bay, 3 Es Canar, 4 Cala san Vicent, 5 Portinatx, 6 Cala Xarraca, 7 Cala Porcs, 8 Cala Bassa, 9 Cala Vedella, 10 Platja d'en Bossa and 11 Cala Talamanca. (**B**) Location of the sampling stations with the presence of dead plants in Formentera, 1 Platja de Migjorn and 2 Cala en Baster. (**C**) Location of sampling stations with presence of dead plant in Mallorca, 1 Sóller Port, 2 Palma Nova, 3 Illetes, 4 South Palma, 5 Es Caló, 6 North Alcúdia (Alcanada), 7 Pollença Port and 8 Platja de Formentor. (**D**) Location of sampling stations with presence of dead plants in Menorca, 1 Cala Pregonda, 2 Cala Mica, 3 Cala Rotja, 4 Cala en Tusqueta, 5 Na Xeringa and 6 Es Grau Bay.

In Eivissa, the Sant Antonio Bay reflects a highly degraded bottom with no *P. oceanica* presence. Formentera exhibits a well-preserved *P. oceanica* meadow, with dead plants identified only in Platja d'es Mitjorn and Cala en Baster (Figure 9A).

Mallorca, along with Eivissa, exhibits the worst *P. oceanica* meadow conservation. Eight zones, including the Pollença Bay and Alcanada area, show the presence of dead plants (Figure 5). The Pollença Port area has a degraded background, making cartography challenging. The Formentor Beach displays medium-low *P. oceanica* conservation, with identified patches of dead plants (Figure 7). In Sóller Port (Figure 8), some patches of dead plants were identified, though more grassland was found than initially thought.

Along the north coast of Menorca (Figure 9D), various locations with dead plants were identified, with Cala en Tusqueta being the most impacted. The presence of *Cymodocea nodosa* competing for space is notable. The importance of preserving areas of dead plants for *P. oceanica* regeneration is observed on this island.

4. Discussion

The comprehensive mapping of *P. oceanica* meadows through remote sensing has been the focus of various studies across the Mediterranean [28–32]. However, none have achieved the level of coverage and precision demonstrated in this research. The marine habitats of Spain, as outlined in the interpretive guide [33], exhibit significant environmental and biological heterogeneity, particularly evident in the Balearic Islands. These islands showcase diverse features, such as maërl beds, expansive seagrass meadows, underwater canyons such as Son Bou, and sedimentary bottoms of coves such as s'Estany des Peix, Fornells, or Addaia [34], hosting formations practically unique in the Mediterranean.

The susceptibility of coastal environments to threats such as climate change, marine pollution, overfishing, and anthropogenic activities leads to dynamic changes, resulting in habitat loss, introduction of non-native species, and degradation of coastlines [33]. Subsequent evaluations will play a crucial role in quantifying the impact of anthropogenic activities and global changes on biodiversity.

The historical backdrop reveals that oceanographic efforts along the Balearic coasts commenced in the 18th century, gaining momentum in the 20th century with comprehensive descriptions of benthic ecosystems [35]. Notably, the commitment to seabed studies was underscored by the Balearic Islands government commissioning the CEAB-CSIC for studies in protected areas [36–38]. Throughout all these centuries, a multitude of studies have been carried out that have allowed us to synthesize and describe the different types of habitats that we find in the entire Balearic Archipelago [39–43]. Despite centuries of research, the precision achieved in the cartography of *P. oceanica* in the Balearic Sea, totaling 553.68 km^2, is unparalleled.

The mapped areas reveal *P. oceanica* as disseminated spots with varying sizes and shapes, often exhibiting elongated or ellipsoid forms. Notably, areas with multiple voids in close proximity suggest potential damage from indiscriminate anchorage or illegal activities, prevalent in regions with high anthropic pressure.

Boat anchorages pose significant threats to *P. oceanica* meadows due to anchor movements and chain interactions. A notable case is the 1170 m long outfall line in Cala Pudent, Mallorca, corresponding to the Torrent Gros submarine outfall [44,45]. The eastern coast of Palma, particularly from the Old Dock to Arenal de Llucmajor, has been extensively modified and impacted by tourism development, affecting coastal dynamics and leading to beach erosion [46].

Regeneration projects, such as the one in Alcanada, further contribute to sediment dynamics that can adversely affect *P. oceanica*. Conservation status varies across islands, with Pollença Bay and Alcanada in Mallorca facing higher anthropic impact and competition from habitats with lower ecological requirements.

In Palma Bay, the *P. oceanica* community faces challenges in areas prone to sediment dynamics, impacting health and coverage. Photophilous zones on dead *P. oceanica* plants, influenced by *Halimeda incrassata*, exhibit compromised health, often characterized by necrosis. The continuous influence of fine sediment dynamics further jeopardizes these photophilic organisms. The easternmost part of Palma Bay hosts *P. oceanica* communities in less shallow areas, struggling against sediment dynamics. The deterioration in coverage and state of conservation may result from various factors, including anthropic impacts, changes in sediment dynamics, and competition with opportunistic species such as *Halimeda incrassata*.

Marine pollution, coastal infrastructure, and the expansion of invasive algae species contribute to the regression of *P. oceanica* meadows. The biological properties of marine phanerogams, such as slow growth and limited recovery capacity, make them particularly vulnerable to human-induced disturbances [45].

5. Conclusions

In conclusion, this study successfully achieved its primary objective of generating a homogeneous, continuous, and high-resolution cartography of *Posidonia oceanica* across the Balearic Islands within the 0 to 35 m depth range. The utilization of various data sources, including aerial photographs, Side-Scan Sonar (SSS) campaigns, and specific sampling using ROV and bathyscope, has resulted in a comprehensive visualization of the precise location and conservation status of *P. oceanica* meadows. The mapped surface area of *P. oceanica* in the Balearic Islands, totaling 553.68 km^2, provides a detailed breakdown with 135.40 km^2 corresponding to the islands of Eivissa and Formentera, 338.92 km^2 to Mallorca, and 79.37 km^2 to Menorca. This comprehensive mapping effort extends to the Serra de Tramuntana region, marking the first-time inclusion of this area, enhancing existing bionomic cartography through Side-Scan Sonar studies and point sampling.

Evaluating the state of conservation reveals significant variations among islands, with Formentera displaying the best-preserved meadows, followed by Menorca. Conversely, the meadows of Eivissa and Mallorca exhibit concerning levels of degradation. Specific areas, such as Sant Antoni Bay in Eivissa and Palma Bay and Pollença Bay in Mallorca, highlight the substantial anthropic pressure and various contributing factors leading to the highly degraded *P. oceanica* meadows. The identified variations and degradation underscore the urgent need for thorough analysis, planning, and management. Furthermore, the data collected emphasize the necessity to expand and enhance existing cartography, not only at a bionomic level but also for the ongoing assessment of habitat conservation status. This study has filled a critical gap by including the Sa Serra de Tramuntana coast in its cartographic scope. It is also evident that there is a compelling need to improve and update our understanding of habitat conservation in vulnerable areas such as the Bays of Pollença and Alcúdia or the Port of Fornells.

Considering these findings, the study concludes with a strong call to action for ongoing research, conservation initiatives, and the development of effective management plans. By addressing the identified challenges and refining our understanding of vulnerable areas, we can actively contribute to the preservation and sustainable management of *Posidonia oceanica* meadows in the Balearic Sea. In this sense, we underscore the urgent need for targeted conservation efforts, considering the diverse challenges posed by anthropogenic activities, climate change, and invasive species. Future research should focus on monitoring the impact of ongoing anthropogenic activities and developing effective conservation strategies to ensure the long-term sustainability of these vital marine ecosystems.

Author Contributions: Methodology, data curation, investigation L.d.V.V., G.X.P. and MB; formal analysis: L.d.V.V., visualization: L.d.V.V. and M.B., writing-original draft–review L.d.V.V. and G.X.P., editing L.d.V.V. and G.X.P., supervision G.X.P. and M.B., funding acquisition: G.X.P., Software: L.d.V.V. and M.B. All authors have read and agreed to the published version of the manuscript.

Funding: This work is a contribution to the research project Chronic Emergencies and Ecosocial Transformations in Touristified Coastal Spaces (PID2022-137648OB-C21). Funded by MCIN/AEI/10.13039/501100011033 and by "ERDF A way of making Europe".

Data Availability Statement: Public data in GOIB.

Conflicts of Interest: The authors declare not to have any conflicts of interest.

References

1. Boudouresque, C.F.; Meinesz, A. Deouverte de l'herbier de Posidonie. Cahier Parc nation. *Port-Cros. France* **1982**, *4*, 1–79.
2. San Félix, M. La Posidònia, el bosc submergit. Bri: Quaderns de Natura de les Balears. *Palma* **2000**, *77*.
3. Campagne, C.S.; Salles, J.-M.; Boissery, P.; Detera, J. The seagrass Posidonia oceanica: Ecosystem services identication and economic evaluation of goods and benets. *Mar. Pollut. Bull.* **2015**, *97*, 391–400. [CrossRef] [PubMed]
4. Boudouresque, C.F.; Pergent, G.; Pergent-Martini, C.; Ruitton, S.; Thibaut, T.; Verlaque, M. The necromass of the Posidonia oceanica seagrass meadow: Fate, role, ecosystem services and vulnerability. *Hydrobiologia* **2016**, *781*, 25–42. [CrossRef]
5. Béthoux, J.P.; Copin-Montégut, G. Biological fixation of atmospheric nitrogen in the Mediterranean Sea. *Limnol. Oceanogr.* **1986**, *31*, 1353–1358. [CrossRef]
6. Mateo, M.; Romero, J.; Pérez, M.; Littler, M.; Littler, D. Dynamics of Millenary Organic Deposits Resulting from the Growth of the Mediterranean Seagrass *Posidonia oceanica*. *Estuarine Coast. Shelf Sci.* **1997**, *44*, 103–110. [CrossRef]
7. Duarte, C.M.; Chiscano, C.L. Seagrass biomass and production: A reassessment. *Aquat. Bot.* **1999**, *65*, 159–174. [CrossRef]
8. Ballesteros, E. Algues bentòniques i fanerògames marines. In *Història Natural de l'Arxipelag de Cabrera. Monografies de la Societat d'Història Natural de les Balears*; Alcover, J.A., Ballesteros, E., Fornós, J.J., Eds.; SHNB: Palma, Spain, 1993; Volume 2, pp. 503–530.
9. Díaz del Río, V.; Somoza, L.; Goy, J.L.; Zazo, C.; Rey, J.; Hernández-Molina, F.J.; Mateu, G. *Mapa Fisiográfico de la Bahía de Palma*; Publicaciones Especiales del Instituto Español de Oceanografía: Madrid, Spain, 1993; Volume 16.
10. Delgado, O.; Ruiz, J.; Pérez, M.; Romero, J.; Ballesteros, E. Effects of fish farming on seagrass (*Posidonia oceanica*) in a Mediterranean bay: Seagrass decline after organic loading cessation. *Oceanol. Acta* **1999**, *22*, 109–117. [CrossRef]
11. Canals, M.; Ballesteros, E. Production of carbonate particles by phytobenthic communities on the Mallorca-Menorca shelf, northwestern Mediterranean Sea. *Deep. Sea Res. Part II Top. Stud. Oceanogr.* **1997**, *44*, 611–629. [CrossRef]

12. Mateu, G.; Viñals, M.J.; Moreiro, M.; Nadal, G. La transgresión flandriense a través de los foraminíferos bentónicos del Mar Balear. In *El Canvi Climàtic: Passat, Present i Future. Monografies de la Societat d'Història Natural de les Balears*; Pons, G.X., Guijarro, J.A., Eds.; SHNB: Palma, Spain, 2001; Volume 9, pp. 13–31.
13. Duarte, C.M. The future of seagrass meadows. *Environ. Conserv.* **2002**, *29*, 192–206. [CrossRef]
14. Marbà, N.; Duarte, C.M.; Holmer, M.; Martínez, R.; Basterretxea, G.; Orfila, A.; Jordi, A.; Tintoré, J. Assessing the effec-tiveness of protection on Posidonia oceanica populations in the Cabrera National Park (Spain). *Environ. Conserv.* **2002**, *29*, 509–518. [CrossRef]
15. Duarte, C.M.; Álvarez, E.; Amengual, J.; Barrón, C.; Basterretxea, G.; Calleja, M.; Deudero, S.; Díaz-Almela, E.; Grau, A.; Massutí, C.; et al. Cap a una millor comprensió de l' estat, ecologia i conservació de les praderies d' angiospermes marines (*Posidonia oceanica* L. Delile) de les Illes Balears/Towards improved understanding of the status, ecology and conservation seagrass (*Posidonia oceanica* L. Delile) meadows in the Balearic Islands. *Boll. Soc. Hist. Nat. Balears* **2003**, *46*, 9–19.
16. European Council. *Council Directive 92/43/EEC of 21 May 1992 on the Conservation of Natural Habitats and of Wild Fauna and Flora*; DOCE n: 206, de 22 de Julio de 1992, 44 pp. DOUE–L-1992-81200; European Council: Brussels, Belgium, 1992.
17. European Council. *Decree 25/2018, of 27 July, on the Conservation of the Posidonia Oceanica in the Balearic Islands*; BOIB n:93 de 28 de Julio de 2018; European Council: Brussels, Belgium, 2018; pp. 25994–26042.
18. Inteview Manu San Félix. Available online: https://www.nationalgeographic.com.es/naturaleza/salvemos-mediterraneo-posidonia-planta-rescate_14324 (accessed on 9 December 2023).
19. Julià, M.; del Valle, L.; Bagur, M.; Marsinyach, E.; Pons, G.X.; Carreras, D. *Cartografía de los Hábitats Marinos de las Islas Baleares: Compilación de Capas y Comunidades Bentónicas*; Observatorio Socioambiental de Menorca (Institut Menorquí d'Estudis): Palma, Spain; Societat d'Història Natural de les Balears: Palma, Spain; Fundació Marilles: Barcelona, Spain, 2019; Volume 157, Available online: https://docs.google.com/file/d/1EhquRA6eD0bKJSeM2jVImuAX1JRHwhus/preview?rm=minimal (accessed on 9 December 2023).
20. Pons, G.X.; del Valle, L. Bionomic Cartography of the Balearic Sea: Evolution of Coastal Marine Habitats of Mallorca. *J. Coast. Res.* **2020**, *95*, 153–157. [CrossRef]
21. European Council. *Decree 75/2005, of 8 July, by which the Balearic Catalog is created Endangered Species and Special Protection, the areas Critical Biology and the Fauna and Flora Advisory Council of the Balearic Islands*; BOIB n. 106, 16 July 2005; European Council: Brussels, Belgium, 2005; p. 11.
22. Government of Spain. *Royal Decree 139/2011, of February 4, for the Development of the List of Wild Species under Special Protection Regime and the Spanish Catalog of Endangered Species*; BOU num. 46, de 4 February 2011, BOE A 2011/3582; Government of Spain: Madrid, Spain, 2011; pp. 20912–90951.
23. Council of Europe. *Convention on the Conservation of European Wildlife and Natural Heritage*; Council of Europe: Bern, Switzerland, 1979; Available online: http://conventions.coe.int/Treaty/EN/Treaties/Html/104.htm (accessed on 9 December 2023).
24. Barcelona Convention—Marine—Environment—European Commission. Available online: https://environment.ec.europa.eu/topics/marine-environment (accessed on 9 December 2023).
25. Telesca, L.; Belluscio, A.; Criscoli, A.; Ardizzone, G.; Apostolaki, E.T.; Fraschetti, S.; Gristina, M.; Knittweis, L.; Martin, C.S.; Pergent, G. Seagrass meadows (*Posidonia oceanica*) distribution and trajectories of change. *Sci. Rep.* **2015**, *5*, 12505. [CrossRef] [PubMed]
26. Gobert, S.; Cambridge, M.T.; Velimirov, B.; Pergent, G.; Lepoint, G.; Bouquegneau, J.M.; Dauby, P.; Pergent-Martini, C.; Walker, D.I. Biology of Posidonia. In *Seagrasses: Biology, Ecology and Conservation*; Larkum, W.D., Orth, R.J., Duarte, C.M., Eds.; Springer: Berlin, Germany, 2006; pp. 387–408.
27. Tragsatec. Atlas Submarí: Cartografía de Posidonia Oceanica a les Illes Balears. In *Informe Descriptivo de los Trabajos y Relación Valorada (Certificación Final)*; Ministerio de Medio Ambiente, y Medio Rural y Marino: Madrid, Spain, 2022; p. 90.
28. Díaz, E.; Marbà, N. *Bases Ecológicas Preliminares para la Conservación de los Tipos de Hábitat de Interés*; 1120 *Posidonion oceanicae*, Praderas de *Posidonia oceanica*; Comunitario en España—Ministerio de Medio Ambiente, y Medio Rural y Marino: Madrid, Spain, 2009; p. 129.
29. Dekker, A.; Brando, V.; Anstee, J.; Fyfe, S.; Malthus, T.; Karpouzli, E. Remote Sensing of Seagrass Ecosystems: Use of Spaceborne and Airborne Sensors. In *Seagrasses: Biology, Ecology and Conservation*; Larkum, W.D., Orth, R.J., Duarte, C.M., Eds.; Springer: Berlin, Germany, 2006; pp. 347–359.
30. Borfecchia, F.; Micheli, C.; Carli, F.; De Martis, S.C.; Gnisci, V.; Piermattei, V.; Belmonte, A.; De Cecco, L.; Martini, S.; Marcelli, M. Mapping Spatial Patterns of *Posidonia oceanica* Meadows by Means of Daedalus ATM Airborne Sensor in the Coastal Area of Civitavecchia (Central Tyrrhenian Sea, Italy). *Remote Sens.* **2013**, *5*, 4877–4899. [CrossRef]
31. Pasqualini, V.; Pergent-Martini, C.; Pergent, G.; Agreil, M.; Skoufas, G.; Sourbes, L.; Tsirika, A. Use of SPOT 5 for mapping seagrasses: An application to Posidonia oceanica. *Remote Sens. Environ.* **2005**, *94*, 39–45. [CrossRef]
32. Fornes, A.; Basterretxea, G.; Orfila, A.; Jordi, A.; Alvarez, A.; Tintore, J. Mapping Posidonia oceanica from IKONOS. *ISPRS J. Photogramm. Remote Sens.* **2006**, *60*, 315–322. [CrossRef]
33. Dimas, X.; Fakiris, E.; Christodoulou, D.; Georgiou, N.; Geraga, M.; Papathanasiou, V.; Orfanidis, S.; Kotomatas, S.; Papatheodorou, G. Marine priority habitat mapping in a Mediterranean conservation area (Gyaros, South Aegean) through multi-platform marine remote sensing techniques. *Front. Mar. Sci.* **2022**, *9*, 953462. [CrossRef]

34. Templado, J.; Ballesteros, E.; Galparsoro, I.; Borja, A.; Serrano, A.; Martín, L.; Brito, A. Inventario español de Hábitats y especies marinos. In *Guía Interpretativa: Inventario Español de Hábitats Marinos*; Gobierno de España; Ministerio de Agricultura, Alimentación y Medio Ambiente: Madrid, Spain, 2012.
35. Ballesteros, E.; Cebrián, E. Llistat preliminar dels hàbitats marins bentònics a les Illes Balears amb alguns comentaris des de la perspectiva de la conservació. In *Llibre Verd de Protecció D'espècies a Les Balears*; Govern de les Illes Balears: Madrid, Spain, 2015; pp. 93–110.
36. Ribera, G.; Coloreu, M.; Rodríguez-Prieto, C.; Ballesteros, E. Phytobenthic Assemblages of Addaia Bay (Menorca, Western Mediterranean): Composition and Distribution. *Bot. Mar.* **1997**, *40*, 523–532. [CrossRef]
37. Ballesteros, E.; Cebrián, E. *Estudi Sobre la Bionomia Bentònica, Biodiversitat i Cartografia de la Reserva dels Freus Entre Formentera i Eivissa*; Informe Final—I; Centre d'Estudis Avançats de Blanes (CEAB-CSIC): Blanes, Spain, 2005; p. 109.
38. Sales, M.; Garcia-Rubies, A.; Cebrian, E.; Ballesteros, E. *Estudi Sobre el Fons Marí de l'illa de l'Aire*; Centre d'Estudis Avançats de Blanes i Conselleria de Medi Ambient del Govern de les Illes Balears: Blanes, Spain, 2004; p. 133.
39. Ballesteros, E. Els herbeis de Caulerpa prolifera (Forsskal) Lamouroux de la badia de Pollença (Mallorca, Mediterrània Occidental). *Boll. Soc. Hist Nat. Balears* **1990**, *33*, 99–116.
40. Ballesteros, E. Els fons rocosos profunds amb *Osmundaria volubilis* (Linné) R.E. Norris a les Balears. *Boll. Soc. Hist Nat. Balears* **1992**, *35*, 33–50.
41. Ballesteros, E. Els fons marins de l'illa de Menorca: Bionomia, estat general de conservació, interès i zones a protegir. In *Jornades Sobre Conservació i Desenvolupament a Menorca*; Vidal, J.M., Rita, J., Eds.; Unesco: London, UK, 1992; pp. 137–141.
42. Ballesteros, E.; López, P.; Garcia, A.; Linares, C.; Cebrián, E. *Els Fons Marins d'Es Vedrà i Es Vedranell: Espècies, Comunitats i Cartografia Bionòmica*; Centre d'Estudis Avançats de Blanes (CEAB-CSIC): Blanes, Spain, 2007.
43. Emaya. *Documento Ambiental del Proyecto de Retirada del Emisario Submarino del Torrent Gros (T.M. Palma de Mallorca, Illes Balears)*; Govern De Les Illes Balears Comunitat Autonoma De Les Illes Balears: Palma, Spain, 2022; p. 67.
44. Emaya. *Proyecto de Retirada del Emisario Submarino del Torrent Gros (Mallorca)*; Emaya: Palma, Spain, 2022; p. 283.
45. Coll, M.À.; Feliu de la Peña, J.F.; Llabrés, N.; Romera, P.; Rullan, E. Evolución de la línea de costa de la playa de Palma-Arenal (1956–2004). *Territoris* **2008**, *7*, 193–202.
46. Ruiz Fernández, J.M.; Bernardeau Esteller, J.; Huete Stauffer, T.; García Muñoz, R.; Marín Guirao, L.; Sandoval Gil, J.M.; Gavilán Alonso, J. *Informe Anual de la Red de Seguimiento Posidonia Oceanica de la Región de Murcia (2004–2009)*; Instituto Español de Oceanografía, Centro Oceanográfico de Murcia: Murcia, Spain, 2009; p. 127.

remote sensing

MDPI

Article

Assessing Shallow Soft Deposits through Near-Surface Geophysics and UAV-SfM: Application in Pocket Beaches Environments

Luciano Galone [1], Sebastiano D'Amico [1,*], Emanuele Colica [1,2], Peter Iregbeyen [1], Pauline Galea [1], Lluís Rivero [3] and Fabio Villani [4]

1 Department of Geosciences, University of Malta, MSD2080 Msida, Malta; luciano.galone@um.edu.mt (L.G.); emanuele.colica@um.edu.mt (E.C.); peter.iregbeyen@um.edu.mt (P.I.); pauline.galea@um.edu.mt (P.G.)
2 Research & Planning Unit, Public Works Department, Ministry for Public Works and Planning, Project House, Triq Francesco Buonamici, FRN1700 Floriana, Malta
3 Department of Mineralogy, Petrology and Applied Geology, Earth Sciences Faculty, University of Barcelona, 08028 Barcelona, Spain; lrivero@ub.edu
4 Istituto Nazionale di Geofisica e Vulcanologia, 00143 Rome, Italy; fabio.villani@ingv.it
* Correspondence: sebastiano.damico@um.edu.mt

Abstract: This study employs a multimethod approach to investigate the sediment distribution in two pocket beaches, Ramla Beach and Mellieha S Beach, in Malta. Both study sites were digitally reconstructed using unmanned aerial vehicle (UAV) photogrammetry. For each case, an ERT and a dense network of ambient seismic noise measurements processed through a horizontal-to-vertical spectral ratio (HVSR) technique were acquired. Electrical resistivity tomography (ERT) analysis enables the estimation of sediment thickness in each beach. HVSR analysis revealed peaks related to beach sediments overlying limestone rocks in both sites and also indicated a deeper stratigraphic contact in Mellieha S Beach. Based on ERT measurements, sediment thickness is calculated for each HVSR measurement. Interpolation of results allows for bedrock surface modelling in each case study, and when combined with digital terrain models (DTMs) derived from photogrammetric models, sediment volumes are estimated for each site. The geometry of this surface is analyzed from a geological perspective, showing structural control of sediment distribution due to a normal fault in Mellieha S Beach and stratigraphic control facilitated by a highly erodible surface in Ramla Beach. The results emphasize the importance of adopting a three-dimensional perspective in coastal studies for precise sediment volume characterization and a deeper understanding of pocket beach dynamics. This practical multimethod approach presented here offers valuable tools for future coastal research and effective coastal management, facilitating informed decision making amidst the growing vulnerability of coastal zones to climate change impacts.

Keywords: horizontal-to-vertical spectral ratio; seismic ambient noise; pocket beach; Malta; near-surface geophysics; electrical resistivity tomography; photogrammetry

Citation: Galone, L.; D'Amico, S.; Colica, E.; Iregbeyen, P.; Galea, P.; Rivero, L.; Villani, F. Assessing Shallow Soft Deposits through Near-Surface Geophysics and UAV-SfM: Application in Pocket Beaches Environments. *Remote Sens.* **2024**, *16*, 40. https://doi.org/10.3390/rs16010040

Academic Editor: José Juan de Sanjosé Blasco

Received: 30 October 2023
Revised: 14 December 2023
Accepted: 18 December 2023
Published: 21 December 2023

1. Introduction

Estimating the thickness of shallow, non-consolidated sedimentary deposits is a common challenge in geology, engineering, and environmental sciences in general. Particularly, understanding sediment thickness in beaches is crucial for scientific research, modelling of erosional and depositional processes, and informed decision making for coastal protection infrastructure. Accurate estimation of sediment volumes in coastal areas is important for a comprehensive understanding of littoral processes, sediment budget studies and geomorphological evolution [1]. However, in these studies, the lower boundary of beach sedimentary deposits is rarely estimated (e.g., [2]) due to the difficulty of data acquisition.

While coring or trenching offers direct measurements of sediment thickness, these methods can be costly, labor-intensive, limited in spatial coverage, and may even, in some cases, be restricted due to ecological or normative reasons that hamper their acquisition.

To address these limitations, geophysical approaches have emerged as promising methods [3]. Nevertheless, the non-uniqueness of results in the inversion process, inherent to the geophysical method, is a limitation to determining the exact sediment thickness solely based on one technique. To ensure the reliability of the results, calibration and validation of sediment thickness estimates by using independent approaches, such as geological data or other geophysical techniques, are necessary.

In this study, we aim to reconstruct the 3D sediment distribution in a dynamic environment by combining surface and subsurface information, utilizing horizontal-to-vertical spectral ratio (HVSR) of ambient seismic noise, supported by electrical resistivity tomography (ERT) and structure from motion (SfM) photogrammetry from unmanned aerial systems (UAV).

The HVSR [4] method is a passive seismic technique that utilizes a three-component seismometer to measure the vertical and horizontal components of ambient seismic noise. The noise from approximately 0.1 to 1 Hz is generated by sources such as ocean waves, extensive regional storms and tectonic activity, while frequencies above 1 Hz typically originate from closer sources, including local storms, wind and human-made activities [5]. HVSR has been widely applied in sediment thickness estimation-related studies in different environments due to its simplicity in field acquisition and processing [6–9], including some applications on beaches [3,10,11]. Its complementarity with ERT, which utilizes active electrical input to model the sub-surface resistivities in 2D or 3D, has been highlighted in several works [12–18]. However, although ERT has also been applied in coastal studies (e.g., [19]), there seems to be a limited number of examples of combined applications of HVSR and ERT in sandy beach environments.

In the last decade, SfM- UAV photogrammetry has emerged as an essential technique in medium-scale geological and geomorphological investigations [20–22] due to the possibility of obtaining high-resolution 3D models, orthomosaics and digital elevation models, complemented by its versatility, cost-effectiveness and rapid data acquisition times.

We selected two pocket beaches in Malta as case studies. Pocket beaches are a common and important feature in the Maltese archipelago, with great ecological and economic significance to the country. A pocket beach is a beach, usually of small dimensions, which is isolated between two headlands [23]. Generally, these are semi-restricted systems, as there is little connection between pocket beaches and headlands tend to regulate and store sediment transport within the beach [24]. However, even if their headlands provide them with some protection, they may be extremely sensitive to low-frequency, high-energy storm events [25] that repeatedly hit the Maltese coasts.

While some works have been carried out concerning coastal studies on the Maltese sandy beaches [26–32], there is a lack of studies that consider a three-dimensional perspective including volumetric computation. In a context where the coastal zone will be among the environments worst affected by projected climate change [33], a precise characterization of those environments is essential to better manage these extremely fragile natural resources.

2. Area of Study

2.1. The Maltese Archipelago

The Maltese archipelago is located in the central Mediterranean area, approximately 100 km south of Sicily. It consists of three main islands: Malta, Comino and Gozo (Figure 1). These islands represent the emerged portion of a wide shallow-water platform. The geology of Malta [34–36] is characterized by a sedimentary sequence comprising four major geological formations (Figure 1). Starting from the oldest formation (Fm), the Lower Coralline Limestone Fm (LCL) is a hard and compact pale grey limestone of the Oligocene age. Globigerina Limestone Fm (GL) is a fine-grained yellowish limestone of the Aquitanian–Langhian age (Lower Miocene). The Blue Clay Fm (BC), consisting of bluish or greyish pelagic clay and limestone, belongs to the Serravallian age (Middle Miocene). Lastly, the Upper Coralline Limestone Fm (UCL) is a coarse-grained pale grey and orange limestone, of the Tortonian–Messinian age (Upper Miocene). Generally, the southeastern part of the island of Malta is shaped by outcrops of LCL and GL, resulting in relatively flat to gently undulating landscapes. The northwestern part of Malta, as well as Comino and Gozo, generally display the complete stratigraphic sequence, where a NE–SW trending horst and graben structure controls flat-topped hills and valleys. Additionally, the islands and the geological sequence have a gentle overall tilt towards the northeast (<10°). These characteristics are the main controlling factors of coastal morphologies: high and sub-vertical cliffs dominate the west coast, while the east side features a smoother coastline. Pocket beaches and embayed areas interrupt the rocky coastline and are typically associated with the NE–SW trending valleys, especially in the northern part of the island of Malta.

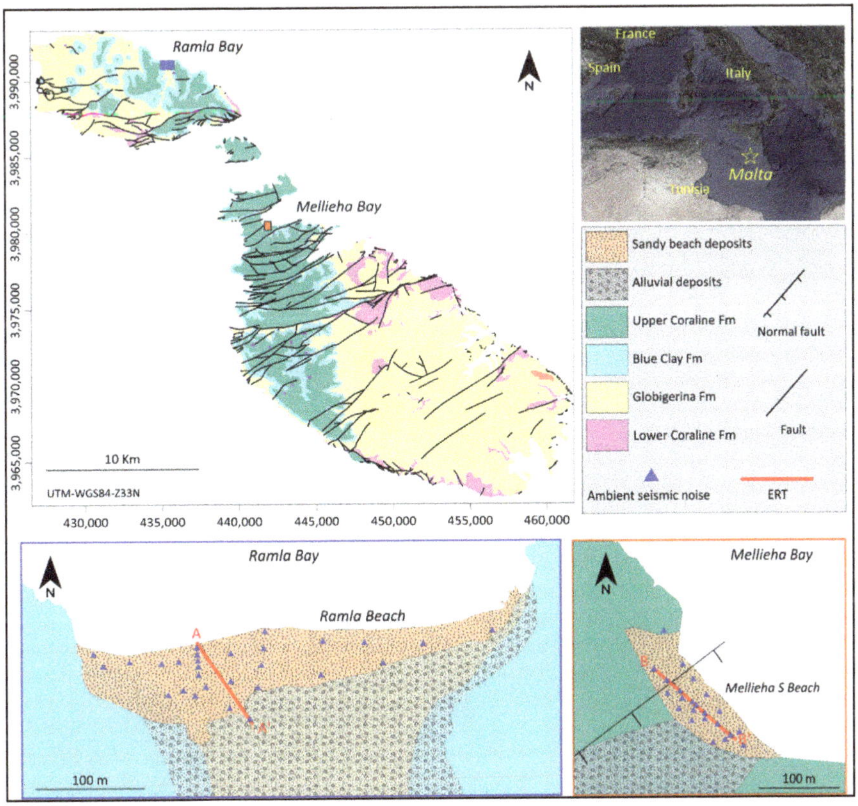

Figure 1. Geological map of the Maltese archipelago. The top right inset shows its position in the Mediterranean Sea. The study areas (Ramla Beach and Mellieha S Beach) are shown in the bottom images with the geophysical measurements presented in this work.

Regarding the recent sedimentary deposits, they are scarce and very localized. Some valleys contain a thin layer of alluvial deposits, typically exhibiting reddish colorations due to iron oxides formed during long-lasting periods of subaerial weathering. The beach deposits represent less than 2.5% of the Maltese coast [37]. These beaches are deposits of gravel and sand, which are predominantly composed of carbonates and subordinately by silts, and originate mainly from the erosion of coastal and shelf Cenozoic limestones and clays and deposited during the Holocene [26,27,30].

2.2. Case Studies

We present two case studies on two iconic beaches of Malta, the one located in Ramla Bay and one on the southern side of Mellieha Bay (Figure 1).

Ramla Bay is located in the northeast of the island of Gozo, at the mouth of a river valley that currently hosts intermittent streams that are functional during storms and runs approximately from south to north. The distance between the headlands that border the bay is approximately 1000 m while the length of the Ramla Beach coastline is nearly 400 m. It is the best-known beach in Gozo, known for its golden-reddish sand, which gives it its name in Maltese (Ramla il-Ħamra) and distinguishes it as unique in the Maltese Islands. The beauty and fragility of this bay fully justify its detailed study.

The slopes that delimit the bay are composed of the BC Fm, and at their highest point, they are crowned by the UCL Fm. Behind the beach deposits, the bay features vegetated dunes and the development of a wetland area, providing it with high ecological importance. The bay holds archaeological and historical interest [38,39] as it houses Roman remains covered by dune sand, a submerged seawall, structures built by the Knights of the Order of St. John in the mid-18th century and the famous Calypso Cave on the western slope.

Mellieha Bay, also known as Għadira bay, is the largest bay in the archipelago, with a width of 1.75 km between the headlands. It contains beach deposits stretching for approximately 850 m. It is a popular beach that is easily accessible from a coastal road that runs along its inland boundary.

This bay represents the northeastern outlet of the Mellieha Valley, which is carved mainly on UCL Fm rocks and has a thin layer of recent sedimentary deposits in its central portion. The valley is bounded on its edges by normal faults forming a graben-like structure. In the centre of the valley, there is a topographic high that crosses it longitudinally, dividing it into two parts. This small hill has higher altitudes at its western boundary; however, it acts as a divide along the entire length of the valley, leading into two small beaches at Mellieha Bay: the northern beach, approximately 500 m long, and the southern beach, about 140 m long, which is the focus of our study and what we call Mellieha S Beach.

3. Materials and Methods

The methodological process employed in this study covers both the surface and subsurface domains of the study sites (Figure 2). To examine the surface of the case studies, field observations were complemented by the generation of 3D photogrammetric models. These models provided orthomosaics and DTMs, allowing for a detailed characterization of the surface. In the subsurface analysis, two near-surface geophysical techniques were employed. Using ambient seismic noise and the HVSR technique, sediment thickness was estimated at different locations along the beaches. These results were interpolated to obtain a continuous interface estimate. Independently obtained ERTs were employed to construct geological sections, providing essential support for determining the average shear wave velocity (Vs) used in thickness estimates derived from the HVSR data. The combination of the surface and subsurface results was ultimately used to assess the sedimentary deposits distribution on the beaches.

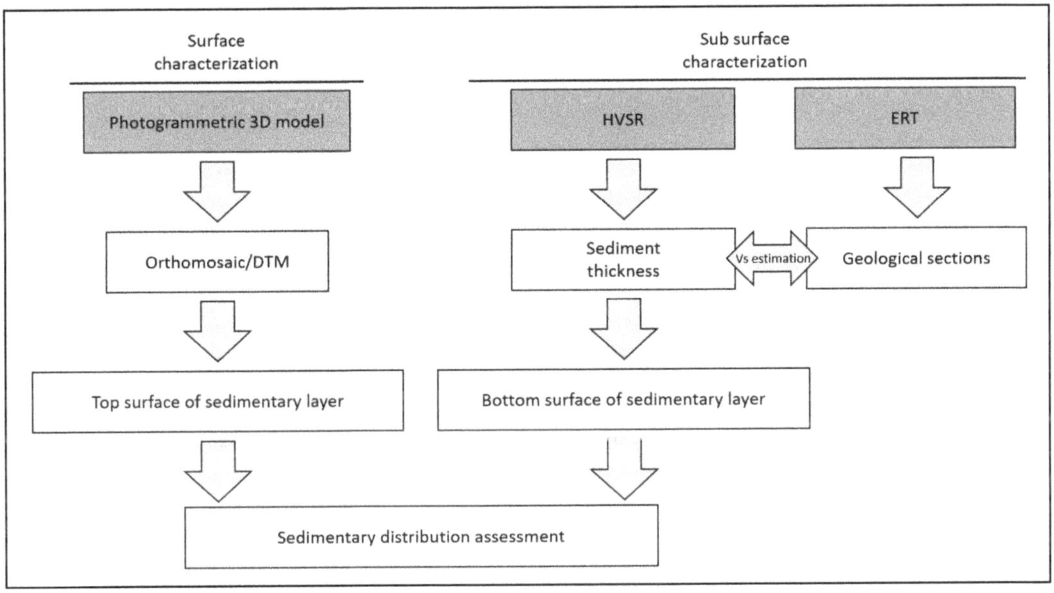

Figure 2. Methodological approach flowchart.

3.1. *SfM Photogrammetric Survey*

Ramla Bay and Mellieha Bay were surveyed using a Phantom 4 Pro UAV in October 2019, employing a pre-programmed flight function that ensured a constant flight altitude and correct overlap between consecutive images (details in Table 1). The acquired images were processed using Agisoft Metashape v. 2.0.0, a digital photogrammetry software that integrates computer vision algorithms such as SfM. This system allows for the estimation of the 3D position of points represented in multiple images, reconstructing the geometry of the object and the camera position, even if internal orientation parameters are not defined. The photogrammetric processing followed a typical workflow involving a series of consecutive steps [20,40].

Table 1. Details of the photogrammetric data acquisition.

Site	Mellieha Bay	Ramla Bay
Date	October 2019	October 2019
Processed images	573	892
Flight altitude	70 m	75 m
Frontal overlap	80%	80%
Lateral overlap	75%	75%
GCPs	21	35

The first step in the photogrammetric workflow was importing the images into the software, avoiding the defective pictures (e.g., out of focus, overexposed). The subsequent step, known as camera alignment, automatically oriented the images in space. This was followed by the creation of a sparse point cloud, forming a 3D point cloud with scattered points. In the next phase, the software used the camera locations and the sparse point cloud to build a dense point cloud. From this dense point cloud, a continuous surface composed of polygons, with the vertices representing the points in the dense cloud, was reconstructed. This step is referred to as mesh reconstruction, and upon applying texture to this model, a textured 3D model was obtained.

The 3D models were accurately scaled and georeferenced using ground control points (GCPs) that were positioned with high precision using a differential Global Navigation Satellite System (GNSS) before image acquisition. Finally, orthomosaics and digital terrain models (DTMs) were extracted from the 3D model.

3.2. Electrical Resistivity Tomography

Two ERT surveys were conducted on the beach deposits (Figure 1). One was carried out at Ramla Beach (May 2020), perpendicular to the coastline and crossing a dune, using 64 electrodes spaced at 1.5 m, with a total array length of 94.5 m, while the other ERT was performed at Mellieha S Beach (February 2022), parallel to the shoreline, using 64 electrodes and 2 m spacing, reaching 126 m of length. The choice of ERT direction was intended to capture maximum geological variability. In the case of Ramla, the ERT was acquired including shoreline beach deposits and dune deposits. At Mellieha S Beach, the ERT covered both sandy beach deposits and an outcrop of the UCL fm. The instrument used was a multichannel digital resistivity meter ELECTRA (Moho srl) featuring a waveform D/A converter with continuous current and voltage control, including feedback. The applied current during measurements was ±10 mA. The Wenner alpha array was employed for both cases [41]. This multi-electrode array is better suited for recovering vertical resistivity contrasts, and thus it is ideal in geological settings with stratified structure and gently dipping bedding, as in our case (sub-horizontal sands overlying a shallow basement).

To assess the quality of electrode–substrate coupling, impedance measurements were performed for each profile. High impedance values (>3000 Ohm) were found for some electrodes in very-dry sandy areas. To improve the electrode–soil contact, the soil on such electrodes was moistened with a saline water solution. To ensure accurate positioning of the electrodes, GNSS equipment was employed. Topographic information was incorporated into the ERT data to calculate apparent resistivity. To convert the pseudo section of apparent resistivity derived from the field data into a 2D model of real resistivity, we employed the Res2DInv software v. 4.08. We employed a robust inversion that usually best resolves sharp conductivity boundaries [42–44]. The absolute error at Ramla beach after 5 iterations was 21.5%, and the absolute error at Mellieha beach after 4 iterations was 17.3%. Despite these relatively high errors, the models developed demonstrated a robust correspondence with the local geology. In complex geological environments, electrical images tend to experience errors over 15% and, in many cases, reducing these errors without introducing artefacts is not possible [45,46].

3.3. Ambient Seismic Noise Measurements

We acquired 28 ambient seismic noise measurements at Ramla Beach and 23 at Mellieha S Beach (Figure 1). We employed 3 triaxial accelerometer devices Tromino to record ambient seismic noise continuously for 16–20 min at each station, using a sampling rate of 128 Hz. Most of the measurements were performed on sandy beach sediments. To obtain a good device–soil coupling and minimize wind effects, the devices were deployed in holes of 20 cm depth where the sand was mechanically compacted. At each beach, one measurement was taken on rock outcrops. For Ramla Beach, in the GL Fm outcrop located at the western limit of the beach, and for Mellieha S Beach, in the northwestern beach limit, on a UCL Fm outcrop (Figure 1).

The acquired ambient seismic noise signals were processed using the HVSR technique [4]. The processing of the ambient seismic noise data followed a standard workflow [47] in the software Grilla v. 9.7.2. First, we divided the signal into 20 s time windows and selected the most stationary ones by applying an anti-triggering algorithm to avoid transient noise during the analysis. Next, we computed and smoothed the Fourier amplitude spectra for each time window. The two horizontal components were then averaged using a quadratic mean to obtain a representative value. Subsequently, the HVSR function was computed for each window by dividing the amplitude of the horizontal component average by the vertical component. Finally, the average HVSR function was calculated by averaging the results obtained from all the selected time windows.

The HVSR curves can exhibit peaks at different frequencies. The fundamental premise for interpreting them is to consider that the substrate can be described as a soft sedimentary layer with low shear wave velocity (Vs) lying over a higher Vs bed-rock. The frequency at which the dominant peak occurs is generally related to the sediment thickness and its Vs. Usually, peaks at higher or lower frequencies of the curve correspond to shallower or deeper subsoil features, respectively.

4. Results

4.1. Photogrammetric Analysis

The SFM analysis conducted at Ramla Bay and Mellieha Bay produced accurate results. The DEMs presented resolutions of 5.66 cm/pixel and 5.9 cm/pixel and the orthophotos a resolution of 1.42 cm/pixel and 1.48 cm/pixel, respectively, and point clouds with an RMS reprojection error of 0.92 and 0.56 pixels. More details are summarized in Table 2.

Table 2. Details of the SfM photogrammetric process.

	Ramla Bay	**Mellieha Bay**
Point Cloud		
Points	638,628 of 745,883	221,216 of 311,141
RMS reprojection error	0.178743 (0.928461 pix)	0.214614 (0.567595 pix)
Max reprojection error	3.86717 (47.1645 pix)	2.86192 (28.6239 pix)
Mean key point size	4.11674 pix	2.19475 pix
Point colors	3 bands, uint8	3 bands, uint8
Average tie point multiplicity	4.39553	7.87079
Dense Point Cloud		
Points	81,101,105	39,735,110
Point colors	3 bands, uint8	3 bands, uint8
Model		
Faces	4,937,852	2,389,756
Vertices	2,531,384	1,229,757
Vertex colors	3 bands, uint8	3 bands, uint8
Texture	8096 × 8096, 4 bands, uint8	9096 × 9096, 4 bands, uint8
Source data	Dense cloud	Dense cloud
DEM		
Size	29,338 × 22,523; 5.66 cm/px	23,579 × 27,507; 5.9 cm/px
Source data	Dense cloud	Dense cloud
Orthomosaic		
Size	94,923 × 54,443; 1.42 cm/px	52,139 × 64,799; 1.48 cm/px
Colors	3 bands, uint8	3 bands, uint8

In Ramla Bay, natural geomorphological features were found to be better preserved due to lower anthropogenic influence. The beach at this site is partially enclosed towards the valley and is limited by a system of fixed dunes covered with vegetation. Small marshy areas develop among these dunes, creating a marshland environment, and some agricultural activity is observed in the area. Steep rocky formations limit the sides of the beach. Towards the western boundary, the beach is delimited by the outcrop of the BC Fm, and at its bottom the transitional contact with the GL Fm has a slight sub horizontal dip to the northeast. GL Fm., with greyish tones in this sector, acts as the bedrock for the beach sediments. As for the eastern boundary of the beach, it is composed of slope deposits and outcrops of the BC Fm. The stratigraphic boundary with the GL Fm is not exposed, presumably due to burial by beach sediments (Figure 3a). The area of sand accumulation was determined to be 14,062 m^2, with 18,344 m^3 of sand above sea level.

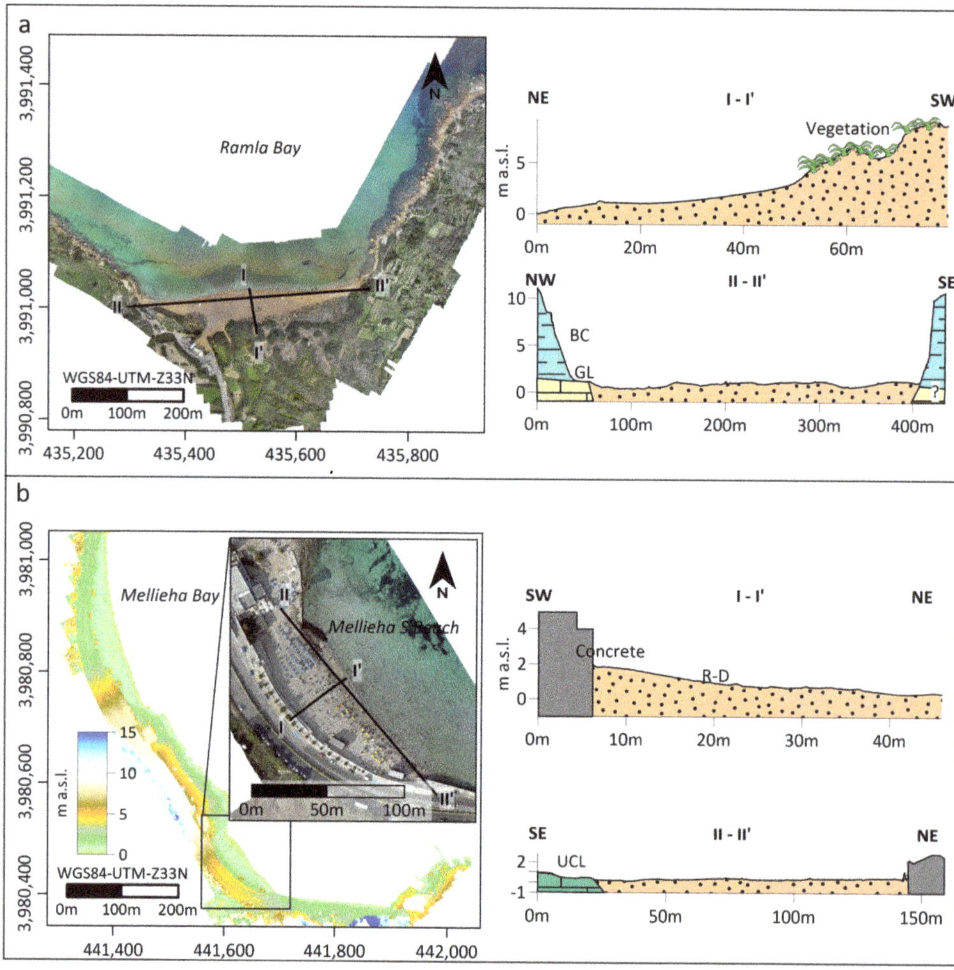

Figure 3. Surface characterization. (**a**) Orthomosaic of Ramla Bay and position of profiles presented on the right: I-I′ orthogonal to the coastline, II-II′ parallel to the coastline. (**b**) DTM of Mellieha Bay and ortho-mosaic of Mellieha S Beach showing the position of the profiles presented: I-I′, orthogonal to the coastline and II-II′, parallel to the coastline. Main surface geological features are labelled and characterized: R-D: recent deposits (mainly sand); UCL: Upper Coralline Limestone Fm; BC: Blue Clay Fm; GL: Globigerina Limestone Fm.

The southern beach of Mellieha Bay, which was the focus of this study, is constrained by a small topographic high to the northwest. This topographic high is composed of UCL Fm and is accompanied by a building that acts as a barrier for sediment in that direction (Figure 3b). Additionally, the beach forms a semi-circular enclosure towards the southwest and southeast and is surrounded by a concrete barrier where a coastal road is located. This configuration traps beach sediments, preventing their movement towards the mainland. The beach lacks dunes or other significant accumulation or erosion landforms, and typically experiences high levels of recreational use. The area covered by beach deposits was estimated to be 4717 m^2, while the volume of sediments above sea level amounts to 4099 m^3.

Regarding the nature of the deposits, both beaches are predominantly composed of sand-sized sediments. However, Ramla Bay stands out for the notable presence of limestone clasts with decimeter dimensions scattered throughout the beach, within a matrix of sandy sediments. These clasts are likely derived from the UCL, which caps the BC Fm and undergoes significant slope erosion processes, especially in the headlands, as described in previous studies [20,40]. These clasts would be transported to the bay by wave action and littoral currents and then deposited onto the beach during storm events.

4.2. ERT Results

In Ramla Beach, ERT A-A′ runs orthogonally to the shoreline (Figure 1). The resistivity model (Figure 4) revealed maximum resistivity values exceeding 3500 Ohm·m in the elevated area corresponding to partially vegetated sandy deposits of a dune. These values were interpreted and supported by field observations, as indicative of unsaturated sand with low moisture content. Beneath this first resistive layer, a continuous layer was observed throughout the entire section, characterized by significantly lower resistivity values of approximately 200 Ohm·m (Figure 4). Within this layer, the minimum resistivity values were found towards the NE limit of the section where the sea is located. This zone of low resistivity was interpreted as saturated beach deposits, where the decrease in resistivity was attributed to an increase in pore water salinity due to the influence of seawater.

At depths close to 4 m b.s.l. and extending to the bottom of the profile, a third domain of resistivities begins on a sub horizontal horizon. This domain exhibits relatively high values that gradually increase with depth, ranging between 500 and 1000 Ohm·m. This domain was interpreted as the limestones of the GL Fm, the rock underlying the sedimentary deposits of the beach. It is worth noting that its surface exhibits a slope towards the sea, likely associated with the marine platform, and a counter-slope inland, probably linked to the archaeological site present in the area [38,39].

The ERT profile B-B′ runs approximately parallel to the shoreline of Mellieha S Beach (Figure 1). The model resistivity section exhibits consistently low resistivity values (mostly below 20 Ohm·m) throughout its extent, but a series of resistivity domains separated by high resistivity gradients can be observed (Figure 3).

From x = 30 to the southeastern limit of the section, the shallowest domain showed medium (around 7 Ohm·m) resistivity values. It extends horizontally with a thickness below one meter, from the surface down to elevation 0, and corresponds to the uppermost layer of unsaturated sand. Below this domain, a zone of very-low resistivity values (below 2 Ohm·m) develops, which we interpret as saturated sedimentary deposits containing saline water with high conductivity. This geoelectric domain exhibits a thinning geometry towards the southeast, with maximum thicknesses around x = 40 of about 5.5–7.5 m. Below this domain and towards the bottom of the profile, a third zone with slightly higher resistivity values was interpreted as the UCL Fm. The anomalously low resistivity values for limestone can be associated with a high degree of chemical weathering of the rock, resulting in a greater percentage of clay minerals and an increase in secondary porosity filled with saline water from the sea.

The overall sedimentary body geometry, combined with prior geological information, supports the interpretation of a normal fault zone characterized by a heterogeneous resistivity values zone around x = 20, where the hanging wall block is located in the southeast and hosts the sedimentary deposits.

In general, the results obtained from ERT measurements for beach deposits align with findings from other studies in similar conditions (e.g., [48]), with saline water-saturated sediments typically exhibiting resistivity values below 3 Ohm·m, while dry sand tends to display resistivity values above 1000 Ohm·m. The low resistivity values of the UCL

Fm under conditions of high chemical weathering agree with an ERT survey conducted on a dissected UCL Fm outcrop (Figure 5) at the northwestern boundary of the Mellieha Valley (Figure 1), and contrast with high UCL Fm values obtained in previous studies on the archipelago, although in different conditions [40].

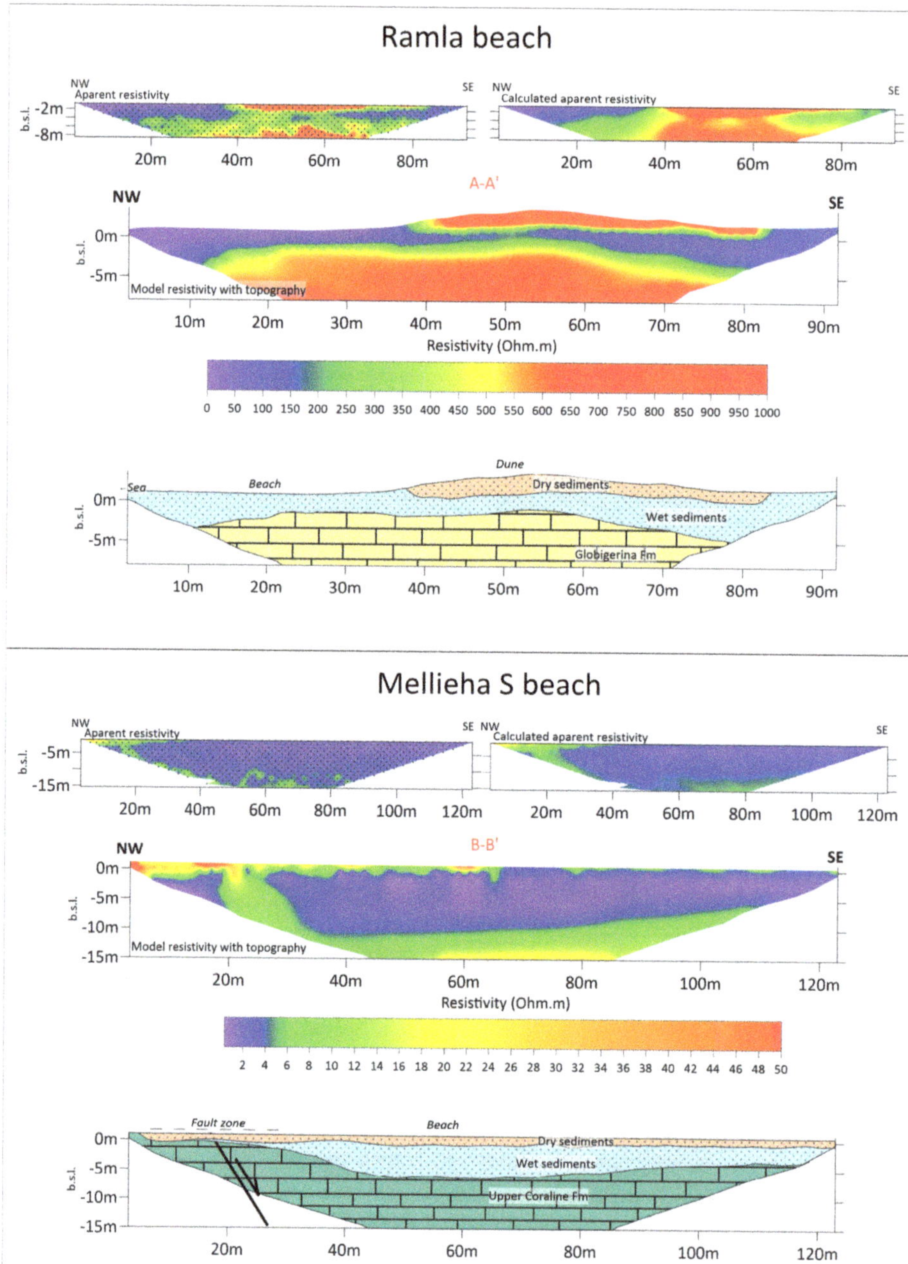

Figure 4. ERT results. For each site, the figure shows the apparent resistivity pseudosection and data distribution, the calculated apparent resistivity pseudosection, the model resistivity with topography and its interpretation. ERTs locations is shown in Figure 1.

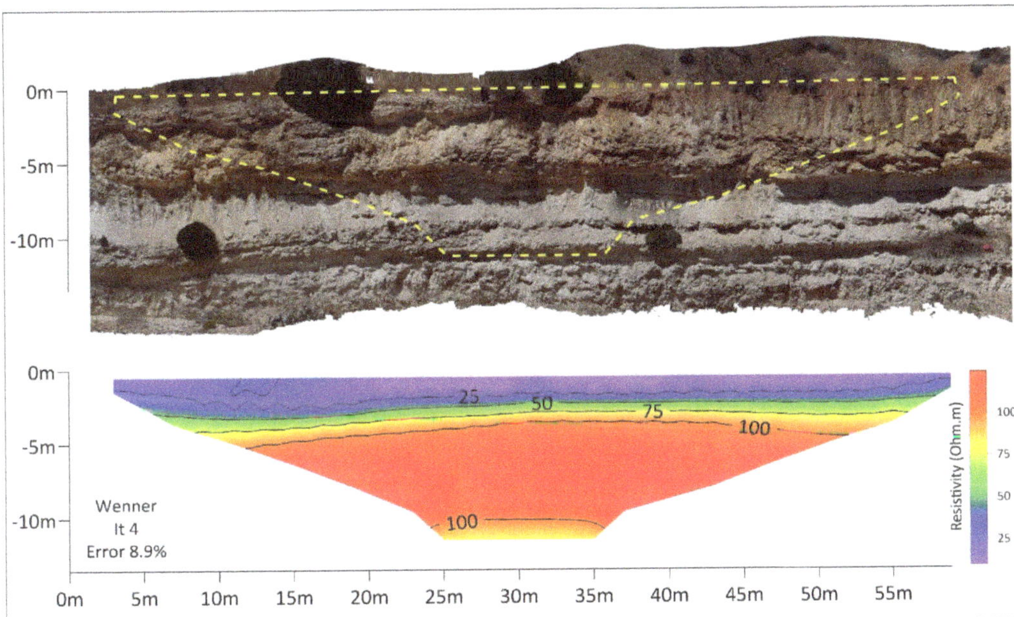

Figure 5. ERT results in a test site in a dissected outcrop in Mellieha Valley, near the Mellieha Bay (Figure 1). The dashed yellow line in the photogrammetric reconstruction shows the ERT results position in the UCL Fm outcrop. Lower resistivity values areas are in the top part of the ERT and coincide with reddish-weathered limestone. Elevations are referred to the ground level.

4.3. HVSR Results

The analysis produced well defined HVSR curves for both sites. To verify the presence of directional effects, the horizontal HVSR were calculated by rotating the NS and EW motion components from 0° (north) to 180° (south) in 10° intervals. Figure 6 displays a typical example of the HVSR and spectral components for each site.

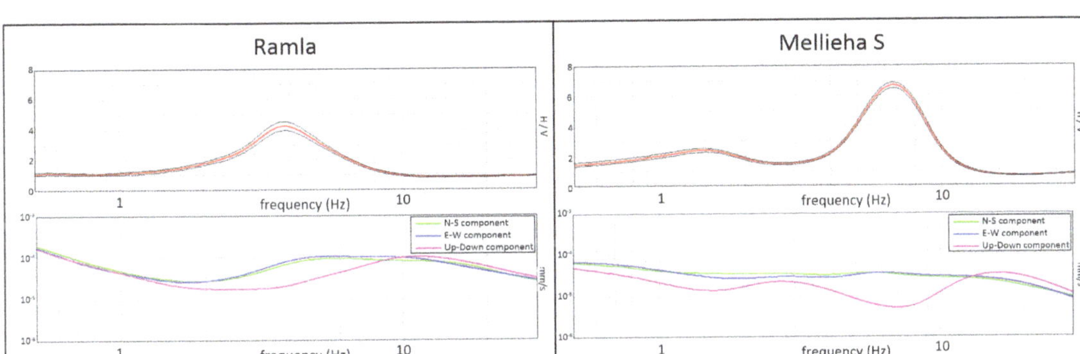

Figure 6. Left: Ramla Beach HVSR example. HVSR peak is present at 3.8 Hz and corresponds with an "eye shape" geometry in the component spectra, where the vertical component presents a fall. Mellieha S Beach HVSR example: Two peaks are present, at 1.5 Hz (F0) and 6.8 Hz (F1). Both peaks correspond with "eye shape" geometry in the component spectra.

The HVSR results from the Ramla Beach exhibit a spectral peak between 2.8 and 7 Hz, with amplitudes ranging from 3 to 7 units, that we call Fs. The results do not show any significant directional amplification effects, as the amplitude of the spectral ratios remained the same for all rotation angles. This and the "eye-shape" of the components spectrum supported a stratigraphic origin of the peaks [47], which we attribute to an impedance contrast resulting from the difference in Vs between the sandy beach deposits with relatively low Vs and the rocky platform of the GL Fm with relatively high Vs. This interpretation is confirmed by the absence of a peak in the HVSR measurement taken at the western boundary of the beach, where sandy sediments were absent, and the measurement was directly performed on the GL Fm.

The HVSR results in Mellieha S Beach exhibit two peaks, one at frequencies between 1 and 2 Hz, that we call F0, and another at frequencies between 6 and 20 Hz, that we call Fs. The measurement taken on the UCL Fm outcrop that bounds the beach on its west side did not exhibit the Fs peak.

The F0 peak, present in all measurements, is interpreted as the result of the impedance contrast between the relatively low Vs BC Fm and the deeper, relatively high Vs GL Fm. This phenomenon has been previously described in various studies conducted on the island (e.g., [49,50]). On the other hand, Fs peaks have been interpreted similarly to those observed in Ramla Bay, although in this case, they are generated by the impedance contrast between the beach sediments and the UCL Fm.

4.4. Sediment Thickness Estimation from HVSR

After identifying the origin of the Fs peaks of the HVSR spectrum generated by the discontinuity between sediments and bedrock, we proceeded to extract the frequency at which each peak occurs at each station. Using these frequencies, it is possible to estimate the depth at which the interface is found at each site.

Two main approaches have been explored in the literature for HVSR peak-to-depth conversion using sediment Vs and peak frequency: a local approach that utilizes borehole data to calculate a local calibration curve (e.g., [7,51]), and a simpler approach, that uses empiric formulas (e.g., [52]). In this study, considering that we do not have borehole data, we decided to approximate the thickness of the sediments by applying the simple formula $H = Vs/4*F$, where H represents the thickness and F is the HVSR peak frequency corresponding to the interfaces between sediments and the underlying bedrock. Even if it can be an oversimplified strategy [7], this approach is valid for obtaining a rough understanding of the subsurface [53].

Vs is a highly variable parameter in sediments, influenced by factors such as texture, packing density, porosity, composition or degree of water saturation. To determine the average Vs of the sediments at each site, we employed an iterative approach. We calculated sediment thickness using the formula $H = Vs/4F$ at HVSR stations located at the same positions as the ERTs and then we evaluated different possible values of Vs to identify the best fit (Figure 7).

At Mellieha S Beach, an average Vs of 180 m/s shows a good fit with the resistivity calculated sedimentary deposits. At Ramla, despite the limited number of HVSR stations, an average Vs of 120 m/s seems to be a reasonable value to avoid overestimation of sediment thickness. The selection of those Vs values are in agreement with previous works and typical Vs on sediments [6,49].

To obtain a model of the geometry of the contact between the sediments and the bedrock at each beach, HVSR-derived thickness values at each station were interpolated by an ordinary kriging algorithm [54] obtaining a grid of thickness that represents the bottom surface of the sedimentary layer.

To obtain a sedimentary depth grid, we subtracted the bottom surface of the sedimentary layer grid from the SfM-derived DTMs, which represent the top surface of the sedimentary layer.

In addition, we calculate the sediment volumes on each beach. For this purpose, three well-established numerical integration methods were used: the Extended Trapezoidal Rule, Extended Simpson's Rule and Extended Simpson's 3/8 Rule [55]. We obtained consistent results, with differences below 0.1 m^3 between methods.

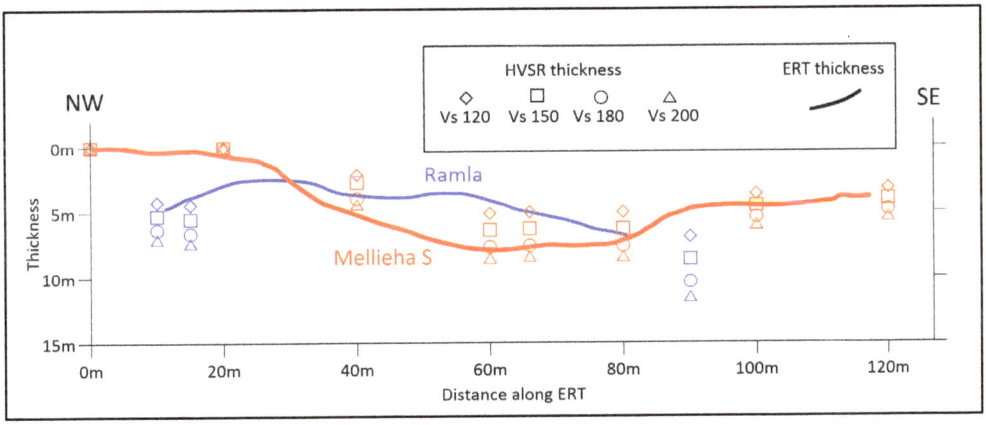

Figure 7. Thickness estimation for Ramla beach (blue) and Mellieha beach (orange) along the ERTs, from the ERTs interpretation and based on the Fs frequencies obtained from HVSR stations utilizing different Vs values.

The results obtained for Ramla Beach (Figure 8) exhibit a total sediment volume of 88,440 m^3 (79% of which is b.s.l.). An ENE trend in the sediment thickness increase is reported, exhibiting a gentle slope (less than 2°) towards the ENE with slight undulations, suggesting that the distribution of the deposits in Ramla Beach is primarily controlled by the stratigraphic contact geometry between the GL Fm, outcropping on the W side of the beach, and the BC Fm, which also exhibits a gentle dip to-wards the ENE (Figure 8b,c). This would have been promoted by the significantly higher erodibility of BC Fm against GL Fm.

In Mellieha S Beach, a total volume of 21,602 m^3 was calculated (72% of which is b.s.l.), which is more than four times lower than Ramla Beach. Maximum thicknesses are observed slightly to the NE of the central sector of the beach, with values close to 7 m (Figure 9a). At the NW boundary, the thickness is minimal and even exposes the UCL, while at the SE boundary of the beach, the thicknesses reach about 3 m. Moving from the center of the beach towards the NW, slope values are about 20°, while the slope towards the SE is less than 5°.

Our interpretation, based both on the ERT and HVSR results, suggests that the geometry of Mellieha S Beach deposits is controlled by structural factors, particularly by the presence of a normal fault (Figure 9b,c). This fault dips to the SW, and the hanging wall block consists of UCL, gently dipping in the opposite direction to the fault.

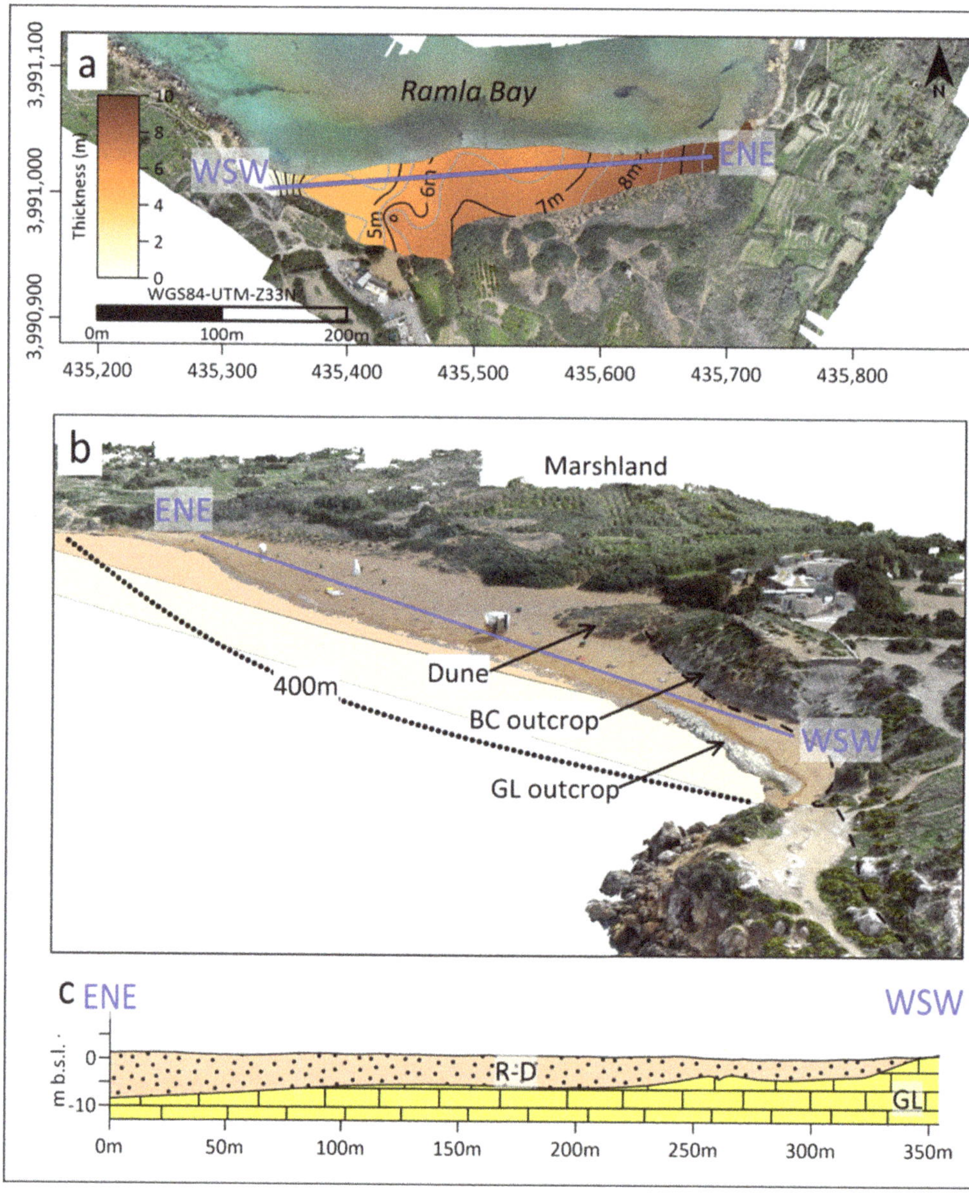

Figure 8. Ramla beach sediment distribution. (**a**) Sediment thickness grid derived from HVSR results. (**b**) 3D photogrammetric model showing main surface features and subsurface representation. (**c**) Cross section of the Ramla Beach, showing the recent deposits resting above the GL Fm.

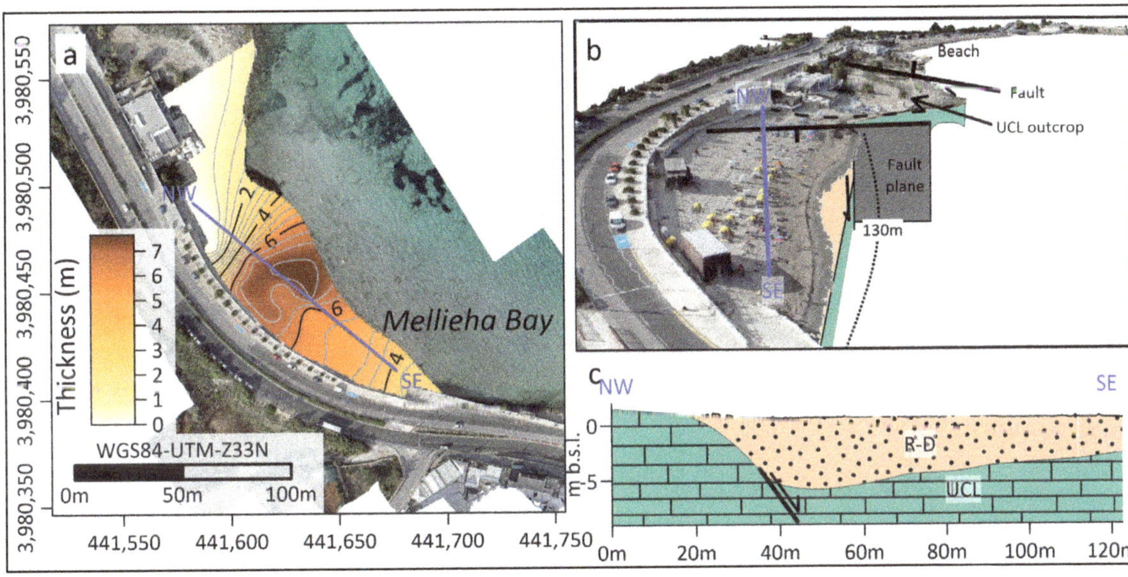

Figure 9. Mellieha S Beach sediment distribution. (**a**) Sediment thickness grid derived from HVSR results. (**b**) 3D photogrammetric model showing main surface features and subsurface representation. (**c**) Cross section of the Mellieha S beach, showing the recent deposits resting above UCL Fm.

5. Discussion

Understanding the sediment distribution in coastal environments is essential for effective coastal management and the preservation of natural resources, especially in a place like Malta where the number of sandy beaches is very low. In this study, we employed a multi-method approach combining ERT, HVSR and SfM photogrammetry techniques to investigate the sediment distribution in two pocket beaches, Ramla Beach and Mellieha S Beach, in Malta.

The results obtained from both ERT and HVSR techniques provided meaningful and consistent findings (Figures 4 and 7–9), enabling the estimation of comparable sediment thickness and morphological characteristics in the studied pocket beaches.

The HVSR maps revealed distinct spatial variations in peak frequencies, which indicated variations in sediment thickness across the study areas (Figures 8 and 9). The observed differences in frequencies were indicative of changes in sediment thickness and provided valuable insights into subsurface properties.

However, it is important to acknowledge the limitations of the study. While efforts were made to obtain comprehensive data coverage, some areas might have been underrepresented, leading to potential gaps in our understanding of sediment distribution in specific regions of the pocket beaches, especially when resolving small-scale features. Furthermore, the geological complexity of the study areas, including the presence of heterogeneities in the bedrock and sediment layers, could have influenced the accuracy of the interpretations. In the absence of borehole data, the direct validation of the geophysical results was not possible. In addition, the selection of the Vs of the sediments, although within the typical ranges, presents a certain degree of uncertainty that is transferred to the HVSR thickness estimation. Despite these limitations, the multi-method approach used in this study provides valuable insights into sediment distribution and subsurface properties of the studied pocket beaches. The findings contribute to our understanding of the main controlling geological factors influencing sediment distribution in the study areas.

6. Conclusions

This study employed a multi-method approach combining HVSR, ERT and SfM photogrammetry from UAV to investigate for first time the volumetric sediment distribution in two pocket beaches in Malta. The results provide valuable insights into the sediment distribution and characteristics of the studied pocket beaches. The HVSR analysis revealed significant variations in sediment thickness, contributing to our understanding of the factors controlling sediment distribution. The ERT data complemented the HVSR results by supporting the modelling of sediment volumes and enhancing the understanding of subsurface resistivities. Additionally, the SfM-UAV photogrammetry technique provided detailed topographic information, enriching our understanding of the beach environments.

This study highlights the importance of considering multi parametric perspectives in coastal studies for accurate characterization of sediment volumes and a better understanding of pocket beach controls. Future research should focus on obtaining borehole data to improve geophysical calibration for sediment thickness estimations.

Given the projected increased vulnerability of coastal zones due to climate change, precise characterizations of coastal environments are even more critical. The multi-method approach used in this study is practical, offering a valuable tool for future coastal studies. The findings can support decision-making processes and the development of sustainable coastal protection strategies to mitigate the impacts of climate change.

Author Contributions: Conceptualization, L.G. and S.D.; methodology, L.G., E.C., P.I., F.V. and S.D.; software, L.G.; validation, L.G. and S.D; formal analysis, L.G., F.V. and S.D.; investigation, L.G., E.C. and P.I., resources, S.D.; data curation, L.G., E.C. and P.I.; writing—L.G. and S.D.; writing—review and editing, L.G., S.D., E.C., P.I., L.R., F.V. and P.G.; visualization, L.G.; supervision, S.D. and F.V.; project administration, S.D., L.G. and E.C.; funding acquisition, L.G., S.D. and F.V. All authors have read and agreed to the published version of the manuscript.

Funding: This work was partially supported by the project Satellite Investigation to study POcket BEach Dynamics (SIPOBED, SRF-2021-2S1, PI: Sebastiano D'Amico),the project Multi-disciplinary monitoring system for a resilient management of coastal areas (REMACO) funded by the INTERREG V A–Italy-Malta Capitalization Programme, the INGV Project "Ricerca 703 Libera" BR2019.23 ("Unveiling silent faults in low strain-rates regions through the integration 704 of high-resolution geophysical and seismological analyses" P.I. Fabio Villani), and by the Internationalisation Partnership & Awards Scheme Plus (IPAS+) supported by the Malta Council for Science and Technology through the project "Near-surface geophysics and geomatic applied to coastal systems" (IPAS-2022-020).

Data Availability Statement: The data presented in this study are available on request from the corresponding author.

Conflicts of Interest: The authors declare no conflict of interest.

References

1. Do, K.; Yoo, J. Morphological Response to Storms in an Embayed Beach Having Limited Sediment Thickness. *Estuar. Coast. Shelf Sci.* **2020**, *234*, 106636. [CrossRef]
2. Cooper, N.J.; Pontee, N.I. Appraisal and Evolution of the Littoral 'Sediment Cell'Concept in Applied Coastal Management: Experiences from England and Wales. *Ocean Coast. Manag.* **2006**, *49*, 498–510. [CrossRef]
3. Morgan, D.; Gunn, D.; Payo, A.; Raines, M. Passive Seismic Surveys for Beach Thickness Evaluation at Different England (UK) Sites. *J. Mar. Sci. Eng.* **2022**, *10*, 667. [CrossRef]
4. Nakamura, Y. A Method for Dynamic Characteristics Estimation of Subsurface Using Microtremor on the Ground Surface. *Railw. Tech. Res. Inst. Q. Rep.* **1989**, *30*, 25–33.
5. Okada, H.; Suto, K. *The Microtremor Survey Method*; Society of Exploration Geophysicists: Houston, TX, USA, 2003; ISBN 978-1-56080-174-0.
6. Villani, F.; D'Amico, S.; Panzera, F.; Vassallo, M.; Bozionelos, G.; Farrugia, D.; Galea, P. Shallow high-resolution geophysical investigation along the western segment of the Victoria Lines Fault (island of Malta). *Tectonophysics* **2018**, *724–725*, 220–233. [CrossRef]
7. Johnson, C.; Lane, J. Statistical Comparison of Methods for Estimating Sediment Thickness from Horizontal-to-Vertical Spectral Ratio (Hvsr) Seismic Methods: An Example from Tylerville, Connecticut, Usa. In Proceedings of the Symposium on the Application of Geophysics to Engineering and Environmental Problems, Denver, CO, USA, 20–24 March 2016; pp. 317–323.

8. Liang, D.; Gan, F.; Zhang, W.; Jia, L. The Application of HVSR Method in Detecting Sediment Thickness in Karst Collapse Area of Pearl River Delta, China. *Environ. Earth Sci.* **2018**, *77*, 259. [CrossRef]
9. Alonso-Pandavenes, O.; Torres, G.; Torrijo, F.J.; Garzón-Roca, J. Basement Tectonic Structure and Sediment Thickness of a Valley Defined Using HVSR Geophysical Investigation, Azuela Valley, Ecuador. *Bull. Eng. Geol. Environ.* **2022**, *81*, 210. [CrossRef]
10. Mohamed, A.M.E.; El-Hussain, I.; Deif, A.; Al-Jabri, K.; Al-Habsi, Z.; El-Hady, S. Near-Surface Site Characterization at Quriyat City, Sultanate of Oman Using HVSR and MASW Techniques. *Arab. J. Geosci.* **2015**, *9*, 23. [CrossRef]
11. Nelson, S.; McBride, J. Application of HVSR to Estimating Thickness of Laterite Weathering Profiles in Basalt. *Earth Surf. Process. Landf.* **2019**, *44*, 1365–1376. [CrossRef]
12. Nguyen, F.; Jongmans, D.; Pirard, E. Near-Surface Geophysical Imaging and Detection of Slow Active Faults. Ph.D. Thesis, Faculty of Applied Sciences, Liège University, Liège, Belgium, 2005.
13. Soupios, P.; Papadopoulos, N.; Papadopoulos, I.; Kouli, M.; Vallianatos, F.; Sarris, A.; Manios, T. Application of Integrated Methods in Mapping Waste Disposal Areas. *Environ. Geol.* **2007**, *53*, 661–675. [CrossRef]
14. Danneels, G.; Bourdeau, C.; Torgoev, I.; Havenith, H.-B. Geophysical Investigation and Dynamic Modelling of Unstable Slopes: Case-Study of Kainama (Kyrgyzstan). *Geophys. J. Int.* **2008**, *175*, 17–34. [CrossRef]
15. Benjumea, B.; Macau, A.; Gabàs, A.; Bellmunt, F.; Figueras, S.; Cirés, J. Integrated Geophysical Profiles and H/V Microtremor Measurements for Subsoil Characterization†. *Near Surf. Geophys.* **2011**, *9*, 413–425. [CrossRef]
16. Gallipoli, M.R.; Giocoli, A.; Piscitelli, S. Joint Application of Low-Cost, Fast Executable and Non-Invasive Geophysical Techniques during Emergency and Microzonation Study: Hints from L'Aquila (Italy) Earthquake. In Proceedings of the Geotechnical and Geophysical Site Characterization: Proceedings of the 4th International Conference on Site Characterization ISC-4, Porto de Galinhas, Brazil, 18–21 September 2012; Volume 1, pp. 1477–1484.
17. Sauret, E.S.G.; Beaujean, J.; Nguyen, F.; Wildemeersch, S.; Brouyere, S. Characterization of Superficial Deposits Using Electrical Resistivity Tomography (ERT) and Horizontal-to-Vertical Spectral Ratio (HVSR) Geophysical Methods: A Case Study. *J. Appl. Geophys.* **2015**, *121*, 140–148. [CrossRef]
18. Abu Zeid, N.; Corradini, E.; Bignardi, S.; Nizzo, V.; Santarato, G. The Passive Seismic Technique 'HVSR' as a Reconnaissance Tool for Mapping Paleo-Soils: The Case of the Pilastri Archaeological Site, Northern Italy. *Archaeol. Prospect.* **2017**, *24*, 245–258. [CrossRef]
19. Zarroca, M.; Bach, J.; Linares, R.; Pellicer, X.M. Electrical Methods (VES and ERT) for Identifying, Mapping and Monitoring Different Saline Domains in a Coastal Plain Region (Alt Empordà, Northern Spain). *J. Hydrol.* **2011**, *409*, 407–422. [CrossRef]
20. Colica, E.; D'Amico, S.; Iannucci, R.; Martino, S.; Gauci, A.; Galone, L.; Galea, P.; Paciello, A. Using Unmanned Aerial Vehicle Photogrammetry for Digital Geological Surveys: Case Study of Selmun Promontory, Northern of Malta. *Environ. Earth Sci.* **2021**, *80*, 551. [CrossRef]
21. Furlani, S.; Antonioli, F.; Colica, E.; D'Amico, S.; Devoto, S.; Grego, P.; Gambin, T. Sea Caves and Other Landforms of the Coastal Scenery on Gozo Island (Malta): Inventory and New Data on Their Formation. *Geosciences* **2023**, *13*, 164. [CrossRef]
22. Leucci, G.; Persico, R.; De Giorgi, L.; Lazzari, M.; Colica, E.; Martino, S.; Iannucci, R.; Galone, L.; D'Amico, S. Stability Assessment and Geomorphological Evolution of Sea Natural Arches by Geophysical Measurement: The Case Study of Wied Il-Mielah Window (Gozo, Malta). *Sustainability* **2021**, *13*, 12538. [CrossRef]
23. Bowman, D.; Guillén, J.; López, L.; Pellegrino, V. Planview Geometry and Morphological Characteristics of Pocket Beaches on the Catalan Coast (Spain). *Geomorphology* **2009**, *108*, 191–199. [CrossRef]
24. Dehouck, A.; Dupuis, H.; Sénéchal, N. Pocket Beach Hydrodynamics: The Example of Four Macrotidal Beaches, Brittany, France. *Mar. Geol.* **2009**, *266*, 1–17. [CrossRef]
25. Pranzini, E.; Rosas, V.; Jackson, N.L.; Nordstrom, K.F. Beach Changes from Sediment Delivered by Streams to Pocket Beaches during a Major Flood. *Geomorphology* **2013**, *199*, 36–47. [CrossRef]
26. Turi, A.; Picollo, M.; Valleri, G. Mineralogy and Origin of the Carbonate Beach Sediments of Malta and Gozo, Maltese Islands. *Boll. Soc. Ceol. Ital.* **1990**, *J09*, 367–374.
27. Gatt, P. Controls on Plio-Quaternary Foreland Sedimentation in the Region of the Maltese Islands. *Boll. Soc. Geol. Ital.* **2007**, *126*, 119–129.
28. Deidun, A.; Gauci, R.; Schembri, J.A.; Šegina, E.; Gauci, A.; Gianni, F.; Gutierrez, J.A.; Sciberras, A.; Sciberras, J. Comparative Median Grain Size Assessment through Three Different Techniques for Sandy Beach Deposits on the Maltese Islands (Central Mediterranean). *J. Coast. Res.* **2013**, *165*, 1757–1761. [CrossRef]
29. Sammut, S.; Gauci, R.; Drago, A.; Gauci, A.; Azzopardi, J. Pocket Beach Sediment: A Field Investigation of the Geodynamic Processes of Coarse-Clastic Beaches on the Maltese Islands (Central Mediterranean). *Mar. Geol.* **2017**, *387*, 58–73. [CrossRef]
30. Gatt, P. Embayment Morphometrics, Granulometry and Carbonate Mineralogy of Sandy Beaches in the Maltese Islands. *Mar. Geol.* **2021**, *432*, 106394. [CrossRef]
31. Randazzo, G.; Italiano, F.; Micallef, A.; Tomasello, A.; Cassetti, F.P.; Zammit, A.; D'Amico, S.; Saliba, O.; Cascio, M.; Cavallaro, F.; et al. WebGIS Implementation for Dynamic Mapping and Visualization of Coastal Geospatial Data: A Case Study of BESS Project. *Appl. Sci.* **2021**, *11*, 8233. [CrossRef]
32. Farrugia, M.T. Public Perceptions on Coastal Erosion in the Maltese Islands: A Case Study of St George's Bay (St Julians) and Pretty Bay (Birżebbuġa). *Nat. Hazards* **2017**, *86*, 587–604. [CrossRef]

33. Cooke, B.C.; Jones, A.R.; Goodwin, I.D.; Bishop, M.J. Nourishment Practices on Australian Sandy Beaches: A Review. *J. Environ. Manag.* **2012**, *113*, 319–327. [CrossRef]
34. Pedley, H.M.; House, M.R.; Waugh, B. The Geology of Malta and Gozo. *Proc. Geol. Assoc.* **1976**, *87*, 325–341. [CrossRef]
35. Pedley, H.M. *The Ghar Lapsi Limestones: Sedimentology of a Miocene Intra-Shelf Graben*; University of Malta Press: Msida, Malta, 1987.
36. Pedley, M. The Calabrian Stage, Pleistocene Highstand in Malta: A New Marker for Unravelling the Late Neogene and Quaternary History of the Islands. *J. Geol. Soc.* **2011**, *168*, 913–926. [CrossRef]
37. Gauci, M.J.; Deidun, A.; Schembri, P.J. Faunistic Diversity of Maltese Pocket Sandy and Shingle Beaches: Are These of Conservation Value? *Oceanologia* **2005**, *47*, 219–241.
38. Ashby, T. Roman Malta. *J. Rom. Stud.* **1915**, *5*, 23–80. [CrossRef]
39. Mifsud, C. Revisiting Ramla L-Ħamra Villa—New Discoveries and Observations on the Roman Villa Complex in Xagħra, Gozo. *Open Archaeol.* **2021**, *7*, 84–97. [CrossRef]
40. Colica, E.; Galone, L.; D'Amico, S.; Gauci, A.; Iannucci, R.; Martino, S.; Pistillo, D.; Iregbeyen, P.; Valentino, G. Evaluating Characteristics of an Active Coastal Spreading Area Combining Geophysical Data with Satellite, Aerial, and Unmanned Aerial Vehicles Images. *Remote Sens.* **2023**, *15*, 1465. [CrossRef]
41. Everett, M.E. *Near-Surface Applied Geophysics*; Cambridge University Press: Cambridge, UK, 2013; ISBN 978-1-107-01877-8.
42. Loke, M.H. Electrical Imaging Surveys for Environmental and Engineering Studies. 1999. Available online: https://pages.mtu.edu/~ctyoung/LOKENOTE.PDF (accessed on 29 October 2023).
43. Loke, M.H.; Dahlin, T. A Comparison of the Gauss–Newton and Quasi-Newton Methods in Resistivity Imaging Inversion. *J. Appl. Geophys.* **2002**, *49*, 149–162. [CrossRef]
44. Loke, M.H.; Acworth, I.; Dahlin, T. A Comparison of Smooth and Blocky Inversion Methods in 2D Electrical Imaging Surveys. *Explor. Geophys.* **2003**, *34*, 182–187. [CrossRef]
45. Naudet, V.; Lazzari, M.; Perrone, A.; Loperte, A.; Piscitelli, S.; Lapenna, V. Integrated Geophysical and Geomorphological Approach to Investigate the Snowmelt-Triggered Landslide of Bosco Piccolo Village (Basilicata, Southern Italy). *Eng. Geol.* **2008**, *98*, 156–167. [CrossRef]
46. Mas-Pla, J.; Rodríguez-Florit, A.; Zamorano, M.; Roqué, C.; Menció, A.; Brusi, D. Anticipating the Effects of Groundwater Withdrawal on Seawater Intrusion and Soil Settlement in Urban Coastal Areas. *Hydrol. Process.* **2013**, *27*, 2352–2366. [CrossRef]
47. Bard, P.-Y. Guidelines for the Implementation of the H/V Spectral Ratio Technique on Ambient Vibrations-Measurements, Processing and Interpretations. Available online: https://scholar.google.com/scholar?cluster=14085203609344808715&hl=en&oi=scholarr (accessed on 29 October 2023).
48. Buquet, D.; Sirieix, C.; Anschutz, P.; Malaurent, P.; Charbonnier, C.; Naessens, F.; Bujan, S.; Lecroart, P. Shape of the Shallow Aquifer at the Fresh Water–Sea Water Interface on a High-Energy Sandy Beach. *Estuar. Coast. Shelf Sci.* **2016**, *179*, 79–89. [CrossRef]
49. Pischiutta, M.; Villani, F.; D'Amico, S.; Vassallo, M.; Cara, F.; Di Naccio, D.; Farrugia, D.; Di Giulio, G.; Amoroso, S.; Cantore, L.; et al. Results from Shallow Geophysical Investigations in the Northwestern Sector of the Island of Malta. *Phys. Chem. Earth Parts A/B/C* **2017**, *98*, 41–48. [CrossRef]
50. Iannucci, R.; Martino, S.; Paciello, A.; D'Amico, S.; Galea, P. Investigation of Cliff Instability at Għajn Ħadid Tower (Selmun Promontory, Malta) by Integrated Passive Seismic Techniques. *J. Seism.* **2020**, *24*, 897–916. [CrossRef]
51. Ibs-von Seht, M.; Wohlenberg, J. Microtremor Measurements Used to Map Thickness of Soft Sediments. *Bull. Seismol. Soc. Am.* **1999**, *89*, 250–259. [CrossRef]
52. Farrugia, D.; Paolucci, E.; D'Amico, S.; Galea, P. Inversion of Surface Wave Data for Subsurface Shear Wave Velocity Profiles Characterized by a Thick Buried Low-Velocity Layer. *Geophys. J. Int.* **2016**, *206*, 1221–1231. [CrossRef]
53. Bignardi, S. The Uncertainty of Estimating the Thickness of Soft Sediments with the HVSR Method: A Computational Point of View on Weak Lateral Variations. *J. Appl. Geophys.* **2017**, *145*, 28–38. [CrossRef]
54. Journel, A.G.; Rossi, M.E. When Do We Need a Trend Model in Kriging? *Math. Geol.* **1989**, *21*, 715–739. [CrossRef]
55. Kalambet, Y.; Kozmin, Y.; Samokhin, A. Comparison of Integration Rules in the Case of Very Narrow Chromatographic Peaks. *Chemom. Intell. Lab. Syst.* **2018**, *179*, 22–30. [CrossRef]

MDPI
St. Alban-Anlage 66
4052 Basel
Switzerland
www.mdpi.com

Remote Sensing Editorial Office
E-mail: remotesensing@mdpi.com
www.mdpi.com/journal/remotesensing

www.ingramcontent.com/pod-product-compliance
Lightning Source LLC
Chambersburg PA
CBHW040805250926
51850CB00007BA/1086
* 9 7 8 3 7 2 5 8 1 4 2 1 3 *